10th Anniversary of Applied Sciences-Invited Papers in Chemistry Section

10th Anniversary of Applied Sciences-Invited Papers in Chemistry Section

Editor

Samuel B. Adeloju

MDPI • Basel • Beijing • Wuhan • Barcelona • Belgrade • Manchester • Tokyo • Cluj • Tianjin

Editor
Samuel B. Adeloju
Charles Sturt University
Australia

Editorial Office
MDPI
St. Alban-Anlage 66
4052 Basel, Switzerland

This is a reprint of articles from the Special Issue published online in the open access journal *Applied Sciences* (ISSN 2076-3417) (available at: https://www.mdpi.com/journal/water/special_issues/hydraulics_numerical_methods).

For citation purposes, cite each article independently as indicated on the article page online and as indicated below:

LastName, A.A.; LastName, B.B.; LastName, C.C. Article Title. *Journal Name* **Year**, *Volume Number*, Page Range.

ISBN 978-3-0365-1114-6 (Hbk)
ISBN 978-3-0365-1115-3 (PDF)

Cover image courtesy of Analytical Chemistry Instrumentation Laboratory, Faculty of Science, Charles Sturt University, NSW, Australia

Contents

About the Editor

Samuel B. Adeloju is a Professor of Chemistry and Deputy Dean of the Faculty of Science at Charles Sturt University, Albury and Editor-in-Chief of the Chemistry Section of the *Applied Sciences* Journal (MDPI). He has made original and innovative contributions to the design, fabrication, and characterization of conducting polymers, biosensors, nanobiosensors, nanomaterials, and microbial fuel cells. In particular, he has made significant contributions to the development of biosensors and nanobiosensors for clinical, environmental, and food analyses, as well as to the development of microbial fuel cells (MFCs) for power generation from wastewaters. Sam has published extensively in international journals, books, and proceedings. He was awarded the Royal Australia Chemical Institute (RACI) awards for Applied Research Medal (2009), R.H. Stokes Medal for Electrochemistry (2011), Doreen Clark Medal for Analytical Chemistry (2013), and Citation Award for Excellence in Leadership and Service (2017). Sam is also on the International Advisory Board for *Sci Journal* (MDPI), Editorial Boards for *Biosensors* (MDPI), *Heliyon* (Elsevier), and other international journals.

Editorial

10th Anniversary of Applied Sciences-Invited Papers in Chemistry Section

Samuel B. Adeloju

Faculty of Science, Charles Sturt University, Albury, NSW 2640, Australia; sadeloju@csu.edu.au; Tel.: +61-2-6051-9681

MDPI´s *Applied Sciences* reached a remarkable milestone in 2020 when the 10th volume of the journal was published and an impact factor of 2.474 was achieved. To celebrate this special occasion, the Chemistry Section has produced a special issue based mainly on invited papers to mark "The 10th Anniversary of Applied Sciences". This Special Issue consists mainly of 12 original research articles and 1 comprehensive review featuring important and recent developments in analytical chemistry, environmental chemistry, nanomaterial chemistry, polymer/coating chemistry, drug delivery chemistry, and flavonol chemistry.

Three of the featured articles are in the broad area of *analytical chemistry*. In one of these papers, Roverso et al. [1] explored the use of a mixed cationic-reverse phase column coupled to high-resolution tandem MS (HR-MS/MS) for the analysis of small highly polar metabolites in biological fluids at low concentrations. They successfully retained and separated trimethylamine N-oxide (TMAO) and the isobaric molecules beta-methylamino-L-alanine (BMAA) and 2,4-diaminobutyric acid (DAB). When applied to plasma and urine samples, they achieved linear concentration ranges of 50–1000 μg/L and 500–10,000 μg/L for TMAO and both BMAA and DAB, respectively. The effectiveness of the method was also investigated for biologically relevant compounds and those with a wider range of polarities. This approach consequently enabled the simultaneous analysis of a larger range of metabolites, from very small and polar compounds to quite lipophilic molecules. In another study, Avino et al. [2] employed ultrasound vortex-assisted dispersive liquid–liquid microextraction and gas chromatography coupled to ion trap mass spectrometry for the determination of non-steroidal anti-inflammatory drugs (NSAIDs) in animal urine samples. The urine samples were initially treated with β-glucuronidase/acrylsulfatase before extracting the NSAIDs with CH_2Cl_2 by the ultrasound vortex-assisted dispersive liquid–liquid microextraction method. They achieved an enrichment factor of about 300–450 times by this approach, with 94.1 to 101.2% recoveries and a relative standard deviation (RSD) of ≤4.1%. The achieved limits of detection (LODs) and limits of quantification (LOQs) were 0.1–0.2 ng mL^{-1} (RSD ≤ 4.5%) and 4.1–4.7 ng mL^{-1} (RSD ≤ 3.5%), respectively. Cordeiro et al. [3] synthesized rapid and cheap MoS_2 nanostructures on paper substrates through microwave-assisted hydrothermal synthesis for the production of low-cost near-infrared photodetectors. The best results were obtained with the interdigital MoS_2 photodetector by using photoactive MoS_2 nanosheets synthesized at 200 °C for 120 min. They achieved a responsivity of 290 mA/W, detectivity of 1.8×10^9 Jones, and an external quantum efficiency of 37% with the photodetector.

Another broad area which is well represented in this special issue is *environmental chemistry*, with four featured articles. Avino et al. [4] reported on the weekly and longitudinal elemental variability in hair samples collected from non-occupationally exposed subjects. Neutron activation analysis was used to determine 30 elements in hair samples collected from different locations on the scalp. Notable differences were observed among samples between the proximal and distal sections. A comparison with other studies was used to establish the relationships and the differences caused by different ethnic origins, lifestyles, diets, and climates among different young populations. In another study, Assi

Citation: Adeloju, S.B. 10th Anniversary of Applied Sciences-Invited Papers in Chemistry Section. *Appl. Sci.* **2021**, *11*, 2831. https://doi.org/10.3390/app11062831

Received: 3 March 2021
Accepted: 18 March 2021
Published: 22 March 2021

et al. [5] explored the use of bottom ash derived from the co-combustion of municipal solid waste and sewage sludge as a stabilizing agent. Two other components used in the stabilization method included flue gas desulfurization residues and coal fly ash. Results obtained by leaching test on the stabilized samples revealed that heavy metal concentrations, particularly for Zn and Pb, were significantly reduced. The three factors responsible for the reduction of the heavy metal concentrations by the stabilization method were the amount of ash used, the Zn and Pb concentrations in the as-received fly ash, and the solution of the final materials. Porto et al. [6] investigated the removal of phosphorus from a secondary-treated effluent of a Portuguese paper company based on the growth and ability of *Chlorella vulgaris*. Results obtained by batch experiments revealed that the undiluted effluent inhibited microalgal growth. Consequently, the dilution of the effluent was necessary to achieve the desired bioremediation. The most diluted effluent enabled the removal of $54 \pm 1\%$ of phosphorus. It was also found that the microalgal growth was dependent on the compounds present in the effluent and on the solution pH. In the last featured article on environmental chemistry, Adeloju et al. [7] reviewed the effects and consequences of arsenic contamination of groundwater with regard to drinking water quality and human health in under-developed countries and remote communities. They discussed the chemistry of arsenic and the factors influencing the form(s) of arsenic present and its fate when introduced into the environment. They also provided a global overview of arsenic contamination of groundwater around the world, and a case study of arsenic contamination in Bangladesh was used to highlight the health-related, agricultural, social, and economic impacts. Furthermore, the available strategies for removal of arsenic and specific examples of the available filter systems for domestic arsenic removal from groundwater for potable water use were discussed.

Polymer/coating chemistry also features very well in this special issue, with three articles. Kanellopoulou et al. [8] reported on their investigation of the effect of the incorporation of superabsorbent polymers (SAPs) into cementitious-based composite materials on their microstructure and self-healing properties. It was found that the compressive strength remained intact for all specimens with the incorporation of the SAPs. This was attributed to the reduction in porosity and the narrower range of pore size distribution of the mortar/SAP specimens. In addition, an up to 60% increase in the self-healing behavior of mortar/SAP specimens was observed when compared to control specimens. In another study, Quintana et al. [9] investigated the use of licorice root extracts produced by ultrasound-assisted extraction in combination with chitosan to produce edible coatings for improving the postharvest quality of fruits. The bioactive extracts were applied to strawberry for the evaluation of their physicochemical and microbiological properties. It was found that the addition of the licorice extract to chitosan improved the rheological properties of the coatings, resulting in reduced rigidity. Furthermore, during storage, good-quality parameters were maintained by the strawberry coated with chitosan and licorice extract. In addition, the coated strawberry achieved the best microbiological preservation when compared with controls. A very interesting use of a combination of polymers for oral drug delivery was presented by Mumuni et al. [10]. They evaluated the use of different ratios of mucin-grafted polyethylene glycol-based micro-particles, both in vitro and in vivo, as carriers for the oral delivery of insulin. The insulin-loaded micro-particles were found to display irregular porosity and shape. The achieved insulin encapsulation efficiency and loading capacity values were >82% and 18%, respectively. The variation in the micro-particle formulation resulted in a variation of the insulin release of between 68% and 92%. More significantly, the oral administration of the insulin-loaded micro-particles achieved a more significant reduction in blood glucose levels than the use of insulin solution. Evidently, the results of this study demonstrate that the use of a combination of polymers for oral delivery of insulin is more effective.

Nanomaterial chemistry also features in this special issue, with two articles. Stefa et al. [11] reported on the use of a hydrothermal method for the synthesis of ZnO-doped ceria nanorods made up of CeO_2/ZnO mixed oxides with different Zn/Ce atomic ratios.

The mixed oxides were found to be superior to CeO_2 and ZnO oxides, which was attributed to the synergistic effect of the ZnO–CeO_2 interaction. It was also established that there was a close correlation between the catalytic activity and oxygen storage capacity of the ZnO-doped ceria nanorods. In another study, Sánchez-Romate et al. [12] produced carbon fiber-reinforced plastic bonded joints with novel carbon nanotube (CNT) adhesive films. By varying the surfactant content used to aid the CNT dispersion, they were able to test the carbon fiber-reinforced plastic bonded joints under different aging conditions. It was found that the electrical response, based on the measurement of the electrical resistance, varied with the aging conditions and after 1 month a higher plasticity region was demonstrated. A sharper increase was also found with 2-month-old samples. Notably, by increasing the surfactant content, the observed changes became more prevalent. However, the presence of more prevalent brittle mechanisms for the CNT-doped joints resulted in a higher established first estimation of damage accumulation for non-aged and 2-month-old samples.

In a final paper on *flavonol chemistry*, Rogozinska and Biesaga [13] utilized LC-MS/MS to investigate the stability of four common types of dietary flavonols, namely: kaempferol, quercetin, isorhamnetin, and myricetin, in the presence of hydrogen peroxide and saliva. They also investigated the influence of saliva on the representative quercetin glycosides rutin, quercitrin, hyperoside, and spiraeoside. They found that flavonol stability generally decreased with increasing B-ring substitution, irrespective of the oxidative agent used. Glycosides were in particular found to be resistant to hydrolysis in the presence of saliva. However, it was proven that saliva is an oxidative agent which often results in the formation of corresponding phenolic acids, thus highlighting the need to give this factor due consideration in flavonol metabolism. The fact that flavonol decomposition starts in the oral cavity suggests that it may not be present in the parent form, but as phenolic acids.

These featured invited papers in this Special Issue for "The 10th Anniversary of Applied Sciences" are good examples of the depth and breadth of chemistry articles published in the Chemistry Section of *Applied Sciences*. The section continues to attract increasing numbers of manuscripts relating to all areas of chemistry for publication in the journal.

Funding: This research received no external funding.

Institutional Review Board Statement: Not Applicable.

Informed Consent Statement: Not Applicable.

Data Availability Statement: Not Applicable.

References

1. Roverso, M.; Di Gangi, I.M.; Favaro, G.; Pastore, P.; Bogialli, S. Use of a Mixed Cationic-Reverse Phase Column for Analyzing Small Highly Polar Metabolic Markers in Biological Fluids for Multiclass LC-HRMS Method. *Appl. Sci.* **2020**, *10*, 7137. [CrossRef]
2. Avino, P.; Notardonato, I.; Passarella, S.; Russo, M.V. Determination of Non-Steroidal Anti-Inflammatory Drugs in Animal Urine Samples by Ultrasound Vortex-Assisted Dispersive Liquid—Liquid Microextraction and Gas Chromatography Coupled to Ion Trap-Mass Spectrometry. *Appl. Sci.* **2020**, *10*, 5411. [CrossRef]
3. Cordeiro, N.J.A.; Gaspar, C.; De Oliveira, M.J.; Nunes, D.; Barquinha, P.; Pereira, L.; Fortunato, E.; Martins, R.; Laureto, E.; Lourenço, S.A. Fast and Low-Cost Synthesis of MoS_2 Nanostructures on Paper Substrates for Near-Infrared Photodetectors. *Appl. Sci.* **2021**, *11*, 1234. [CrossRef]
4. Avino, P.; Lammardo, M.; Petrucci, A.; Rosada, A. Weekly and Longitudinal Element Variability in Hair Samples of Subjects Non-Occupationally Exposed. *Appl. Sci.* **2021**, *11*, 1236. [CrossRef]
5. Assi, A.; Bilo, F.; Zanoletti, A.; Borgese, L.; Depero, L.E.; Nenci, M.; Bontempi, E. Stabilization of Municipal Solid Waste Fly Ash, Obtained by Co-Combustion with Sewage Sludge, Mixed with Bottom Ash Derived by the Same Plant. *Appl. Sci.* **2020**, *10*, 6075. [CrossRef]
6. Porto, B.; Gonçalves, A.L.; Esteves, A.F.; De Souza, S.M.A.G.U.; De Souza, A.A.U.; Vilar, V.J.P.; Pires, J.C.M. Microalgal Growth in Paper Industry Effluent: Coupling Biomass Production with Nutrients Removal. *Appl. Sci.* **2020**, *10*, 3009. [CrossRef]
7. Adeloju, S.; Khan, S.; Patti, A. Arsenic Contamination of Groundwater and Its Implications for Drinking Water Quality and Human Health in Under-Developed Countries and Remote Communities—A Review. *Appl. Sci.* **2021**, *11*, 1926. [CrossRef]
8. Kanellopoulou, I.A.; Kartsonakis, I.A.; Charitidis, C.A. The Effect of Superabsorbent Polymers on the Microstructure and Self-Healing Properties of Cementitious-Based Composite Materials. *Appl. Sci.* **2021**, *11*, 700. [CrossRef]

9. Quintana, S.E.; Llalla, O.; García-Zapateiro, L.A.; García-Risco, M.R.; Fornari, T. Preparation and Characterization of Lico-rice-Chitosan Coatings for Postharvest Treatment of Fresh Strawberries. *Appl. Sci.* **2020**, *10*, 8431. [CrossRef]
10. Mumuni, M.A.; Calister, U.E.; Aminu, N.; Franklin, K.C.; Oluseun, A.M.; Usman, M.; Abdulmumuni, B.; James, O.Y.; Ofokansi, K.C.; Anthony, A.A.; et al. Mucin-Grafted Polyethylene Glycol Microparticles Enable Oral Insulin Delivery for Improving Diabetic Treatment. *Appl. Sci.* **2020**, *10*, 2649. [CrossRef]
11. Stefa, S.; Lykaki, M.; Binas, V.; Pandis, P.K.; Stathopoulos, V.N.; Konsolakis, M. Hydrothermal Synthesis of ZnO–Doped Ceria Nanorods: Effect of ZnO Content on the Redox Properties and the CO Oxidation Performance. *Appl. Sci.* **2020**, *10*, 7605. [CrossRef]
12. Sánchez-Romate, X.F.; Jiménez-Suárez, A.; Sánchez, M.; Prolongo, S.G.; Güemes, A.; Ureña, A. Electrical Monitoring as a Novel Route to Understanding the Aging Mechanisms of Carbon Nanotube-Doped Adhesive Film Joints. *Appl. Sci.* **2020**, *10*, 2566. [CrossRef]
13. Malgorzata, R.; Magdalena, B. Decomposition of Flavonols in the Presence of Saliva. *Appl. Sci.* **2020**, *10*, 7511. [CrossRef]

Communication

Use of a Mixed Cationic-Reverse Phase Column for Analyzing Small Highly Polar Metabolic Markers in Biological Fluids for Multiclass LC-HRMS Method

Marco Roverso *, Iole Maria Di Gangi, Gabriella Favaro, Paolo Pastore and Sara Bogialli

Department of Chemical Sciences, University of Padua, Via Marzolo, 1, 35131 Padova, Italy; iole.digangi@gmail.com (I.M.D.G.); gabriella.favaro@unipd.it (G.F.); paolo.pastore@unipd.it (P.P.); sara.bogialli@unipd.it (S.B.)

* Correspondence: marco.roverso@unipd.it; Tel.: +39-049-8275181

Received: 17 September 2020; Accepted: 12 October 2020; Published: 14 October 2020

Abstract: The determination of small highly polar metabolites at low concentrations is challenging when reverse-phase (RP) chromatography is used for multiclass analysis. A mixed cationic-RP column coupled to high-resolution tandem MS (HR-MS/MS) was tested for highly polar compounds in biological fluids, i.e., trimethylamine N-oxide (TMAO) and the isobaric molecules beta-methylamino-L-alanine (BMAA) and 2,4-diaminobutyric acid (DAB). The efficient retention and separation of the above compounds were obtained with common and MS-friendly RP conditions, reaching high selectivity and sensitivity. The method was firstly assessed in plasma and urine, showing good linearity in the range 50–1000 µg/L and 500–10,000 µg/L for TMAO and both BMAA and DAB, respectively. Excellent precision (RDS < 3%) and good accuracies (71–85%) were observed except for BMAA in plasma, whose experimental conditions should be specifically optimized. Preliminary tests performed on compounds with biological relevance and a wider range of polarities proved the effectiveness of this chromatographic solution, allowing the simultaneous analysis of a larger panel of metabolites, from very small and polar compounds, like trimethylamine, to quite lipophilic molecules, such as corticosterone. The proposed LC-HRMS protocol is an excellent alternative to hydrophilic interaction liquid chromatography and ion-pairing RP chromatography, thus providing another friendly analytical tool for metabolomics.

Keywords: polar amino acids; mixed cationic-RP column; LC-HRMS; BMAA; TMAO

1. Introduction

The separation of small strongly polar compounds is challenging for reverse-phase (RP) chromatography. The weak interaction between the hydrophobic stationary phase and analytes usually leads to short retention times, matrix interferences and poor selectivity [1]. Derivatization steps and ion-pairing chromatography are possible solutions. However, derivatization procedures are usually time-consuming, and both reagents and secondary reactions may interfere with the analysis. Ion pairing is another useful trick but may result in low sensitivity with LC-MS. Hydrophilic interaction liquid chromatography (HILIC) is nowadays a quite widespread approach for the analysis of highly polar molecules such as amino acids, saccharides, nucleic acids, and phosphate-containing molecules [2]. In this case, very common drawbacks are peak distortion, large consumption of organic solvents, long equilibration times, and lack of solubility of some polar compounds in organic solvents. Mixed-mode liquid chromatography is an interesting alternative to the RP and HILIC stationary phases. These phases, firstly described in 1984 by Bischoff and McLaughlin [3], can be very effective for the simultaneous analysis of polar and non-polar compounds, due to the combined effects of different interaction mechanisms [4].

Beta-N-methylamino-L-alanine (BMAA) is a polar amino acid with a supposed neurotoxic activity. BMAA was described as a metabolite of several cyanobacterial strains, which can be bioaccumulated and biomagnified across the food chain [5]. Chronic exposure to BMAA has been associated with degenerative neurological conditions such as amyotrophic sclerosis, Parkinson's disease and dementia [6]. The risk assessment of BMAA is highly debated in the scientific literature since results are difficult to compare, because of the possible false-positives obtained by using different analytical techniques [7]. The main issue is the presence of several isomers of BMAA, such as 2,4-diaminobutyruc acid (DAB), so that the chromatographic separation and the correct quantification in biological and food samples remain challenging [8,9]. Most of the existing analytical procedures are based on the derivatization of the analytes with 6-aminoquinolonyl-N-hydroxysuccinimdyl prior to RP chromatography, or HILIC and ion-pairing RP separation without derivatization [5,8–11]. Solid-phase extraction performed using strong cation exchange sorbents is also reported for improving the sensitivity of the method and for samples clean up [12]. Recent works take advantage of tandem MS to improve the selectivity of the method and avoid false positives [13].

Trimethylamine N-oxide (TMAO) is a highly polar metabolite whose production is controlled by gut microbiota and liver enzymes. The production of TMAO is strictly related to the dietary consumption of L-carnitine and lectin rich food. Recently, the alteration of TMAO concentration in plasma was positively associated with an increased risk for cardiovascular diseases, heart failure, obesity, impaired glucose tolerance, diabetes, and colorectal cancer [14,15]. The quantitative analysis of TMAO is usually performed with RP or HILIC chromatography coupled to electrospray (ESI) mass spectrometry [16–18]. BMAA and TMAO, such as other polar metabolites, are potentially useful biomarkers to be monitored in various biological fluids, but the experimental conditions conventionally used for RP mode can inhibit the simultaneous multiclass determination of very different compounds, which is a goal of the metabolomic approach.

In this preliminary research work, a new separation approach, based on the use of a mixed cation-RP stationary phase, is evaluated to obtain the best retention and selectivity in the direct analysis of highly polar compounds such as TMAO, BMAA, and DAB. The analytical method is based on LC coupled to high-resolution tandem mass spectrometry (LC-HR-MS/MS) and applied to various biological matrices, such as plasma and urine, for clinical applications. Furthermore, the present study outlines future applications of this protocol to a wider panel of polar and non-polar analytes in the framework of metabolomic investigations.

2. Materials and Methods

2.1. Chemicals

Analytical grade BMAA, DAB, TMAO, deuterated trimethylamine N-oxide (D9, 98%) (TMAO-d9), trimethylamine, kynurenic acid, dopamine, homocysteine, carbidopa, picolinic acid, L-DOPA, 3-hydroxykynurenine, corticosterone and formic acid (FA) were purchased from Sigma-Aldrich Italy (Milan, Italy). LC-MS solvent grade acetonitrile was purchased from Carlo Erba Reagents (Milano, Italy). Ultrapure-grade water was produced by a Pure-Lab Option Q apparatus (Elga Lab Water, High Wycombe, UK). Standard stock solutions of the analytes under study were prepared in water/acetonitrile 50:50 at 1000 mg/L and stored at −20 °C until use. A mixed working standard solution was obtained by suitable dilutions in mobile phases. TMAO-d9 was used as an internal standard (IS).

2.2. Instrumentation

The LC-MS/MS system was an Ultimate 3000 UHPLC coupled to a Q-Exactive hybrid quadrupole-Orbitrap mass spectrometer (Thermo Fisher Scientific, Waltham, MA, USA).

A Luna HILIC (100 mm × 2.0 mm i.d. × 3 µm, Phenomenex, Bologna, Italy) and an Acclaim Mixed-Mode WCX-1 (150 mm × 2.1 mm i.d. × 3 µm, Thermo Fisher Scientific, Waltham, MA, USA) were used as columns thermostated at 15 °C. The elution was performed in a gradient mode (0–6 min 0% B,

6–15 min 0% B to 100% B, 15–20 min 100% B, equilibration time 10 min) at a flow rate of 0.25 mL/min, using water (eluent A for Acclaim, B for HILIC) and acetonitrile (eluent B for Acclaim, A for HILIC) both 10 mM FA. This flow rate was a compromise between the chromatographic selectivity and the ESI-MS responses. The injection volume was 5 μL. The MS conditions were the following: electrospray (ESI) ionization in positive mode, resolution 35,000 in MS and 17,500 in MS/MS (at *m/z* 200), AGC target 3×10^6 and 2×10^5 in MS and tandem MS, respectively; max injection time 200 ms, scan range 50–750 Da, isolation window 4.0 *m/z*, normalized collision energy 35 in HCD mode. The capillary voltage was 3.5 kV, the capillary temperature was 320 °C, auxiliary gas, and sheath gas was nitrogen at 40 and 20 a.u., respectively, while sweep gas was not used. Calibration was performed with Pierce™ ESI Positive Ion Calibration Solution (Thermo Fisher Scientific, Waltham, MA, USA). The MS data were analyzed with the Xcalibur 4.0™ software (Thermo Fisher Scientific, Waltham, MA, USA).

2.3. Sample Preparation and Calibrations

A five-point external calibration curve was prepared from stock solutions of TMAO, BMAA and DAB. The concentrations of BMAA and DAB in the five solutions were 50, 100, 250, 750 and 1000 μg/L, while TMAO was present at 5, 10, 25, 75 and 100 μg/L. IS was added in all the calibration solutions to a final concentration of 250 μg/L. The peak area ratio (A/A_{IS}) related to the selected fragment ions acquired in tandem MS mode for each compound was plotted against concentrations for calibration purposes. Linearity was assessed using the least-squares regression.

Mouse plasma and urine samples were collected as previously reported [19]. For the matrix-matched calibration, plasma samples were pooled from ten different animals and split into 100 μL aliquots. Five aliquots were spiked with both BMAA and DAB at concentrations of 500, 1000, 2500, 7500 and 10,000 μg/L and TMAO at concentrations of 50, 100, 250, 750 and 1000 μg/L, respectively; one aliquot was treated as blank. The basal concentration of TMAO, which is a metabolite physiologically present in plasma and urine, was preliminarily estimated by the standard addition method. Plasma aliquots were extracted with 400 μL of ice-cold acetonitrile added with 10 mM FA and 625 μg/L of IS, vortexed and centrifuged at 14,000× *g* and 4 °C for 10 min. 100 μL of the supernatant was further diluted 1:1 with water 10 mM FA and injected for the analysis. The same protocol was used for urine samples. The selected analytes were analyzed in their free form, i.e., not bound to proteins, which were removed during the sample preparation.

3. Results and Discussion

The chromatographic separation of BMAA, DAB and TMAO is hindered by poor retention in RP mode, and HILIC conditions have been used to overcome these problems. Figure 1A shows the extracted ion chromatogram (EIC, mass accuracy 10 ppm) of $[M+H]^+$ precursor ions at *m/z* 119.0815, 76.0757 and 85.1322 for isobaric BMAA and DAB, TMAO and TMAO-d9, respectively, obtained from the analysis of a working standard solution at 2500 μg/L in HILIC mode. Although the isobaric BMAA and DAB are greatly retained by this stationary phase (rt = 7.03 and 7.17 min, respectively), their separation is not effective. Conversely, TMAO and TMAO-d9 are not retained and are detectable at the column dead time. Although several attempts were made to increase the efficiency of the BMAA and DAB separation and the TMAO retention by varying both additive type and concentration (in particular, by changing FA concentration from 10 mM to 25 mM and by substituting FA with ammonium acetate, from 1 mM to 10 mM in both water and acetonitrile), no significant improvements were obtained. The effective separation of TMAO, and/or BMAA and DAB using HILIC chromatography are anyway reported in the literature, but the proposed methods are based on stationary phases that are quite different from the one herein described, e.g., ZIC-HILIC [8], ethylene-bridged hybrid (BEH) particles [12] or amide-based stationary phase [17].

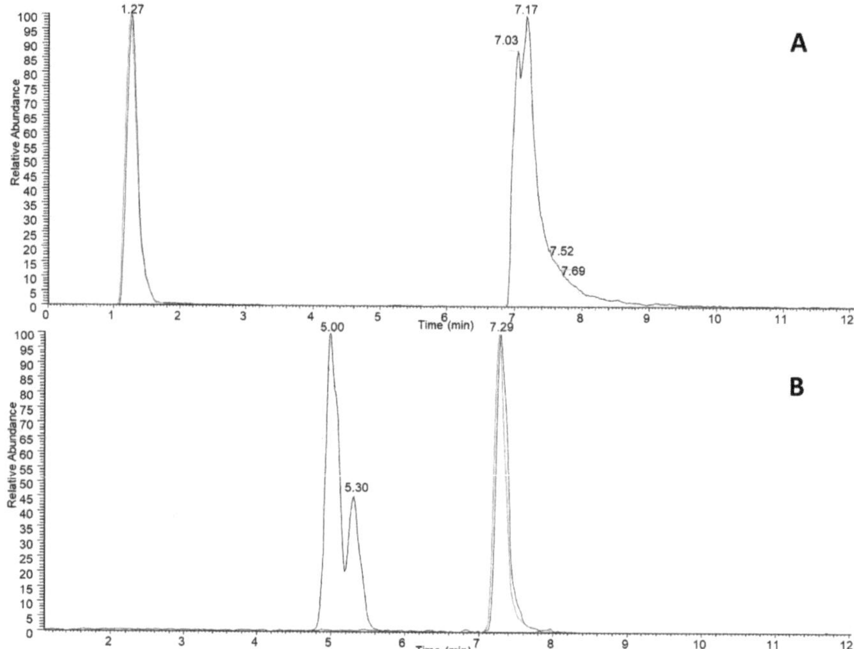

Figure 1. (**A**) Overlapped EIC chromatograms (accuracy 10 ppm) for BMAA ($[M+H]^+$, *m/z* 119.0815, black, rt = 7.03 min), DAB ($[M+H]^+$, *m/z* 119.0815, black, rt = 7.17 min), TMAO ($[M+H]^+$, *m/z* 76.0757.0815, red, rt = 1.27 min) and TMAO-d9 (IS) ($[M+H]^+$, *m/z* 85.1322, green, rt = 1.27 min) obtained with Luna HILIC. (**B**) Overlapped EIC chromatograms (accuracy 10 ppm) for BMAA ($[M+H]^+$, *m/z* 119.0815, black, rt = 5.00 min), DAB ($[M+H]^+$, *m/z* 119.0815, black, rt = 5.30 min), TMAO ($[M+H]^+$, *m/z* 76.0757.0815, red, rt = 7.32 min) and TMAO-d9 (IS) ($[M+H]^+$, *m/z* 85.1322, green, rt = 7.29 min) obtained with and Acclaim Mixed-Mode WCX-1. The concentration is 2500 μg/L for each analyte.

Better results, in terms of peaks separation and retention, were obtained with a mixed cationic-RP column in reverse-phase conditions, as reported in Figure 1B. In this case, the isobars BMAA (rt = 5.00 min) and DAB (rt = 5.30 min) were retained and satisfactorily separated. Even for TMAO and TMAO-d9 (rt = 7.29 min) retention was considerably improved.

Greater sensitivity and selectivity can be achieved by acquiring data in high-resolution MS/MS mode, in particular for BMAA and DAB. The comparison between the MS/MS spectra of BMAA and DAB (Figure 2, panel A), shows the presence of two significantly different fragment ions for DAB and BMAA: signal at *m/z* 88.0399 was related to the loss of substituted amino moiety of BMAA, and signal a *m/z* 101.0714 due to the loss of the terminal NH_3 of DAB. Once selected the precursor ions, the EIC (mass accuracy: 10 ppm) related to the fragment ions at *m/z* 58.0880, *m/z* 68.1302, *m/z* 101.0714 and *m/z* 88.0399 for TMAO, TMAO-d9, DAB and BMAA, respectively, were acquired in parallel reaction monitoring mode (Figure 2, panel B), thus improving the MS sensitivity and the selectivity obtained by the chromatographic separation. The obtained separation is anyway necessary when lower selective detection systems, such as LC coupled to single-stage or low-resolution MS are used. In the last cases, the shared fragment ions can produce inaccuracies or false-positive results.

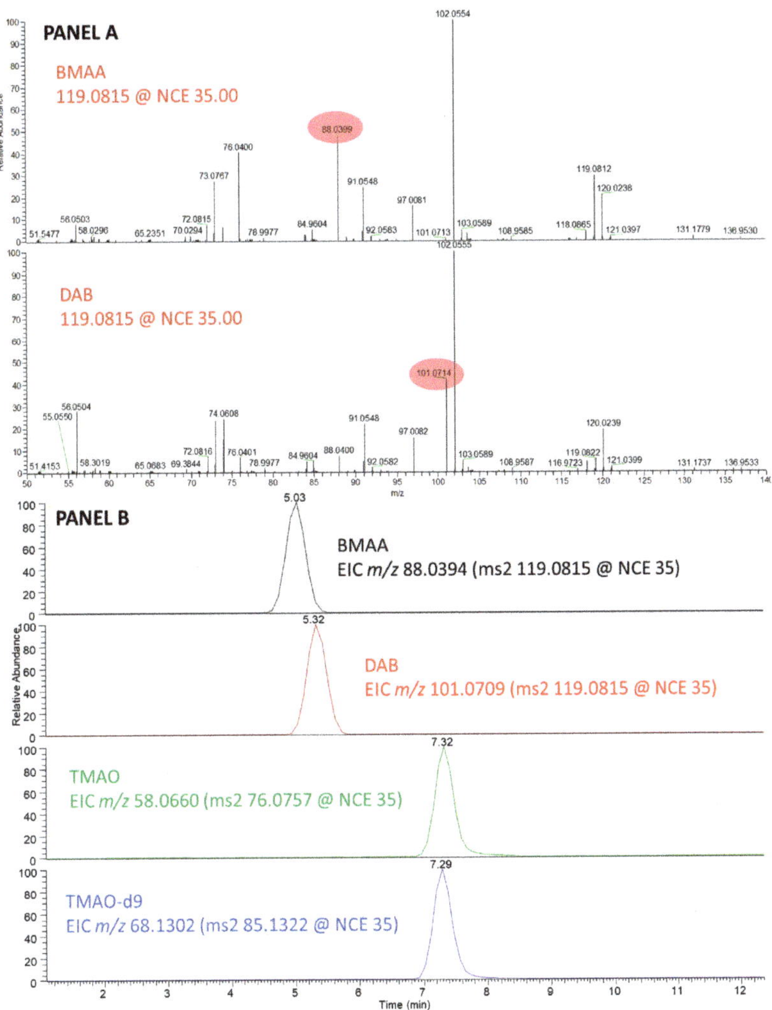

Figure 2. *Cont.*

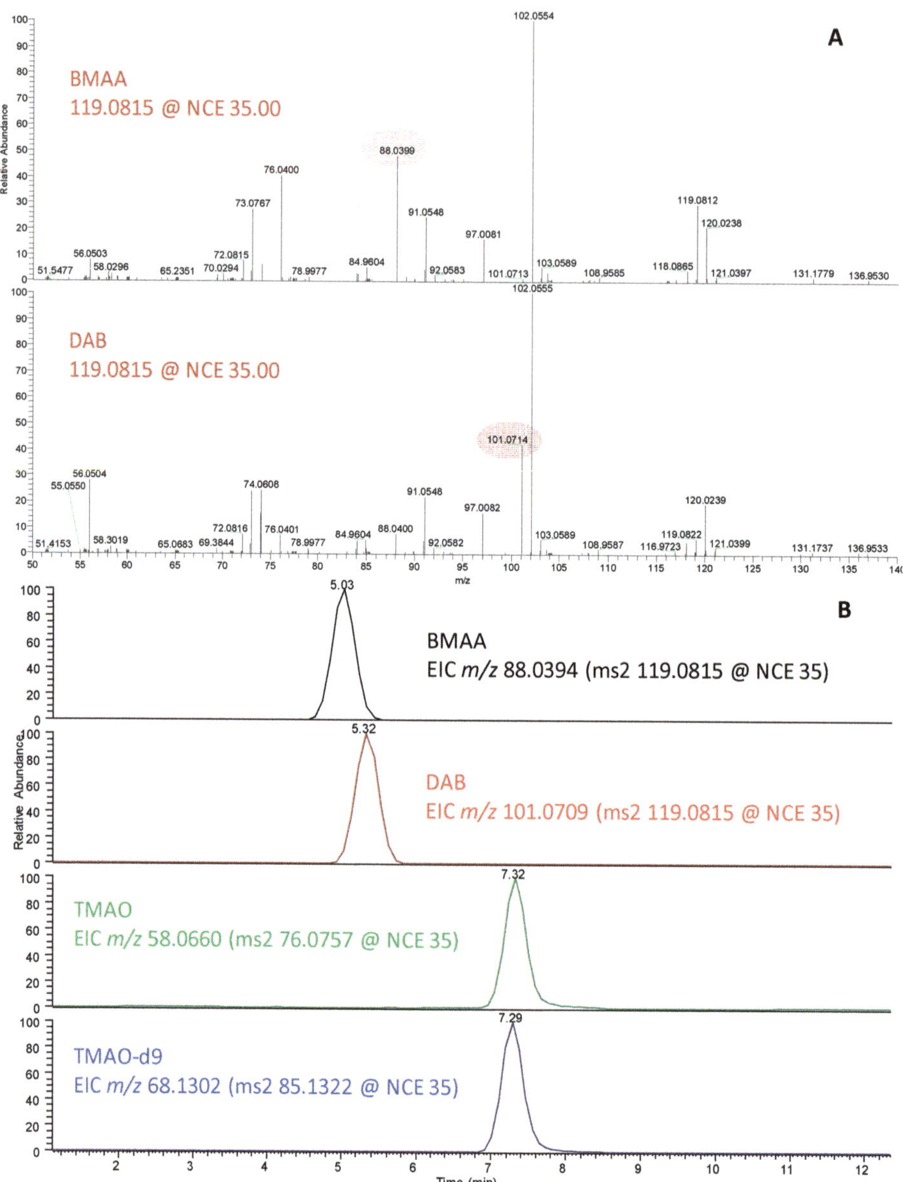

Figure 2. (**A**) Comparison between the MS/MS spectra of BMAA and DAB, both obtained in HCD mode with a normalized collision energy (NCE) of 35. (**B**) EIC MS/MS chromatograms (accuracy: 10 ppm) obtained with Acclaim Mixed-Mode WCX-1: BMAA: EIC for *m/z* 88.0394 (parent ion: *m/z* 119.0815, NCE:35); DAB: EIC for *m/z* 101.0709 (parent ion: *m/z* 119.0815, NCE:35); TMAO: EIC for *m/z* 58.0660 (parent ion: *m/z* 85.1322, NCE:35); TMAO-d9 (IS): EIC for *m/z* 68.1302 (parent ion: *m/z* 85.1322, NCE:35). The concentration is 2500 μg/L for each analyte.

The method linearity, precision and accuracy, the last intended as the combined contribution of matrix effects and recoveries, were evaluated in plasma and urine to perform a preliminary evaluation of the performances. Matrix-matched calibration curves were performed in triplicate and were obtained by spiking blank samples at the same nominal concentrations of the external calibration curves. A very good linearity ($R^2 > 0.99$) was observed for all the selected analytes in water, plasma and urine matrices. Precision, evaluated from the standard deviations of the regression slopes, was excellent and showed relative standard deviations (RSD%) < 3% for all the matrices taken into consideration. Matrix effects and recoveries were evaluated from the percent slope ratio of the matrix-matched and external standard calibration curves. Values within the range 70–110% are generally acceptable, as strong matrix effects and poor recoveries can be excluded. Considering the excellent precision of the method, it is possible to assume good recoveries and a limited matrix effect for BMAA and DAB in urine, as the values obtained for the percent slope ratio were 71% and 85%, respectively. Results obtained for plasma are acceptable for DAB (73%) and quite good for TMAO (81%). Such method performances were already proved to be reliable for the quantification of TMAO in mouse plasma (RSD of the IS < 15% for N = 77 samples) [19]. Nevertheless, the combined matrix effect, in terms of ESI signals variation and recovery, was not acceptable in plasma for BMAA, as values are lower than 50%. In this last case, probably due to the interfering compounds co-extracted from this complex matrix, it will be necessary to increase the dilution of the sample before analysis or modify the extraction procedure or the chromatographic conditions. The choice of a suitable IS could anyway improve its accuracy. LODs, assessed from the lowest point of the matrix-matched calibration curve, and corresponding to an S/N value of 3, were estimated to be 10 μg/L for BMAA and DAB and 2 μg/L for TMAO.

Preliminary quantitative data in plasma and urine samples were obtained only for TMAO, which was present as an average ($n = 5$) basal concentration of 121 ± 8 μg/L in plasma and 193 ± 9 μg/L in urine, respectively. BMAA and DAB were always lower than LODs in both plasma and urine samples, but the collection of specimens suspected to be positively correlated to BMAA has to be specifically planned, and it is beyond the focus of this work.

Further tests are in progress in our laboratory using the chromatographic method herein reported in order to expand the panel of analytes potentially quantifiable using this mixed cationic-RP column. The novel set of analytes included: other isomers of BMAA, such as beta-amino-*N*-methylalanine and *N*-(2aminoethyl) glycine; levodopa, carbidopa and dopamine, which are compounds involved in the Parkinson's disease, and may be useful for the assessment of the possible adverse effects of BMAA; trimethylamine, short-chain fatty acids, e.g., butyric, isobutyric, valeric, isovaleric, hexanoic and acetic acid are key compounds, together with TMAO, linked to the gut microbiota, whose alteration was recently associated with the development of type 2 diabetes and obesity [20]; metabolites such as picolinic and nicotinic acid, tryptophan, kynurenic acid, 3-hydroxykinurenine were selected as representative of other specific metabolic pathways, e.g., the tryptophan metabolism. Corticosterone, an important intermediate for the synthesis of glucocorticoid hormones in humans, was selected in order to evaluate the chromatographic interactions of the mixed-RP stationary phase with lipophilic substances. Preliminary results regarding the chromatographic separation of some of the selected analytes obtained by the Acclaim Mixed-Mode WCX-1 column are reported in Figure 3. As shown, this column was suitable for retaining compounds with a wide range of polarities, from trimethylamine to corticosterone.

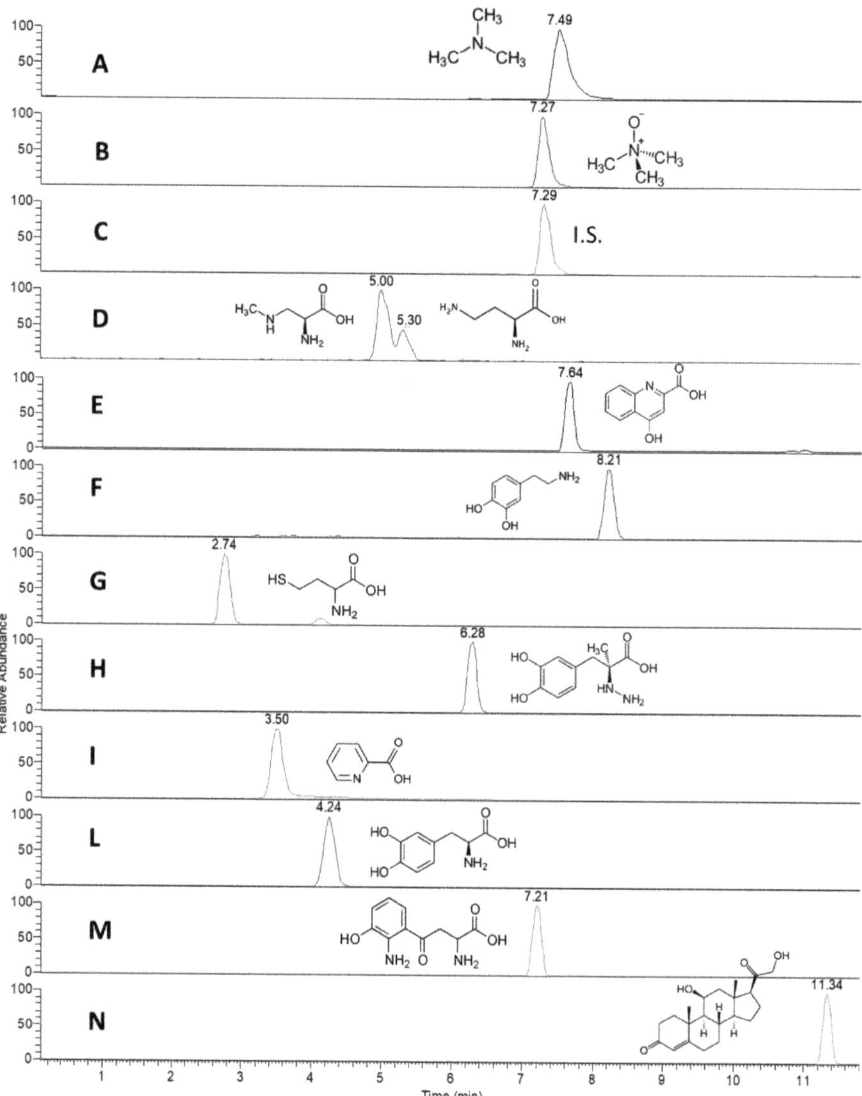

Figure 3. EIC chromatograms obtained with an Acclaim Mixed-Mode WCX-1 in ESI(+)-HRMS. (**A**) Trimethylamine, $[M+H]^+$, EIC at *m/z* 60.0808. (**B**) TMAO, $[M+H]^+$, EIC at *m/z* 76.0757. (**C**) TMAO-d9 (IS), $[M+H]^+$, EIC at *m/z* 85.1322. (**D**) BMAA (left peak) and DAB (right peak), $[M+H]^+$, EIC at *m/z* 119.0815. (**E**) Kynurenic acid, $[M+H]^+$, EIC at *m/z* 190.0499. (**F**) Dopamine, $[M+H]^+$, EIC at *m/z* 154.0863. (**G**) Homocysteine, $[M+H]^+$, EIC at *m/z* 136.0427. (**H**) Carbidopa, $[M+H]^+$, EIC at *m/z* 227.1026. (**I**) Picolinic acid, $[M+H]^+$, EIC at *m/z* 124.0393. (**L**) L-DOPA, $[M+H]^+$, EIC at *m/z* 198.0761. (**M**) 3-hydroxykynurenine, $[M+H]^+$, EIC at *m/z* 225.0870. (**N**) Corticosterone, $[M+H]^+$, EIC at *m/z* 347.2217. Mass accuracy: 10 ppm. The concentration is 2500 µg/L for each analyte.

4. Conclusions

In this work, the effective chromatographic retention of selected highly polar metabolites was carried out by using a mixed cationic-RP column, simultaneously obtaining an efficient separation of the isobaric BMAA and DAB without derivatization and ion pairing. The selectivity of the method was increased by HR tandem MS, avoiding the contribution of the partial co-eluted peaks. A preliminary evaluation of the method performances showed good linearity, acceptable recoveries and matrix effects for all the analytes in urine, and DAB and TMAO in plasma. The full validation of the method, including the assessment of the LOD, LOQ, repeatability and reproducibility, is in progress. Further evaluation of the column retention and selectivity started on a larger panel of analytes with different chemical properties and related to the metabolism of tryptophan, the short-chain fatty acids, and other isomers of BMAA and molecules related to the Parkinson disease. The versatility of this alternative chromatographic method is of particular interest in the field of metabolomics, where it is essential to analyze simultaneously various classes of molecules with very different chemical properties, in terms of polarity and molecular weight.

Author Contributions: Conceptualization, M.R., S.B. and I.M.D.G.; methodology, M.R., S.B. and I.M.D.G.; validation, M.R. and I.M.D.G.; formal analysis, M.R. and I.M.D.G.; investigation, M.R., S.B. and I.M.D.G.; resources, S.B. and P.P.; data curation, M.R., G.F. and I.M.D.G.; writing—original draft preparation, M.R., S.B. and G.F.; writing—review and editing, M.R., S.B. and P.P.; visualization, S.B. and P.P.; supervision, S.B.; funding acquisition, S.B. and P.P. All authors have read and agreed to the published version of the manuscript.

Funding: This research received no external funding.

Acknowledgments: The authors are grateful to the research group of Gian Paolo Fadini (University of Padua) for collaboration with the TMAO project.

Conflicts of Interest: The authors declare they have no conflicts of interest.

References

1. Rojo, D.; Barbas, C.; Ruperez, F.J. LC-MS metabolomics of polar compounds. *Bioanalysis* **2012**, *4*, 1235–1243. [CrossRef] [PubMed]
2. Jandera, P. Stationary and mobile phases in hydrophilic interaction chromatography: A review. *Anal. Chim. Acta* **2011**, *692*, 1–25. [CrossRef] [PubMed]
3. Bischoff, R.; McLaughlin, L.W. Nucleic acid resolution by mixed-mode chromatography. *J. Chromatogr. A* **1984**, *296*, 329–337. [CrossRef]
4. Lemasson, E.; Richer, Y.; Bertin, S.; Hennig, P.; West, C. Characterization of retention mechanisms in mixed-mode HPLC with a bimodal reversed-phase/cation-exchange stationary phase. *Chromatographia* **2018**, *81*, 387–399. [CrossRef]
5. Manolidi, K.; Triantis, T.M.; Kaloudis, T.; Hiskia, A. Neurotoxin BMAA and its isomeric amino acids in cyanobacteria and cyanobacteria-based food supplements. *J. Hazard. Mater.* **2019**, *365*, 346–365. [CrossRef] [PubMed]
6. Kurland, L.T. Amyotrophic lateral sclerosis and parkinson's disease complex on Guam linked to an environmental neurotoxin. *Trends Neurosci.* **1988**, *11*, 51–54. [CrossRef]
7. Faassen, E.J. Presence of the neurotoxin BMAA in aquatic ecosystems: What do we really know? *Toxins* **2014**, *6*, 1109–1138. [CrossRef] [PubMed]
8. Faassen, E.J.; Antoniou, M.G.; Beekman-Lukassen, W.; Blahova, L.; Chernova, E.; Christophoridis, C.; Combes, A.; Edwards, C.; Fastner, J.; Harmsen, J.; et al. A collaborative evaluation of LC-MS/MS based methods for BMAA analysis: Soluble bound BMAA found to be an important fraction. *Mar. Drugs* **2016**, *14*, 45. [CrossRef] [PubMed]
9. Banack, S.A.; Downing, T.G.; Spácil, Z.; Purdie, E.L.; Metcalf, J.S.; Downing, S.; Esterhuizen, M.; Codd, G.A.; Cox, P.A. Distinguishing the cyanobacterial neurotoxin β-*N*-methylamino-l-alanine (BMAA) from its structural isomer 2,4-diaminobutyric acid (2,4-DAB). *Toxicon* **2010**, *56*, 868–879. [CrossRef]
10. Jiang, L.; Aigret, B.; De Borggraeve, W.M.; Spacil, Z.; Ilag, L.L. Selective LC-MS/MS method for the identification of BMAA from its isomers in biological samples. *Anal. Bioanal. Chem.* **2012**, *403*, 1719–1730. [CrossRef] [PubMed]

11. Bhandari, D.; Bowman, B.A.; Patel, A.B.; Chambers, D.M.; De Jesús, V.R.; Blount, B.C. UPLC-ESI-MS/MS method for the quantitative measurement of aliphatic diamines, trimethylamine *N*-oxide, and β-methylamino-L-alanine in human urine. *J. Chromatogr. B Anal. Technol. Biomed. Life Sci.* **2018**, *1083*, 86–92. [CrossRef] [PubMed]
12. Yan, B.; Liu, Z.; Huang, R.; Xu, Y.; Liu, D.; Lin, T.F.; Cui, F. Optimization of the determination method for dissolved cyanobacterial toxin BMAA in natural water. *Anal. Chem.* **2017**, *89*, 10991–10998. [CrossRef] [PubMed]
13. Roy-Lachapelle, A.; Solliec, M.; Sauvé, S. Determination of BMAA and three alkaloid cyanotoxins in lake water using dansyl chloride derivatization and high-resolution mass spectrometry. *Anal. Bioanal. Chem.* **2015**, *407*, 5487–5501. [CrossRef] [PubMed]
14. Janeiro, M.H.; Ramírez, M.J.; Milagro, F.I.; Martínez, J.A.; Solas, M. Implication of trimethylamine n-oxide (TMAO) in disease: Potential biomarker or new therapeutic target. *Nutrients* **2018**, *10*, 1398. [CrossRef] [PubMed]
15. Subramaniam, S.; Fletcher, C. Trimethylamine N-oxide: Breathe new life. *Br. J. Pharmacol.* **2018**, *175*, 1344–1353. [CrossRef] [PubMed]
16. Zhao, X.; Zeisel, S.H.; Zhang, S. Rapid LC-MRM-MS assay for simultaneous quantification of choline, betaine, trimethylamine, trimethylamine *N*-oxide, and creatinine in human plasma and urine. *Electrophoresis* **2015**, *36*, 2207–2214. [CrossRef] [PubMed]
17. Liu, J.; Zhao, M.; Zhou, J.; Liu, C.; Zheng, L.; Yin, Y. Simultaneous targeted analysis of trimethylamine-*N*-oxide, choline, betaine, and carnitine by high performance liquid chromatography tandem mass spectrometry. *J. Chromatogr. B Anal. Technol. Biomed. Life Sci.* **2016**, *1035*, 42–48. [CrossRef] [PubMed]
18. Wang, Z.; Levison, B.S.; Hazen, J.E.; Donahue, L.; Li, X.M.; Hazen, S.L. Measurement of trimethylamine-*N*-oxide by stable isotope dilution liquid chromatography tandem mass spectrometry. *Anal. Biochem.* **2014**, *455*, 35–40. [CrossRef] [PubMed]
19. Scattolini, V.; Dalla Costa, F.; Cignarella, A.; Vettore, M.; Di Gangi, I.M.; Bogialli, S.; Avogaro, A.; Fadini, G.P. Interplay between gut microbiota and p66Shc affects obesity-associated insulin resistance. *FASEB* **2018**, *32*, 4004–4015.
20. Sanna, S.; van Zuydam, N.R.; Mahajan, A.; Kurilshikov, A.; Vich Vila, A.; Võsa, U.; Mujagic, Z.; Masclee, A.A.M.; Jonkers, D.M.A.E.; Oosting, M.; et al. Causal relationships among the gut microbiome, short-chain fatty acids and metabolic diseases. *Nat. Genet.* **2019**, *51*, 600–605. [CrossRef] [PubMed]

Article

Determination of Non-Steroidal Anti-Inflammatory Drugs in Animal Urine Samples by Ultrasound Vortex-Assisted Dispersive Liquid–Liquid Microextraction and Gas Chromatography Coupled to Ion Trap-Mass Spectrometry

Pasquale Avino, Ivan Notardonato, Sergio Passarella and Mario Vincenzo Russo *

Department of Agricultural, Environmental and Food Sciences, University of Molise, via De Sanctis, I-86100 Campobasso, Italy; avino@unimol.it (P.A.); ivan.notardonato@unimol.it (I.N.); sergio.passarella@studenti.unimol.it (S.P.)

* Correspondence: mvrusso@unimol.it; Tel.: +39-0874-404-717

Received: 26 May 2020; Accepted: 29 June 2020; Published: 6 August 2020

Featured Application: The paper would like to show an easy, rapid, and affordable protocol to be used for determining four non-steroidal anti-inflammatory drugs (NSAIDs) (i.e., acetylsalicylic acid, ibuprofen, naproxen, and ketoprofen) in urine samples at trace levels. The method could be routinely used in several situations, from medicine and veterinary to doping issues.

Abstract: A low solvent consumption method for the determination of non-steroidal anti-inflammatory drugs (NSAIDs) in animal urine samples is studied. The NSAIDs were extracted with CH_2Cl_2 by the ultrasound vortex assisted dispersive liquid–liquid microextraction (USVA-DLLME) method from urine samples, previously treated with β-glucuronidase/acrylsulfatase. After centrifugation, the bottom phase of the chlorinated solvent was separated from the liquid matrix, dried with Na_2SO_4, and derivatized with N,O-bis(trimethylsilyl)trifluoroacetamide (BSTFA) + trimethylchlorosilane (TMCS) (99 + 1). After cooling at room temperature, the solution was concentrated under nitrogen flow, and 1 μL of solution was analyzed in gas chromatography/ion trap-mass spectrometry (GC-IT-MS). The enrichment factor was about 300–450 times and recoveries ranged from 94.1 to 101.2% with a relative standard deviation (RSD) of ≤4.1%. The USVA-DLLME process efficiency was not influenced by the characteristics of the real urine matrix; therefore, the analytical method characteristics were evaluated in the range 1–100 ng mL^{-1} ($R^2 \geq 0.9950$). The limits of detection (LODs) and limits of quantification (LOQs) were between 0.1 and 0.2 ng mL^{-1} with RSD ≤4.5% and between 4.1 and 4.7 ng mL^{-1} with RSD ≤3.5%, respectively, whereas inter- and intra-day precision was 3.8% and 4.5%, respectively. The proposed analytical method is reproducible, sensitive, and simple.

Keywords: non-steroidal anti-inflammatory drug (NSAID); urine; doping analysis; dispersive liquid–liquid microextraction (DLLME); gas chromatography mass spectrometry (GC-MS)

1. Introduction

Anti-inflammatory drugs, used for reducing inflammation, are of two types, i.e., cortisone-based and non-steroidal anti-inflammatory drugs (NSAIDs). The latter are, in all likelihood, the best known and most used category of anti-inflammatory drugs in therapy [1]. NSAIDs are a wide class of drugs showing anti-inflammatory, analgesic, and antipyretic action and include some of the best-known molecules used to fight pain [2]: ibuprofen, nimesulide, ketoprofen, naproxen, and diclofenac. They are able to stop the inflammation process by their mechanism of action, i.e., interfering with the synthesis of prostanoids; molecules that play a fundamental role in these processes [3]. To do this, the NSAIDs block one or more passages of the metabolism of arachidonic acid, which is the precursor of prostaglandins [4]. Further, NSAIDs can also be used as pain relievers and antipyretics [5,6].

NSAIDs are associated with a small increase in the risk of a heart attack, stroke, or heart failure [7]. However, even in this case, the real danger depends on the type of molecule taken, the duration of the treatment, and the doses taken. Short-term use can instead trigger less serious but sometimes serious adverse effects, such as ulcers, gastric bleeding, and kidney damage [8–10]. In addition, NSAIDs can trigger allergic reactions and interfere with the activity of antihypertensive drugs [11].

Furthermore, NSAIDs are commonly used in animal medicine in different inflammatory situations (e.g., for curing musculoskeletal problems in equines) [12–14]. On the other hand, these drugs are improperly used for masking inflammation and pain of an animal, especially before horse racing. NSAIDs are substances prohibited in horse competitions and are considered one of the main doping agents [15–18]. For instance, salicylic acid, a NSAID used for the treatment of pain and fever, has an allowed threshold of 750 $\mu g\ mL^{-1}$ in urine, or 6.5 $\mu g\ mL^{-1}$ in plasma, for equines [19].

NSAIDs are considered safe drugs, but acute overdose or chronic abuse can give serious toxic effects [20,21]. They are weak in acid (pK_a 3–5) and some of them show short half-lives (e.g., ibuprofen 2–3 h [22]), whereas others show long half-lives (e.g., phenylbutazone residual can also be detected after 24 h [23]). A screening procedure is necessary for detecting such drugs in urine samples. Different analytical methods are present in literature, mainly based on liquid–liquid extraction (LLE) or solid-phase extraction (SPE), followed by chromatographic methods (i.e., HPLC with fluorescence detector HPLC-FLD, HPLC-diode array detection (DAD), gas chromatography mass spectrometry (GC-MS), GC-MS/MS, UHPLC-MS/MS, capillary electrophoresis CE-DAD, and CE-MS) [20,24–33]. Further, a derivatization step is necessary before the GC-MS analysis [30,31,34].

Recently, Rezaee et al. introduced the dispersive liquid–liquid microextraction (DLLME) [35]. The extraction is based on the addition of both an immiscible solvent with higher density to the aqueous sample and a dispersant solvent for increasing the contact between the two immiscible solvents. For many years, researchers have deepened this method by applying it to different matrices [36–38], especially for avoiding (at least, for reducing) the use of highly toxic chloro-solvents [39]. In this way, several protocols based on ultrasound vortex assisted DLLME (USVA-DLLME) for determining toxic compounds in foodstuffs have been investigated and set up [40–44].

The aim of this study was to develop a simple method for the simultaneous screening and confirmation of four NSAIDs, i.e., acetylsalicylic acid (ASA), ibuprofen (IBP), naproxen (NAP), and ketoprofen (KPF), in animal urine samples. The entire procedure, not previously reported in literature, starts with the extraction procedure, i.e., the USVA-DLLME method, followed by the NSAID derivatization step with N,O-bis(trimethylsilyl)trifluoroacetamide (BSTFA)-trimethylchlorosilane (TMCS) to form the relative trimethylsilyl (TMS) derivates: gas chromatography coupled with an ion trap-mass spectrometry detector (GC-IT-MS) has allowed us to detect the NSAID residues in real samples.

2. Materials and Methods

2.1. Materials

Ethanol, C_2Cl_2, $CHCl_3$, $C_2H_4Cl_2$, $C_2H_2Cl_4$, and acetone were of pesticide grade (Carlo Erba, Milan, Italy), whereas NaCl, acetic acid, NaOH, HCl, and Anhydrous Na_2SO_4 were of analytical grade (Carlo Erba). Standards of acetylsalicylic acid, ibuprofen, naproxen, and ketoprofen were purchased as powder from Sigma–Aldrich (Milan, Italy), whereas anthracene, used as the internal standard (IS), was provided by LabService Analytical (Anzola Emilia, Bologna, Italy). Beta-glucuronidase/arylsulfatase and BSTFA-TMCS (99 + 1) solutions were given by Sigma–Aldrich.

The solutions (1 mg mL^{-1}) of each analyte, i.e., acetylsalicylic acid, ibuprofen, naproxen, and ketoprofen (Table 1), were prepared in acetone. These solutions were further diluted for preparing final working standard solutions for spiking both the blank solutions (simulated urine samples) and real samples.

Table 1. The non-steroidal anti-inflammatory drugs (NSAIDs) investigated in this paper, with their corresponding abbreviations, Chemical Abstracts Service (CAS) number, chemical structure, molecular weight (MW), target, and qualifier ions (selected ion monitoring (SIM), abundance 100%).

Compound [a]	# CAS	Formula	MW	Target Ion [b]	Qualifier Ion [b]	II Ion [b]
ASA	50-78-2	$C_9H_8O_4$	180.16	252 $[C_{12}H_{16}O_4Si]^+$	209 $[C_{10}H_{13}O_3Si]^+$	149 $[C_8H_5O_3]^+$
IBP	15687-27-1	$C_{13}H_{18}O_2$	206.29	278 $[C_{15}H_{26}O_2Si]^+$	160 $[C_{12}H_{16}]^+$	263 $[C_{15}H_{18}O_2Si]^+$
NAP	22204-53-1	$C_{14}H_{14}O_3$	230.26	302 $[C_{17}H_{22}O_3Si]^+$	185 $[C_{13}H_{13}O]^+$	243 $[C_{14}H_{12}O_3]^+$
KPF	22071-15-4	$C_{16}H_{14}O_3$	254.28	325 $[C_{19}H_{22}O_3Si]^+$	282 $[C_{15}H_{14}O_3Si]^+$	295 $[C_{17}H_{15}O_3Si]^+$

[a] Abbreviations: acetylsalicylic acid (ASA), ibuprofen (IBP), naproxen (NAP), ketoprofen (KPF); [b] target and qualifier ions of the trimethylsilyl derivates.

The anthracene solution (1 mg mL^{-1}) was prepared in ethanol, and by further dilution the working solution was obtained. NaOH 1 M, HCl 1 M, and CH_3COOH 1 M was used to adjust the pH of the blank, and real samples were prepared with ultrapure water (resistivity 18.2 MΩ cm^{-1}) and obtained by means of a Milli-Q purification system (Millipore, Bradford, MA, USA).

2.2. Sample Preparation

2.2.1. Preparation of Simulated Urine Samples

For simulating a urine sample, an aqueous solution containing the most present components was prepared as follows: urea 14 g L^{-1}, creatinine 0.4 g L^{-1}, uric acid 0.05 g L^{-1}, glucose 0.06 g L^{-1}, mono potassium phosphate 0.2 g L^{-1}, and sodium chloride 13 g L^{-1}.

2.2.2. Preparation of Animal Urine Samples

Animal urine samples were provided by small farm owners near Campobasso (Molise, Italy). Each sample was filtered through a 0.45 μm pore size cellulose acetate filter and buffered at pH 5 with few drops of acetic acid, with the addition of a few μL of NaOH 1 M. Before performing the extraction and derivatization procedures, the animal urine samples were subjected to enzymatic hydrolysis. With total of 9 mL of sample and 100 μL of β-glucuronidase/arylsulfatase [45], the IS (5 μL of anthracene, 60 ng μL^{-1}) were incubated overnight at 37 °C.

2.2.3. USVA-DLLME and Derivatization Procedure

The extraction procedure was performed as follows: the mixture of dispersive (1 mL of acetone) and extraction (250 μL of CH_2Cl_2) solvent was injected above the sample level of the solution previously kept at room temperature at pH 3 with a few μL of HCl [46]. The solution was subjected to vortex for 1 min and ultrasounds for 2 min: This occurrence was repeated three times, followed by centrifugation for 10 min at 4000 rpm at room temperature. The organic phase was withdrawn with a micro-syringe and placed in a vial with the addition of a few grains of anhydrous sodium sulphate. A total of 50 μL of BSTFA + TMCS (99 + 1, v + v) were added [47] and the vial was closed and heated up to 50 °C for 30 min. Afterwards, the vial was cooled at room temperature and the organic phase was concentrated to a final volume of 20–50 μL under a slight nitrogen flow and 1 μL were analyzed in GC-IT-MS.

2.3. GC-IT-MS Apparatus

Analysis and data acquisition were performed using a gas chromatograph Finnigan Trace GC Ultra, equipped with an ion trap mass-spectrometry detector Polaris Q (Thermo Fisher Scientific, Waltham, MA, USA), a programmed temperature vaporizer (PTV) injector, and a PC with a chromatography station Xcalibur 1.2.4 (Thermo Fisher Scientific).

A fused-silica capillary column with a chemically bonded phase (SE-54, 5% phenyl-95% dimethylpolysiloxane) was prepared in our laboratory [48–50] with the following characteristics: 30 m × 250 µm i.d.; N (theoretical plate number) 132,000 for *n*-dodecane at 90 °C; K′, capacity factor, 7.0; d_f, (film thickness) 0.246 µm; u_{opt} (optimum linear velocity of carrier gas, hydrogen) 39.5 cm s^{-1}; utilization of theoretical efficiency (UTE) 95%. A 1 µL sample was injected into the PTV injector in the splitless mode. A total of 10 s after, the injection the vaporizer was heated from 110 °C to 290 °C at 800 °C min^{-1}; the splitter valve was opened after 120 s (split ratio 1:50). The transfer line and ion source were held at 270 °C and 250 °C, respectively. Helium (IP 5.5) was used as a carrier gas at a flow rate of 10 mL min^{-1}. The oven temperature program was as follows: 100 °C for 60 s, 10 °C min^{-1} up to 290 °C, and held for 120 s. The IT/MS was operated in the electron ionization mode (70 eV), and the analytes were qualitatively identified in the full-scan mode (*m*/*z* 100–500) and quantified in the selected ion monitoring (SIM) mode (Table 1). The quantitative analysis was performed by calibration graphs of ratio $Area_{(NSAID)}/Area_{(IS)}$ plotted versus each NSAID concentration (ng mL^{-1}). All the samples were determined in triplicate.

3. Results and Discussion

For USVA-DLLME extraction of the four investigated NSAIDs from animal urine samples, several parameters that control the optimal extraction performance were investigated and optimized using the one variable at a time method. It should be highlighted that the entire analytical methodology has been studied by means of simulated urine samples, prepared according to what reported in Section 2.2.1 and after applied to real urine samples. Simultaneously, the use of β-glucuronidase was welcome because it increased the IBP detection [33].

3.1. Parameter Optimization

The parameter optimization was addressed to find out the best analytical conditions for achieving high recoveries and accurate and precise determinations of the NSAIDs in animal urine samples. In this way, extraction solvent and volume, sample pH, and NaCl effect were deeply investigated.

First, the study dedicated its attention on the choice of organic extraction solvent. This issue plays a key role in the extraction efficiency. Chlorinated solvents are generally used because they show characteristics (higher density than water, low solubility in water) appropriate to obtaining high extraction efficiency and worthy gas chromatographic performance. Following these considerations, our attention was focused to five solvents: dichloromethane (CH_2Cl_2; d = 1.3255 g mL^{-1}), chloroform ($CHCl_3$; d = 1.4788 g mL^{-1}), carbon tetrachloride (CCl_4; d = 1.5940 g mL^{-1}), 1,2-dichloroethane ($C_2H_4Cl_2$; d = 1.2454 g mL^{-1}), and 1,1,2,2-tetrachloroethane ($C_2H_2Cl_4$; 1.5953 g mL^{-1}). Table 2 reports the results of the performance of a 300 µL volume of each solvent on simulated urine samples spiked with 20 ng mL^{-1} of each NSAID: Dichloromethane shows the best recoveries, ranging between 94.6% and 98.5% for IBP, NAP, and KPF, respectively, and 82.5% for ASA with a relative standard deviation (RSD, %) below 3.0. The recoveries are calculated as the accuracy (IS added before the extraction) [51].

The extraction recovery, defined as the percentage of the total analyte (n_0), that was extracted to the sediment phase (n_{sed}) has been determined according to the formula reported in a previous paper [36]. Over the extraction solvent choice, another quite important parameter is its volume, used to achieve the highest recoveries. The strength of the DLLME regards an extraction solvent volume as low as possible for obtaining good performance. Leong and Huang [39] highlighted that an extraction solvent volume leads to a change in the sediment phase volume and therefore in the enrichment factors (EFs). For these reasons, the effect of different dichloromethane volumes (200, 250, 300 µL) were investigated (Table 3): a volume of 250 µL is sufficient to obtain good recoveries for all the NSAIDs, i.e., 94.2% for ASA, 100.1 for IBP, 99.8 for NAP, and 101.2 for KPF with RSDs ≤3.1.

Table 2. Effect of different extraction solvents on the NSAID recovery accuracy (%). The conditions were as follows: 9 mL of simulated urine samples spiked with NSAIDs (20 ng mL^{-1} of each), 1 mL of acetone, 300 µL of extraction solvent, and 5 µL of anthracene (I.S.; 60 ng μL^{-1}). In brackets are reported the relative standard deviations (RSDs, %); each analysis was in triplicate.

Compound	Accuracy (%)				
	CH_2Cl_2	$CHCl_3$	CCl_4	CH_2ClCH_2Cl	$CHCl_2CHCl_2$
ASA	82.5 (2.5)	79.2 (3.2)	74.1 (2.7)	76.1 (2.1)	74.2 (3.0)
IBP	94.6 (3.0)	81.7 (2.6)	82.3 (3.0)	83.2 (2.7)	85.2 (2.9)
NAP	96.8 (2.9)	83.4 (3.1)	84.5 (3.1)	85.4 (3.0)	86.3 (3.1)
KPF	98.5 (2.8)	85.2 (3.0)	86.5 (2.9)	89.1 (2.9)	86.2 (2.9)

Table 3. Effect of different volumes of CH_2Cl_2 on the NSAID recoveries (%). The conditions were as follows: 9 mL of simulated urine samples spiked with NSAIDs (20 ng mL^{-1} of each), 1 mL of acetone, different volumes of CH_2Cl_2 as extraction solvent, and 5 µL of IS (60 ng μL^{-1}). In brackets are reported the RSDs (%); each analysis was in triplicate.

Compound	Recovery (%)		
	200 µL	250 µL	300 µL
ASA	82.3 (3.1)	94.2 (2.8)	82.5 (2.5)
IBP	95.2 (2.7)	100.1 (3.1)	94.6 (3.0)
NAP	96.8 (2.9)	99.8 (2.5)	96.8 (2.9)
KPF	95.1 (3.2)	101.2 (3.0)	98.5 (2.8)

Another parameter influencing the extraction is the pH of the solution. In fact, it should be remembered that NSAIDs are weak acids. Particularly, ASA shows a pK_a of 3.5 [52], IBP of 5.3 [53], NAP of 4.14, and KPF of 4.45 [54]. Solutions of simulated urine samples at different pH were tested for studying the best acidic conditions. Table 4 evidences that the best recoveries and RSDs are obtained at pH 3: in fact, they range between 93.5 and 100.1% and between 3.4 and 4%, respectively.

Table 4. The effect of pH on the NSAID recoveries (%). The conditions were as follows: 9 mL of simulated urine samples spiked with NSAIDs (20 ng mL^{-1} of each), 1 mL of acetone, 250 µL of CH_2Cl_2, and 5 µL of IS (60 ng μL^{-1}). In brackets are reported the RSDs (%); each analysis was in triplicate.

Compound	Recovery (%)				
	pH 2	pH 3	pH 4	pH 5	pH 6
ASA	94.1 (5.6)	93.5 (3.5)	86.1 (4.0)	82.0 (4.2)	70.0 (4.3)
IBP	100.2 (3.3)	99.7 (3.4)	97.2 (3.9)	92.5 (4.1)	82.2 (4.1)
NAP	99.8 (3.8)	100.1 (3.9)	96.8 (4.1)	91.0 (4.0)	81.5 (4.0)
KPF	101.0 (3.7)	99.7(4.0)	97.2 (4.2)	90.2 (3.9)	84.7 (4.3)

Finally, the effect of different NaCl quantities on the NSAID recoveries was evaluated. Table 5 shows that the salt decreased the NSAID solubility (salting out) below and above 13 g L^{-1} concentration. Further, the decision to perform the whole study at NaCl concentration of 13 g L^{-1} was essentially due to two considerations: (1) this concentration was the average of those reported in the real urine samples, which was between 10 and 16 g L^{-1} of NaCl [55]; (2) the percentage NSAID recoveries obtained and reported in Table 5 were very similar to each other for NaCl concentrations between 10 and 15 g L^{-1}.

Finally, it should be highlighted that two other interesting parameters, such as vortex time and ultrasonication time, were extensively studied in previous papers by this group [41–44].

Table 5. Effect of different NaCl amounts on the NSAID recoveries (%). The conditions were as follows: 9 mL of simulated urine samples spiked with NSAIDs (20 ng mL^{-1} of each), pH 3, 1 mL of acetone, 250 μL of CH_2Cl_2, and 5 μL of IS (60 ng μL^{-1}). In brackets are reported the RSDs (%); each analysis was in triplicate.

Compound	Recovery (%)					
	5 g L^{-1}	10 g L^{-1}	13 g L^{-1}	15 g L^{-1}	20 g L^{-1}	25 g L^{-1}
ASA	80.1 (4.0)	89.5 (3.9)	93.2 (3.6)	93.8 (3.5)	85.2 (4.1)	83.5(3.9)
IBP	92.5 (3.7)	97.2 (3.5)	99.4 (3.5)	99.5 (3.7)	92.5 (3.8)	90.1 (4.0)
NAP	94.2 (4.1)	97.1 (4.2)	100.6 (3.9)	101.0 (4.0)	91.6 (4.2)	89.2 (4.1)
KPF	95.2 (3.9)	97.5 (4.0)	99.8 (3.9)	100.1 (7.8)	93.2 (4.0)	90.2 (3.8)

3.2. GC-IT-MS Method Validation

Using optimized parameters, all the analytical data were investigated. Table 6 shows the correlation coefficients (R^2) in the range 1–100 μg L^{-1}, along with the limits of detection (LODs) and limits of quantification (LOQs), repeatability (as intra-day precision) and reproducibility (as inter-day precision), and EFs of each NSAID considered. LODs and LOQs were determined according to Knoll's definition [56,57], i.e., an analyte concentration that produces a chromatographic peak equal to three times (LOD) and seven times (LOQ) the standard deviation of the baseline noise. All the compounds show a good linearity in the investigated range (≥0.995) and LODs and LOQs between 0.1–0.2 μg L^{-1} and 4.1–4.7 μg L^{-1}, respectively, with high intra- and inter-day precision (≤3.8 and ≤4.5, respectively). The EFs, defined as the ratio between the analyte concentration in the sediment phase (C_{sed}) and the initial analyte concentration (C_0) in the sample (EF = C_{sed}/C_0) [35], were also studied, ranging between 350–450.

Table 6. Correlation coefficients (R^2) calculated in the range 1–100 μg L^{-1}, limit of detection (LOD; μg L^{-1}) and limit of quantification (LOQ; μg L^{-1}) and inter- and intra-day precision (expressed as RSD, %) of each NSAID determined by GC-IT-MS.

Compound	R^2	LOD	LOQ	Intra-day	Inter-day	EF
ASA	0.9950	0.2	4.1	3.8	4.5	350
IBP	0.9972	0.1	4.7	3.2	4.0	450
NAP	0.9987	0.1	4.7	3.5	4.3	385
KPF	0.9981	0.1	4.5	3.3	4.2	412

Finally, for a complete analytical methodology evaluation, the recoveries have been studied in the investigated matrices, i.e., animal urine, at two different spiked NSAID concentrations (20 ng mL^{-1} and 50 ng mL^{-1}). Table 7 shows these data: recoveries in animal urine samples between 93.8 and 102 with RSDs ≤3.2.

Table 7. Average NSAID recoveries (%) obtained at different spiking concentrations on real urine samples. In brackets are reported the RSDs (%); each analysis was in triplicate.

Compound	Recovery (%)	
	Animal Urine [a]	
	20 ng mL^{-1}	50 ng mL^{-1}
ASA	93.8 (3.2)	94.3 (2.9)
IBP	99.8 (3.0)	100.2 (3.0)
NAP	101.0 (3.2)	102.0 (3.1)
KPF	100.2 (2.9)	99.8 (3.2)

[a] Goat urine sample.

Finally, Table 8 shows a comparison among different methods present in literature [58–63] for analyzing NSAIDs. The extraction methods were different: three papers were based on hollow-fiber liquid microextraction [59,60,62], whereas two papers were on rotating disk sorptive [63] and liquid–liquid extraction [61]. According to the analytical techniques, three studies used HPLC with ultraviolet (UV) [58,61,63] and one the diode array detection (DAD) [59], one used the ultra-performance liquid chromatography coupled with tandem mass spectrometry (UPLC-MS/MS) [62], and one used GC equipped with a flame ionization detector (FID) [60]. Looking at the comparison among the different studies with parameters developed in this study, the main advantages regard LODs and LOQs, recoveries, and RSDs, whereas EFs are good except those reported by Payan et al. [59]. On the other hand, the whole procedure can be routinely applied and does not require particular technology, such as the use of rotating disks or hollow fiber.

Table 8. Linear range (LR, µg mL^{-1}), volume of extraction solvent (Ex. Solv., µL), extraction time (Ex. Tm., min), enrichment factor (EF), limit of detection (LOD, µg L^{-1}), limit of quantification (LOQ, µg L^{-1}), recovery (%), and relative standard deviation (RSD, %) of each NSAID investigated in this study and relative comparison with similar studies present in literature.

Method	Matrix	Analyte	LR	Ex. Solv.	Ex. Tm.	EF	LOD	LOQ	Recovery	RSD	Ref.
LLE-HPLC-UV [1]	human plasma	KPF, NAP, FPC [7], IBP, DIC [7]	100–100,000	600	n.r. [8]	n.r. [8]	11.5–75	n.r. [8]	97.1–146.6	<10.1	[58]
HF-LPME-HPLC-DAD (or FLD) [2]	human urine	DIC [7], ASA, IBP	41–10,000	50	15	70–1010	12.3–52.9	n.r. [8]	82.3–99	<1.8	[59]
HFLM-SPME-GC-FID [3]	human urine	IBP, NAP, DIC [7]	0.08–400	6	80	46.5–60.5	0.03–0.07	0.08–0.1	80.2–98.5	<12.1	[60]
SPE-SUPRASF-HPLC-UV [4]	human urine, water	DIC [7], MFA [7]	10–300	1500	25	431–489	0.4–7.0	n.r. [8]	90.4–103.8	<6.2	[61]
HF-LPME-UPLC-(MS/MS) [5]	water, juice, soda, energy drink	ASA, IBP, NAP, DIC [7]	1–5000	15	30	195–350	0.5–1.25	2.0–5.0	87.9–115.2	<12.0	[62]
RDSE-HPLC-UV [6]	human urine	DIC [7], IBP, KPF, NAP	200–2000	200	20	15–18	21.7–44.0	72.4–146.6	100–110	<12.0	[63]
USVA-DLLME-GC-IT-MS	animal urine	ASA, IBP, NAP, KPF	1–100	250	19	300–450	0.1–0.2	4.1–4.7	94.1–101	<4.1	This study

[1] Liquid–liquid extraction-high performance liquid chromatography-ultraviolet (LLE-HPLC-UV) detector; [2] hollow-fiber liquid-phase microextraction-high performance liquid chromatography-diode-array detection; [3] hollow-fiber liquid membrane-protected solid-phase microextraction-gas chromatography flame ionization detector; [4] solid phase extraction combined with supramolecular solvent-high performance liquid chromatography-UV; [5] hollow-fiber liquid-phase microextraction-ultra performance liquid chromatography-tandem mass spectrometry; [6] rotating disk sorptive extraction HPLC-UV; [7] diclofenac (DIC), fenoprofen (FPC), mefenamic acid (MFA); [8] not reported.

3.3. Application to Real Animal Urine Samples

Using the entire analytical USVA-DLLME-GC-IT-MS protocol previously developed (briefly resuming: 9 mL of simulated urine sample solution at pH 3 containing 5 µL of I.S., 60 ng mL^{-1}, addition of 1 mL of acetone and of 250 µL dichloromethane as extraction solvent, three times of 1 min vortex and 2 min ultrasounds, centrifugation for 10 min at 4000 rpm, 1 µL injection into GC-IT-MS), some animal urine samples have been analyzed, particularly three animal urine samples, i.e., two from goats and one from a sheep. All the subjects were healthy. No residues (i.e., levels below the LODs) were found in all the samples. The analysis allows us to investigate the presence of such compounds at trace levels in these matrices, but it does not furnish evidence as to whether there was a previous assumption of such molecules. As an example are shown in Figure 1, the gas chromatograms in SIM mode of a simulated sample of urine (a) and one of goat urine sample (b) both additions with 30 ng mL−1 of each NSAID. The peaks are well-solved and the determinations are precise and accurate.

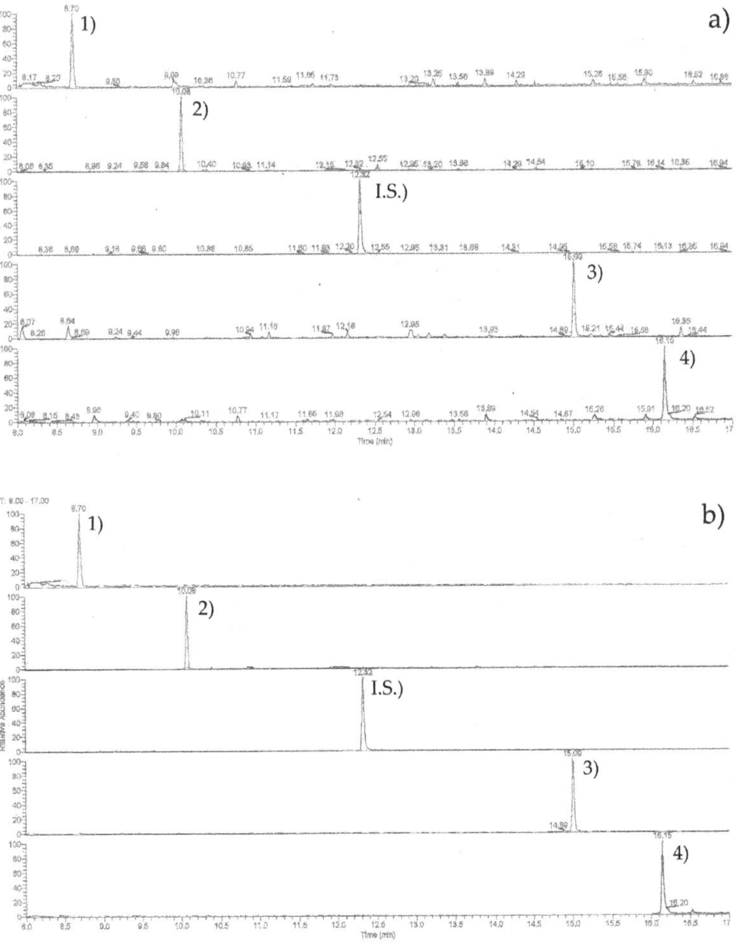

Figure 1. Gas chromatography/ion trap-mass spectrometry (GC-IT-MS) chromatograms in SIM mode of (**a**) simulated urine and (**b**) goat urine samples, both spiked with 30 ng mL^{-1} of each NSAID. For experimental conditions, see text. Peak list: 1. acetylsalicylic acid; 2. ibuprofen; internal standard (IS); 3. naproxen; 4. ketoprofen.

4. Conclusions

This paper highlights an affordable method for analyzing NSAIDs in animal urine samples. The method used for the animal urine samples in this study can also be applied to human urine samples, as a lead on to the discussion about athletes. In fact, athletes often make excessive use of anti-inflammatories in order to compete, even in less than optimal physical conditions. Many athletes take NSAIDs to compete or even simply train, even in the presence of pain, joint inflammation, trauma etc. Incorrect use of these drugs can lead to serious damage to health. Further, with regards to "premedication" in the sports field, it should be highlighted that the NSAIDs are not among the substances prohibited by the anti-doping measures and are therefore only drugs at risk of easy inappropriate abuse. Equine doping can also be defined as "the use of any exogenous agent (pharmacological, endocrinological, hematological, etc.) or clinical manipulation, which, in the absence of suitable and necessary therapeutic indications, is aimed to improve performance, outside the adjustments induced by training. In this view, this paper shows a simple, rapid, and sensitive method for determining four NSAIDs in animal urine samples. The very low LODs and LOQs and the high precision reached by means of a modified DLLME method coupled with GC-IT-MS allow us to apply the entire procedure to routine screening and monitoring of such compounds in doping cases or other similar situations.

Author Contributions: Conceptualization, M.V.R.; methodology, M.V.R. and P.A.; validation, I.N. and S.P.; formal analysis, S.P.; investigation, I.N.; resources, M.V.R.; data curation, M.V.R. and P.A.; writing—original draft preparation, P.A.; writing—review and editing, M.V.R. and P.A.; supervision, M.V.R.; project administration, M.V.R. All authors have read and agreed to the published version of the manuscript.

Funding: This research received no external funding.

Acknowledgments: The authors would like to thank Alessandro Ubaldi for his helpful suggestions in the data interpretation.

Conflicts of Interest: The authors declare no conflict of interest.

References

1. Baigent, C.; Bhala, N.; Emberson, J.; Merhi, A.; Abramson, S.; Arber, N.; Baron, J.A.; Bombardier, C.; Cannon, C.; Farkouh, M.E.; et al. Vascular and upper gastrointestinal effects of non-steroidal anti-inflammatory drugs: Meta-analyses of individual participant data from randomised trials. *Lancet* **2013**, *382*, 769–779.
2. Atkinson, T.J.; Fudin, J. Nonsteroidal antiinflammatory drugs for acute and chronic pain. *Phys. Med. Rehabil. Clin. N. Am.* **2020**, *31*, 219–231. [CrossRef]
3. Gunaydin, C.; Sirri Bilge, S. Effects of nonsteroidal anti-inflammatory drugs at the molecular level. *Eurasian J. Med.* **2018**, *50*, 116–121. [CrossRef] [PubMed]
4. Samad, T.A.; Sapirstein, A.; Woolf, C.J. Prostanoids and pain: Unraveling mechanisms and revealing therapeutic targets. *Trends Mol. Med.* **2002**, *8*, 390–396. [CrossRef]
5. Wasfi, I.A.; Boni, N.S.; Abdel Hadi, A.A.; ElGhazali, M.; Zorob, O.; Alkatheeri, N.A.; Barezaiq, I.M. Pharmacokinetics, metabolism and urinary detection time of flunixin after intravenous administration in camels. *J. Vet. Pharmacol. Therap.* **1998**, *21*, 203–208. [CrossRef] [PubMed]
6. Greisman, L.A.; Mackowiak, P.A. Fever: Beneficial and detrimental effects of antipyretics. *Curr. Opin. Infect. Dis.* **2002**, *15*, 241–245. [CrossRef]
7. Varas-Lorenzo, C.; Maguire, A.; Castellsague, J.; Perez-Gutthann, S. Quantitative assessment of the gastrointestinal and cardiovascular risk-benefit of celecoxib compared to individual NSAIDs at the population level. *Pharmacoepidemiol. Drug Saf.* **2007**, *16*, 366–376. [CrossRef]
8. Pountos, I.; Georgouli, T.; Calori, G.M.; Giannoudis, P.V. Do nonsteroidal anti-inflammatory drugs affect bone healing? A critical analysis. *Sci. World J.* **2012**, *2012*, 606404. [CrossRef]
9. Maniar, K.H.; Jones, I.A.; Gopalakrishna, R.; Vangsness, C.T. Lowering side effects of NSAID usage in osteoarthritis: Recent attempts at minimizing dosage. *Expert Opin. Pharmacother.* **2018**, *19*, 93–102. [CrossRef]

10. Bjarnason, I.; Scarpignato, C.; Holmgren, E.; Olszewski, M.; Rainsford, K.D.; Lanas, A. Mechanisms of damage to the gastrointestinal tract from nonsteroidal anti-inflammatory drugs. *Gastroenterology* **2018**, *154*, 500–514. [CrossRef]
11. Szeto, C.C.; Sugano, K.; Wang, J.G.; Fujimoto, K.; Whittle, S.; Modi, G.K.; Chen, C.H.; Park, J.B.; Tam, L.S.; Vareesangthip, K.; et al. Non-steroidal anti-inflammatory drug (NSAID) therapy in patients with hypertension, cardiovascular, renal or gastrointestinal comorbidities: Joint APAGE/APLAR/APSDE/APSH/APSN/PoA recommendations. *Gut* **2020**, *69*, 617–629. [CrossRef] [PubMed]
12. González, G.; Ventura, R.; Smith, A.K.; de la Torre, R.; Segura, J. Detection of non-steroidal anti-inflammatory drugs in equine plasma and urine by gas chromatography-mass spectrometry. *J. Chromatogr. A* **1996**, *719*, 251–264. [CrossRef]
13. Thevis, M.; Fußhöller, G.; Schänzer, W. Zeranol: Doping offence or mycotoxin? A case-related study. *Drug Test Anal.* **2011**, *3*, 777–783. [CrossRef] [PubMed]
14. Conaghan, P.G. A turbulent decade for NSAIDs: Update on current concepts of classification, epidemiology, comparative efficacy, and toxicity. *Rheumatol. Int.* **2012**, *32*, 1491–1502. [CrossRef] [PubMed]
15. Jaussaud, P.; Guieu, D.; Courtot, D.; Barbier, B.; Bonnaire, Y. Identification of a tolfenamic acid metabolite in the horse by gas chromatography-mass spectrometry. *J. Chromatogr.* **1992**, *3*, 136–140. [CrossRef]
16. Cortout, D.; Jaussaud, P. Anti-inflammatoires et contrôle antidopage chez le cheval. *Sci. Sports* **1993**, *8*, 53–54. [CrossRef]
17. Tsitsimpikou, C.; Spyridaki, M.H.E.; Georgoulakis, I.; Kouretas, D.; Konstantinidou, M.; Georgokopulous, C.G. Elimination profiles of flurbiprofen and its metabolites in equine urine for doping analysis. *Talanta* **2001**, *55*, 1173–1180. [CrossRef]
18. Fédération Équestre Internationale (FEI). Equine Prohibited Substances Database. Available online: http://prohibitedsubstancesdatabase.feicleansport.org/search/ (accessed on 23 May 2020).
19. International Federation of Horseracing Authorities (IFHA). International Agreement: Art. 6A-Prohibited Substances. Available online: https://www.ifhaonline.org/default.asp?section=IABRW&AREA=2#article6a (accessed on 23 May 2020).
20. Maurer, H.H.; Tauvel, F.X.; Kraemer, T. Screening procedure for detection of non-steroidal anti-inflammatory drugs and their metabolites in urine as part of a systematic toxicological analysis procedure for acidic drugs and poisons by gas chromatography-mass spectrometry after extractive methylation. *J. Anal. Toxicol.* **2001**, *25*, 237–244.
21. Oliva, A.; De Giorgio, F.; Arena, V.; Fucci, N.; Pascali, V.L.; Navarra, P. Death due to anaphylactic shock secondary to intravenous self-injection of Toradol®: A case report and review of the literature. *Clin. Toxicol.* **2007**, *45*, 709–713. [CrossRef]
22. Albert, K.S.; Gernaat, C.M. Pharmacokinetics of ibuprofen. *Am. J. Med.* **1984**, *77*, 40–46. [CrossRef]
23. Soma, L.R.; Uboh, C.E.; Maylin, G.M. The use of phenylbutazone in the horse. *J. Vet. Pharmacol. Therap.* **2011**, *35*, 1–12. [CrossRef] [PubMed]
24. El Haj, B.M.; Al Ainri, A.M.; Hassan, M.H.; Bin Khadem, R.K.; Marzouq, M.S. The GC/MS analysis of some commonly used non-steriodal anti-inflammatory drugs (NSAIDs) in pharmaceutical dosage forms and in urine. *Forensic Sci. Int.* **1999**, *105*, 141–153. [CrossRef]
25. Kim, J.Y.; Kim, S.J.; Paeng, K.J.; Chung, B.C. Measurement of ketoprofen in horse urine using gas chromatography-mass spectrometry. *J. Vet. Pharmacol. Ther.* **2001**, *24*, 315–319. [CrossRef]
26. Wasfi, I.A.; Hussain, M.M.; Elghazali, M.; Alkatheeri, N.A.; Abdel Hadi, A.A. The disposition of diclofenac in camels after intravenous administration. *Vet. J.* **2003**, *166*, 277–283. [CrossRef]
27. Levreri, I.; Caruso, U.; Deiana, F.; Buoncompagni, A.; De Bernardi, B.; Marchese, N.; Melioli, G. The secretion of ibuprofen metabolites interferes with the capillary chromatography of urinary homovanillic acid and 4-hydroxy-3-methoxymandelic acid in neuroblastoma diagnosis. *Clin. Chem. Lab. Med.* **2005**, *43*, 173–177. [CrossRef]
28. Mareck, U.; Sigmund, G.; Opfermann, G.; Geyer, H.; Thevis, M.; Schänzer, W. Identification of the aromatase inhibitor letrozole in urine by gas chromatography/mass spectrometry. *Rapid Commun. Mass Spectrom.* **2005**, *19*, 3689–3693. [CrossRef]
29. Główka, F.K.; Karaźniewicz, M. High performance capillary electrophoresis method for determination of ibuprofen in human serum and urine. *Anal. Chim. Acta* **2005**, *540*, 95–102. [CrossRef]

30. Azzouz, A.; Ballesteros, E. Gas chromatography–mass spectrometry determination of pharmacologically active substances in urine and blood samples by use of a continuous solid-phase extraction system and microwave-assisted derivatization. *J. Chromatogr. B* **2012**, *891*, 12–19. [CrossRef]
31. Yilmaz, B.; Erdem, A.F. Determination of ibuprofen in human plasma and urine by gas chromatography/mass spectrometry. *J. AOAC Int.* **2014**, *97*, 415–420. [CrossRef]
32. Waraksa, E.; Woźniak, M.K.; Kłodzińska, E.; Wrzesień, R.; Bobrowska-Korczak, B.; Namieśnik, J. A rapid and sensitive method for the quantitative analysis of ibuprofen and its metabolites in equine urine samples by gas chromatography with tandem mass spectrometry. *J. Sep. Sci.* **2018**, *41*, 3881–3891. [CrossRef]
33. Waraksa, E.; Woźniak, M.K.; Banaszkiewicz, L.; Kłodzińska, E.; Ozimek, M.; Wrzesień, R.; Bobrowska-Korczak, B.; Namieśnik, J. Quantification of unconjugated and total ibuprofen and its metabolites in equine urine samples by gas chromatography-tandem mass spectrometry: Application to the excretion study. *Microchem. J.* **2019**, *150*, 104129. [CrossRef]
34. Waraksa, E.; Wójtowicz-Zawadka, M.; Kwiatkowska, D.; Jarek, A.; Małkowska, A.; Wrzesien, R.; Namiesnika, J. Simultaneous determination of ibuprofen and its metabolites in complex equine urine matrices by GC-EI-MS in excretion study in view of doping control. *J. Pharm. Biomed. Anal.* **2018**, *152*, 279–288. [CrossRef] [PubMed]
35. Rezaee, M.; Assadi, Y.; Hosseini, M.R.M.; Aghaee, E.; Ahmadi, F.; Berijani, S. Determination of organic compounds in water using dispersive liquid-liquid microextraction. *J. Chromatogr. A* **2006**, *1116*, 1–9. [CrossRef]
36. Cinelli, G.; Avino, P.; Notardonato, I.; Centola, A.; Russo, M.V. Rapid analysis of six phthalate esters in wine by ultrasound-vortex-assisted dispersive liquid–liquid micro-extraction coupled with gas chromatography-flame ionization detector or gas chromatography-ion trap mass spectrometry. *Anal. Chim. Acta* **2013**, *769*, 72–78. [CrossRef] [PubMed]
37. Avino, P.; Notardonato, I.; Perugini, L.; Russo, M.V. New protocol based on high-volume sampling followed by DLLME-GC-IT/MS for determining PAHs at ultra-trace levels in surface water samples. *Microchem. J.* **2017**, *133*, 251–257. [CrossRef]
38. Notardonato, I.; Russo, M.V.; Vitali, M.; Protano, C.; Avino, P. Analytical method validation for determining organophosphorus pesticides in baby foods by a modified liquid-liquid microextraction method and gas chromatography-ion trap/mass spectrometry analysis. *Food Anal. Methods* **2019**, *12*, 41–50. [CrossRef]
39. Leong, M.I.; Huang, S.D. Determination of volatile organic compounds in water using ultrasound-assisted emulsification microextraction followed by gas chromatography. *J. Sep. Sci.* **2012**, *35*, 688–694. [CrossRef] [PubMed]
40. Russo, M.V.; Notardonato, I.; Avino, P.; Cinelli, G. Fast determination of phthalate ester residues in soft drinks and light alcoholic beverages by ultrasound/vortex assisted dispersive liquid-liquid microextraction followed by gas chromatography-ion trap mass spectrometry. *RSC Adv.* **2014**, *4*, 59655–59663. [CrossRef]
41. Russo, M.V.; Avino, P.; Notardonato, I. Fast analysis of phthalates in freeze-dried baby foods by ultrasound-vortex-assisted liquid-liquid microextraction coupled with gas chromatography-ion trap/mass spectrometry. *J. Chromatogr. A* **2016**, *1474*, 1–7. [CrossRef]
42. Russo, M.V.; Avino, P.; Perugini, L.; Notardonato, I. Fast analysis of nine PAHs in beer by ultrasound-vortex-assisted dispersive liquid-liquid micro-extraction coupled with gas chromatography-ion trap mass spectrometry. *RSC Adv.* **2016**, *6*, 13920–13927. [CrossRef]
43. Notardonato, I.; Salimei, E.; Russo, M.V.; Avino, P. Simultaneous determination of organophosphorus pesticides and phthalates in baby food samples by ultrasound-vortex-assisted liquid-liquid microextraction and GC-IT/MS. *Anal. Bioanal. Chem.* **2018**, *410*, 3285–3296. [CrossRef] [PubMed]
44. Notardonato, I.; Passarella, S.; Ianiri, G.; Di Fiore, C.; Russo, M.V.; Avino, P. Analytical method development and chemometric approach for evidencing presence of plasticizer residues in nectar honey samples. *Int. J. Environ. Res. Public Health* **2020**, *17*, 1692. [CrossRef] [PubMed]
45. Croes, K.; Goeyens, L.; Baeyens, W.; Van Loco, J.; Impens, S. Optimization and validation of a liquid chromatography tandem mass spectrometry (LC/MS^n) method for analysis of corticosteroids in bovine liver: Evaluation of Keyhole Limpet beta-glucuronidase/sulfatase enzyme extract. *J. Chromatogr. B* **2009**, *877*, 635–644. [CrossRef] [PubMed]
46. Cinelli, G.; Avino, P.; Notardonato, I.; Russo, M.V. Ultrasound-vortex-assisted dispersive liquid–liquid microextraction coupled with gas chromatography with a nitrogen-phosphorus detector for simultaneous and rapid determination of organophosphorus pesticides and triazines in wine. *Anal. Methods* **2014**, *6*, 782–790. [CrossRef]

47. Sebők, Á.; Vasanits-Zsigrai, A.; Palkó, G.; Záray, G.; Molnár-Perl, I. Identification and quantification of ibuprofen, naproxen, ketoprofen and diclofenac present in waste-waters, as their trimethylsilyl derivatives, by gas chromatography mass spectrometry. *Talanta* **2008**, *76*, 642–650. [CrossRef]
48. Russo, M.V.; Goretti, G.C.; Liberti, A. A fast procedure to immobilize polyethylene glycols in glass capillary columns. *J. High Resolut. Chromatogr.* **1985**, *8*, 535–538. [CrossRef]
49. Cartoni, G.P.; Castellani, L.; Goretti, G.; Russo, M.V.; Zacchei, P. Gas-liquid microcapillary columns precoated with graphitized carbon black. *J. Chromatogr.* **1991**, *552*, 197–204. [CrossRef]
50. Russo, M.V.; Goretti, G.; Soriero, A. Preparation and application of fused-silica capillary microcolumns (25–50 μm ID) in gas chromatography. *Ann. Chim.* **1996**, *86*, 115–124.
51. Matuszewski, B.K.; Constanzer, M.L.; Chavez-Eng, C.M. Strategies for the assessment of matrix effect in quantitative bioanalytical methods based on HPLC–MS/MS. *Anal. Chem.* **2003**, *75*, 3019–3030. [CrossRef]
52. Dressman, J.B.; Nair, A.; Abrahamsson, B.; Barends, D.M.; Groot, D.W.; Kopp, S.; Langguth, P.; Polli, J.E.; Shah, V.P.; Zimmer, M. Biowaiver monograph for immediate-release solid oral dosage forms: Acetylsalicylic acid. *J. Pharm. Sci.* **2012**, *101*, 2653–2667. [CrossRef]
53. Bushra, R.; Aslam, N. An overview of clinical pharmacology of ibuprofen. *Oman Med. J.* **2010**, *25*, 155–161. [CrossRef] [PubMed]
54. Sangster, J. LogKOW Database. A Databank of Evaluated Octanol-Water Partition Coefficients (Log P). Available online: http://logkow.cisti.nrc.ca/logkow/search.html (accessed on 23 May 2020).
55. Spek, J.W.; Bannink, A.; Gort, G.; Hendriks, W.H.; Dijkstra, J. Effect of sodium chloride intake on urine volume, urinary urea excretion, and milk urea concentration in lactating dairy cattle. *J. Dairy Sci.* **2012**, *95*, 7288–7298. [CrossRef] [PubMed]
56. Knoll, J.E. Estimation of the limit of detection in chromatography. *J. Chromatogr. Sci.* **1985**, *23*, 422–425. [CrossRef]
57. Russo, M.V.; Avino, P.; Cinelli, G.; Notardonato, I. Sampling of organophosphorus pesticides at trace levels in the atmosphere using XAD-2 adsorbent and analysis by gas chromatography coupled with nitrogen-phosphorus and ion-trap mass spectrometry detectors. *Anal. Bioanal. Chem.* **2012**, *404*, 1517–1527. [CrossRef] [PubMed]
58. Sun, Y.; Takaba, K.; Kido, H.; Nakashima, M.N.; Nakashima, K. Simultaneous determination of arylpropionic acidic non-steroidal anti-inflammatory drugs in pharmaceutical formulations and human plasma by HPLC with UV detection. *J. Pharm. Biomed. Anal.* **2003**, *30*, 1611–1619. [CrossRef]
59. Payán, M.R.; López, M.Á.B.; Fernández-Torres, R.; Bernal, J.L.P.; Mochón, M.C. HPLC determination of ibuprofen, diclofenac and salicylic acid using hollow fiber-based liquid phase microextraction (HF-LPME). *Anal. Chim. Acta* **2009**, *653*, 184–190. [CrossRef]
60. Sarafraz-Yazdia, A.; Amiri, A.; Rounaghia, G.; Eshtiagh-Hosseini, H. Determination of non-steroidal anti-inflammatory drugs in urine by hollow-fiber liquid membrane-protected solid-phase microextraction based on sol-gel fiber coating. *J. Chromatogr. B* **2012**, *908*, 67–75. [CrossRef]
61. Rezaei, F.; Yamini, Y.; Moradi, M.; Ebrahimpour, B. Solid phase extraction as a cleanup step before microextraction of diclofenac and mefenamic acid using nanostructured solvent. *Talanta* **2013**, *105*, 173–1178. [CrossRef]
62. Zhang, H.; Du, Z.; Ji, Y.; Mei, M. Simultaneous trace determination of acidic non-steroidal anti-inflammatory drugs in purified water, tap water, juice, soda and energy drink by hollowfiber-based liquid-phase microextraction and ultra-high pressure liquid chromatography coupled to tandem mass spectrometry. *Talanta* **2013**, *109*, 177–184.
63. Manzo, V.; Miró, M.; Richtera, P. Programmable flow-based dynamic sorptive microextraction exploiting an octadecyl chemically modified rotating disk extraction system for the determination of acidic drugs in urine. *J. Chromatogr. A* **2014**, *1368*, 64–69. [CrossRef]

Article

Fast and Low-Cost Synthesis of MoS_2 Nanostructures on Paper Substrates for Near-Infrared Photodetectors

Neusmar J. A. Cordeiro [1,*], Cristina Gaspar [2], Maria J. de Oliveira [2], Daniela Nunes [2], Pedro Barquinha [2], Luís Pereira [2], Elvira Fortunato [2], Rodrigo Martins [2], Edson Laureto [1] and Sidney A. Lourenço [3,*]

1 Laboratório de Óptica e Optoeletrônica, Departamento de Física, Universidade Estadual de Londrina (UEL), CP6001, Londrina, Paraná CEP 86051-970, Brazil; laureto@uel.br
2 CENIMAT | i3N, Departamento de Ciência dos Materiais, Faculdade de Ciências e Tecnologia, Universidade Nova de Lisboa (UNINOVA), 1099-085 Lisboa, Portugal; chg12706@fct.unl.pt (C.G.); mj.oliveira@campus.fct.unl.pt (M.J.d.O.); daniela.gomes@fct.unl.pt (D.N.); pmcb@fct.unl.pt (P.B.); lmnp@fct.unl.pt (L.P.); emf@fct.unl.pt (E.F.); rm@uninova.pt (R.M.)
3 Laboratório de Fotônica e Materiais Nanoestruturados, Departamento de Física, Universidade Tecnológica Federal do Paraná (UTFPR), Londrina, Paraná CEP 86036-370, Brazil
* Correspondence: neusmar.jr@gmail.com (N.J.A.C.); sidneylourenco@utfpr.edu.br (S.A.L.)

Featured Application: This work presents an analysis of time and temperature influence of microwave-assisted hydrothermal synthesis on the direct growth of 2D-MoS2 nanostructures on cellulose paper substrates, and the production of MoS2-based low-cost photosensors with high responsivity and detectivity values.

Abstract: Recent advances in the production and development of two-dimensional transition metal dichalcogenides (2D TMDs) allow applications of these materials, with a structure similar to that of graphene, in a series of devices as promising technologies for optoelectronic applications. In this work, molybdenum disulfide (MoS_2) nanostructures were grown directly on paper substrates through a microwave-assisted hydrothermal synthesis. The synthesized samples were subjected to morphological, structural, and optical analysis, using techniques such as scanning electron microscopy (SEM), X-ray diffraction (XRD), and Raman. The variation of synthesis parameters, as temperature and synthesis time, allowed the manipulation of these nanostructures during the growth process, with alteration of the metallic (1T) and semiconductor (2H) phases. By using this synthesis method, two-dimensional MoS_2 nanostructures were directly grown on paper substrates. The MoS_2 nanostructures were used as the active layer, to produce low-cost near-infrared photodetectors. The set of results indicates that the interdigital MoS_2 photodetector with the best characteristics (responsivity of 290 mA/W, detectivity of 1.8×10^9 Jones and external quantum efficiency of 37%) was obtained using photoactive MoS2 nanosheets synthesized at 200 °C for 120 min.

Keywords: MoS_2; microwave-assisted hydrothermal synthesis; low-cost photosensors

Citation: Cordeiro, N.J.A.; Gaspar, C.; Oliveira, M.J.d.; Nunes, D.; Barquinha, P.; Pereira, L.; Fortunato, E.; Martins, R.; Laureto, E.; Lourenço, S.A. Fast and Low-Cost Synthesis of MoS_2 Nanostructures on Paper Substrates for Near-Infrared Photodetectors. *Appl. Sci.* **2021**, *11*, 1234. https://doi.org/10.3390/app11031234

Academic Editor: Samuel B. Adeloju
Received: 30 December 2020
Accepted: 25 January 2021
Published: 29 January 2021

1. Introduction

Since its discovery in 2004, graphene has become one of the nanomaterials of great interest in the construction of devices, due to its high electronic conductivity, mechanical flexibility, and low production cost [1]. Despite the good results obtained with graphene [2,3], the absence of energy bandgap restricts its application in some devices, such as photodetectors, mostly due to low intrinsic responsivity. This led to the development of a series of other two-dimensional materials with different characteristics, such as the hexagonal boron nitride [4], silicene [5], borophene [6], black phosphourous [7], and two-dimensional transition metal dichalcogenides (2D TMDs) [8]. The latter have been thoroughly explored in recent years for several applications [8].

Among the 2D TMDs materials, composed of a transition metal (M) and a chalcogen (X) with generalized form MX_2 in which M = Mo, W, Nb, Ta, Hf, Pt, and so on, and

X = S, Se, Te, the MoS_2 nanostructures has generated great interest in the scientific community [8–10]. Featuring a metastable metallic phase (1T), which can be stabilized or converted to a stable semiconductor phase (2H) by the appropriate heat treatment [11] or use of microwave radiation [12], MoS_2 has unique characteristics, such as high carrier mobility, strong electron-hole confinement and variable bandgap (1.20 to 1.89 eV) [13,14]. The bandgap of the MoS_2 increases with the decrease of the crystal thickness, to below 100 nm, due to the quantum confinement effect [15], and reaches 1.89 eV for a single monolayer [13]. Thus, it can cover an extent NIR (near-infrared) electromagnetic spectrum (6560 to 10,332 nm) by changing the monolayer number. Additionally, the material changes its electronic structure from an indirect bandgap, in its bulk form, to a direct bandgap in the MoS_2 monolayer [16], enabling optical applications. The relationship between its optical and electrical characteristics with the number of stacked layers and the control of the 1T and 2H phases turns the MoS_2 into a material of great interest for applications in chemical sensors [17], hydrogen evolution reaction [18], electronic devices [19], sodium-ion battery [20], and photodetectors [21,22].

The most commonly used synthesis methods of the two-dimensional MoS_2 are exfoliation (liquid [23,24] and mechanical [24]), chemical vapor deposition (CVD) [25], and hydrothermal [26]. Due to a better relationship between uniformity, the quantity of produced material and production cost, the hydrothermal method has been the most applied in the synthesis of MoS_2 nanostructures, usually presenting synthesis times between 20 and 24 h [27,28]. The microwave-assisted hydrothermal method has proved to be a good alternative to the conventional hydrothermal method, offering shorter synthesis times due to "molecular heating", consequently lower energy consumption, enhanced reaction selectivity, and homogeneous volumetric heating [29–31]. This method also allows the direct growth of the material on a substrate, being an alternative to conventional deposition or costly transfer methods [28,32]. Synthesis parameters, such as time and temperature, as well as the chemical processes involved in the production of the material, allow the direct growth of the MoS_2 nanostructures on flexible substrates [28], like cellulose-based substrates [33–35]. These advantages will have impact in the development of paper-based electronics that emerge as a future alternative to traditional electronics, seeking low-cost electronic systems and components which can be environmentally friendly [36–38]. To meet the requirements imposed on this new generation of devices, different methods of producing materials and printing technologies are employed for the production of electronic devices [36,38–40], like solar cells [41], thin film transistors [42], light emitting devices [43], electrochromic devices [44], and photosensors [28], among others.

In this work the authors adopted the microwave-assisted hydrothermal method, with different synthesis parameters, for direct growth of MoS_2 nanostructures on tracing paper substrates. Structural and morphological characterizations of the MoS_2 nanosheets were performed, where it was possible to observe the dispersion of the 2H and 1T phases of the nanostructures grown on paper, making these samples potentially interesting for applications in optoelectronic devices. Thus, near infrared photodetectors of MoS_2 nanostructures, grown on cellulose substrates, were built with different synthesis parameters. These interdigital MoS_2 photodetectors (with detection at 980 nm) were obtained with high responsivity (290 mA/W) and detectivity of 1.8×10^9 Jones. These devices presented higher responsivity values than that reported in the literature, as compared with MoS_2 photodetectors produced by the conventional hydrothermal method.

2. Materials and Methods

Molybdenum disulfide (MoS_2) nanosheets were grown on cellulose paper substrates by a microwave-assisted hydrothermal method optimizing the hydrothermal temperature and reaction time. Sodium molybdate dihydrate ($Na_2MoO_4.2H_2O$) and thiourea ($CS(NH_2)_2$) were used as precursors of molybdenum and sulfur, respectively. Cellulose paper was used as a substrate, which was previously washed by sonication in deionized water and by acetone and isopropyl alcohol, both for 15 min.

For direct growth on the cellulose fibers, the substrates were previously immersed in a seed solution composed of sodium molybdate and thiourea (1:4) for 60 min. After that time, the substrates were dried on a hotplate at 80 °C for 60 min. Substrates with deposited seeds were placed in a hydrothermal reactor with nutrition solution also composed of sodium molybdate and thiourea (1:4). Then, they were taken to the microwave oven for synthesis that was carried out with a power of 100 W, maximum pressure of 280 psi, and with different values of temperature and time of synthesis. Three different synthesis temperatures were used: 190 °C, 200 °C, and 220 °C. These were chosen to apply the shortest possible time for the growth of the nanostructures, corresponding to each temperature used, until the maximum time of 120 min. Thus, the different temperatures and synthesis times used were: 190 °C for 30, 45 and 120 min, 200 °C for 15, 30, 45, 60 and 120 min, 220 °C for 05, 15, 30, 45, 60, and 120 min. After the growth of the MoS_2 on the cellulose fibers, the paper with the grown nanostructures were washed with ethanol in ultrasound for 15 min to remove the nanostructures not fixed on the sample's surface. Finally, the samples were dried in a nitrogen flow.

Interdigital contacts were deposited by the screen-printing technique using commercial conductive ink CRSN2442 Suntronic, composed of silver nanoparticles dispersed on the solvent (solvent-based), acquired from Coats Screen Inks GmbH. The interdigital electrodes have a total dimension of 10.0 × 6.0 mm, 500 μm electrode wide and a 500 μm active channel between the electrodes.

Structural and optical properties of the MoS_2 nanostructures, grown on cellulose paper substrate, were studied by X-ray diffraction (XRD), scanning electron microscopy (SEM), and micro-Raman spectroscopy. The crystal phase was obtained by X-ray diffraction (X'Pert PRO MPD), using Cu-kα radiation (λ = 1.540598 Å). The XRD patterns were collected in symmetric configuration in the 2θ range of 5−70°, using a step angle of 0.05°, step time of 2 s, operating at 40 kV tension and 30 mA current. The morphology was studied by scanning electron microscopy (Carl Zeiss AURIGA CrossBeam workstation). Raman measurements were performed in a Renishaw inVia Reflex micro-Raman spectrometer equipped with an air-cooled CCD detector and a HeNe laser operating at 50 mW of 532 nm laser excitation. The spectral resolution of the spectroscopic system is 0.3 cm^{-1}. The laser beam was focused with a 50 × Leica objective lens (N Plan EPI) with a numerical aperture of 0.75. An integration time of 3 scans of 1.5 s each was used for all single-scan measurements to reduce the random background noise induced by the detector, without significantly increasing the acquisition time. The intensity of the incident laser was 2.5 mW. Triplicates were taken for all spectra. The spectrograph was calibrated between different Raman sessions using the Raman line at 521 cm^{-1} of an internal Si wafer for reducing possible fluctuations of the Raman system. All spectra were recorded at room temperature. Raw data were collected digitally with the Wire 5.0 software for processing. The characterization of the photodetectors was performed with the use of an AUTOLAB potentiostat (PGSTAT204) for ixV electrical measurements and chronoamperometry and a 980 nm diode laser for irradiation of the sensors, under ambient conditions. The power of the laser light was recorded using a handheld digital power meter console (PM100D) from THORLABS. Neutral density filters were used to vary the excitation power incident on the devices.

3. Results and Discussion

The microwave-assisted hydrothermal synthesis method was used for the direct growth of MoS_2 nanostructures on cellulose paper substrates due to the ease of the technique's application, as well as its ability to change synthesis parameters to optimize the produced structures [29–31]. The growth was carried out using a two-step process, where the synthesis is performed on a substrate with a thin layer of deposited seeds.

For construction of NIR photodetectors, a layer of interdigital silver electrodes was deposited on the layer of the MoS_2 nanostructures using the screen-printing technique. Further details on the growth of nanostructures and the construction of the devices can be found in the Experimental Section. Figure 1 shows a diagram of the MoS_2 nanostructures synthesis and the photodetector production.

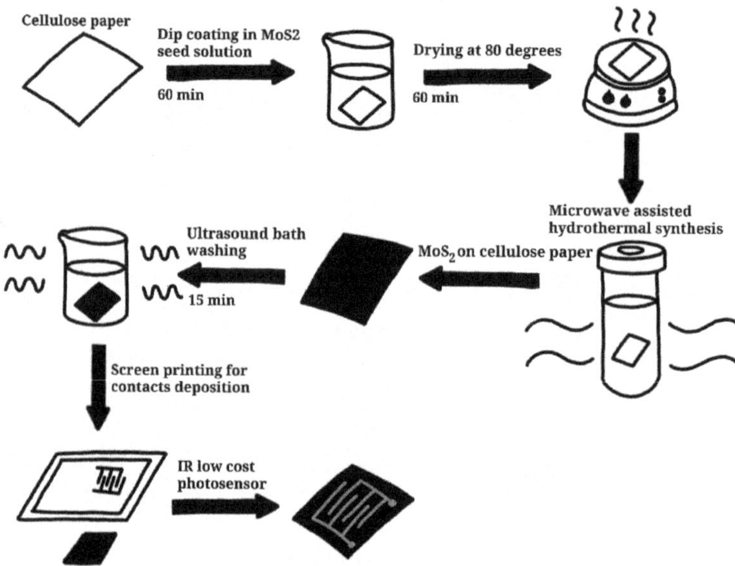

Figure 1. Scheme of the synthesis process of the molybdenum disulfide (MoS_2) nanostructures in paper substrates by the microwave-assisted hydrothermal method, and the production of infrared radiation photodetectors.

Figure 2 shows SEM images of the samples synthesized with a synthesis temperature of 220 °C at different times.

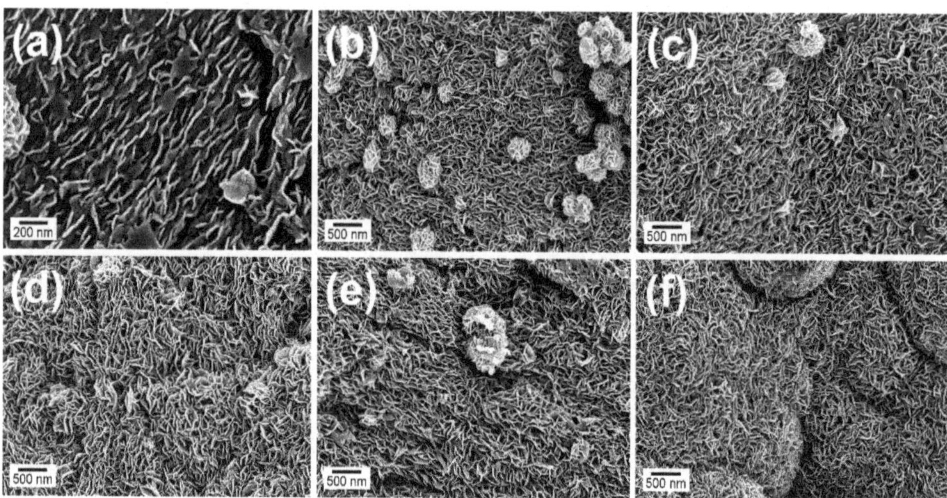

Figure 2. Scanning electron microscopy (SEM) images of the MoS_2 nanostructures grown directly on cellulose paper substrates with a synthesis temperature of 220 °C and time of: (**a**) 05 min, (**b**) 15 min, (**c**) 30 min, (**d**) 45 min, (**e**) 60 min, and (**f**) 120 min.

Microscopy images show that the beginning of the growth process is characterized by the formation of vertically aligned MoS_2 nanosheets on the paper surface (see also Figure S1 in the Supplementary File), while the paper fibers are not fully covered. For longer synthesis times, the surface of the paper becomes completely covered by vertically aligned MoS_2 nanosheets, regardless the temperature. The effect of temperature is noticed on the time required to observe the first nanosheets on the paper surface. Such a result is expected since, for lower temperatures, the solution is exposed to lower energies during the synthesis process, requiring a longer time for nanomaterial formation. At 190 °C, it is possible to observe that, even for the longest synthesis time (120 min), the profile of the paper fibers is still perceived, indicating that it was not possible to form a thick and uniform nanostructure layer. On the other hand, at 200 °C the fibers are uniformly coated with a thick layer of MoS_2 nanosheets. It is important to note that the samples synthesized at this temperature had a more uniform surface profile than the samples produced by the conventional hydrothermal method reported in the literature [28,33,35], which generally have agglomerations of MoS_2 nanosheets in microspherical profile. A surface totally covered with nanostructures is also observed in the samples synthesized at 220 °C. However, at this temperature, it is possible to observe that, for longer synthesis times, the sample surface has regions with a spherical profile very similar to those synthesized by the conventional hydrothermal method [28,32]. This indicates that there was the production of a large number of nanostructures during the synthesis process, where they tend to agglomerate in spherical shape, microflowers, formed by nanosheets. The growth kinetics of MoS_2 nanostructures on cellulose paper and other flexible substrates was recently discussed by Sahatiya et al. [28] through classical nucleation and growth theory. These spherical structures are formed when there is excessive nanosheets growth, which generally occurs for longer synthesis times or for MoS_2 synthesis without the presence of the substrate [26]. In the synthesis with direct growth on substrate, the excess of grown nanostructures ends up detaching from the paper surface and agglomerating into spherical particles to decrease the surface energy. Note the early formation of these small flowers for samples synthesized at 200 °C (Supplementary Figure S1).

Figure 3 shows XRD diffractograms at different temperatures and synthesis times of the MoS_2 nanostructures, in addition to the XRD diffractogram of cellulose paper and the JCPDS n° 37-1492 card pattern of 2H-MoS_2 for comparison purposes [28,45]. Dashed vertical lines are used to facilitate the identification of phases in the diffractograms obtained at different temperatures and synthesis times. It is possible to identify regions, close do 15° and 22°, with the characteristic peaks of the cellulose diffractograms, the 2H-MoS_2 phases with XRD peaks at ~32.8°, ~49.8°, and 57.8° relative to (100), (105), and (110) planes, respectively. For $2\theta > 30°$ the XRD peaks fit well with JCPDS card n° 37-1492, where the presence of (100) and (110) peaks, ate ~32.8° and 57.3° respectively, show that the samples present similar atomic arrangement along the basal planes with the bulk MoS_2.

According to the standard 2H-MoS_2 diffractogram pattern, represented by the JCPDS card n° 37-1492 shown in Figure 3, there is a diffraction peak at 14.4° regarding the plane (002) of the layered material. However, two other peaks appear in the diffractogram, at 9.3° and 18.6°, identified in Figure 3 by # symbols. MoS_2 is a representative 2D layered material and the weak van der Waals interaction between MoS_2 layers favors the intercalation of molecules and ions in this space [32,46]. The interlayer distance of the (002) plane, referring to the 14.4° peak, of intercalation-free MoS_2 is 0.62 nm, while the corresponding interlayer distances of 9.4° and 18.6° peaks, present in the diffractograms of the synthesizes samples, were 0.93 nm and 0.49 nm, respectively, calculated using the Bragg equation. Several groups attribute the interlayer distances of 0.93 nm and 0.49 nm to the distance between two adjacent MoS_2 expand layers—resulting from the intercalation of different materials as CTAB [47], mesoporous carbon (MoS_2/m-C) [48], NH_4^+ [49], and NH_3 molecules [46]—and that distance between intercalated molecules and the adjacent MoS_2 layers, respectively [32,47,48,50]. Other authors have attributed the 18.6° peak to a second-order diffraction [49] It is important to note that, while some authors have attributed the shifted 9.4° and 18.6° peaks to the presence of 1T-MoS_2 phase [49,51], Lei et al. [52] have alerted about the difficult to use the expanded interlayer spacing of MoS_2 as direct evidence to confirm the existence of the 1T-MoS_2 phase.

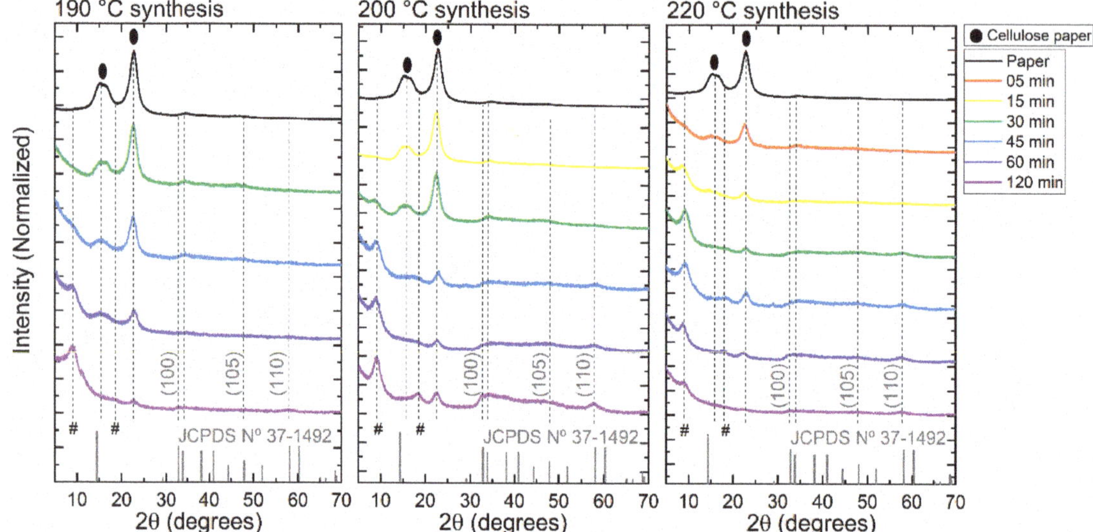

Figure 3. X-ray diffraction patterns of MoS_2 nanosheets growth on cellulose paper substrate at different synthesis temperatures and times and the JCPDS 37-1492 card pattern of 2H-MoS_2. The peaks identified by circles are related to the cellulose paper. The # symbol is used to identify the peaks at 9.3° and 18.6°.

Liu et al. have shown that the use of the $Na_2MoO_4.2H_2O$ and $CS(NH_2)_2$ with high thiourea concentration induce the formation of ammonium ion intercaled into MoS_2 layer [49]. In the synthesis procedure presented in this work, $Na_2MoO_4.2H_2O$ and thiourea were used as precursors of molybdenum and sulfur, respectively, in the proportion of (1:4). Thus, we believe that the MoS_2 nanosheets grown on cellulose paper substrates by a microwave-assisted hydrothermal method present NH_4^+ intercaled between layers of MoS_2. It should also be noted that there is an decrease in the intensity of the cellulose diffraction peaks when subjected to heat treatments (probably associated with crystallinity of the cellulose) [53]; even so, the diffraction peaks of MoS_2 stand out in the diffractogram for high synthesis times, due to the greater amount of nanomaterial grown on the paper surface, for (100), (105) and (110) planes, or the possible increase of the intercalation with the synthesis times, for the 9.4° and 18.6° peaks.

It can be noted that there is a continuous increase in peak intensity at 9.2° up to 30 min, in samples synthesized at 220 °C. Then the intensity of the entire diffractogram, including those peaks related to cellulose paper, starts to decrease for higher times. The high temperature leads to degrading the paper structure and can affecting the growth process for longer synthesis times, leading to less production of nanostructures on the substrate surface. This result is in line with what is reported in the literature, which leads to cellulose degradation processes starting above 200 °C [53]. In summary, the growth process of the MoS_2 on paper substrate seems to be better at 200 °C.

For a complementary analysis of the presence of 1T and 2H-MoS_2 phases, Raman spectroscopy measurements were performed, looking for the presence of characteristic peaks of these phases on the samples. After confirming the presence of both phases in the synthesized samples by Raman spectroscopy, an analysis of the spatial distribution of 1T and 2H-MoS_2 was performed by the micro-Raman technique. By mapping an area of 225 μm [2] and using steps of 1 μm, the regions of the spectra between 146 to 148 cm^{-1}

(referring to the main vibrational mode of the 1T phase), and between 378 to 385 cm^{-1} (referring to the $E^1{}_{2g}$ vibrational mode of the 2H phase) [12,46], were highlighted. The Raman spectrum and the micro-Raman mapping of the samples synthesized at 220 °C, for times of 05 and 120 min, are shown in Figure 4.

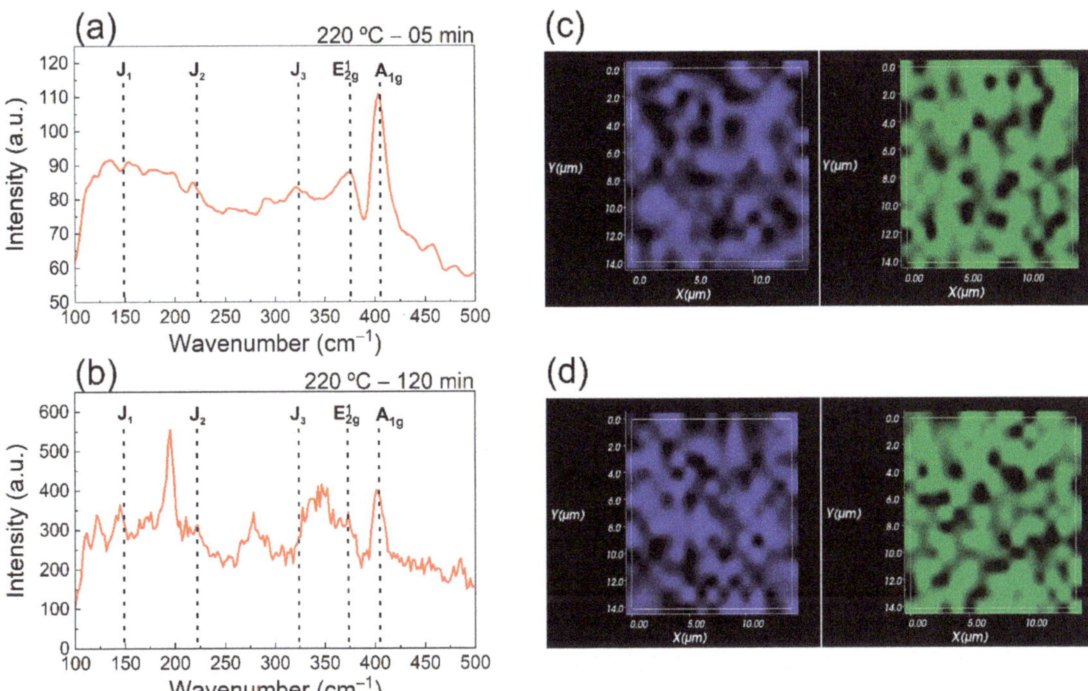

Figure 4. Raman spectra and micro-Raman mapping of the MoS_2 nanosheets synthetized at 220 °C with synthesis times of (**a**,**c**) for 05 min and (**b**,**d**) for 120 min. The blue and green mappings represent 1T- and 2H-MoS2 phases in the samples.

As can be seen in Figure 4a,b, it was possible to identify characteristic Raman peaks of the 1T (vibrational modes J_1, J_2 and J_3) and 2H (vibrational modes $E^1{}_{2g}$ and A_{1g}) phases in the sample spectrum [46], thus confirming the presence of both phases in the synthesized nanostructure. Figure 4c,d shows the micro-Raman mapping, confirming the presence of both phases, the 1T (blue) and the 2H (green), in both samples and in the same space region. These phases are not occupying well-defined regions, but they are dispersed throughout the mapping area. This behavior is observed for all samples (regardless synthesis time and temperature), as shown in Figure 4 and Figure S2 of the supplementary information. The dispersion of the metallic and semiconductor phases of MoS_2 in the samples may be associated with the continuous production of the metallic phase during synthesis. Over time, the 1T phase is converted to a 2H phase, due to the low stability of the metallic phase and presence of microwave radiation during the synthesis process [12]. Thus, the 1T and 2H phases are present in the samples, as can be seen in the diffractogram of the sample synthesized at 200 °C for 120 min. The greater dispersion of phases on the surface of the substrate can be good for certain applications, since it increases the metal/semiconductor interface (1T/2H) which leads to greater carrier mobility due to the presence of the 1T phase.

As shown in Figure 1, low-cost interdigital photodetectors were built on cellulose paper substrate with a MoS_2 active layer synthesized by the microwave-assisted hydrothermal method at different temperatures and times. Figure 5a–c show the current x voltage curves under laser illumination (980 nm) of the photodetectors with a MoS_2 photosensitive layer synthesized at 190, 200, and 220 °C, respectively. The curves are linear and symmetric for the small bias voltage, indicating a ohmic-like contact as observed for the single-layer MoS_2 phototransistor synthesized using the CVD technique [54], or by photodetector built with a multilayer MoS_2 synthesized by the conventional hydrotermal method, using in this case Ag NPs (nanoparticles) as contact [28]. The small bandgap energy of the multilayer MoS_2 may generate a very small Schottky barrier between Ag NPs electrodes and MoS_2. After illumination, carriers are injected into the small conduction band of the multilayer MoS_2 generating photocurrent. The photocurrent increase can be associated with a bias voltage through to the reduction of the carrier transit time ($\tau_{transit} = l^2/\mu V$), where l is the length of the channel, μ is the carrier mobility, and V the bias potential [22].

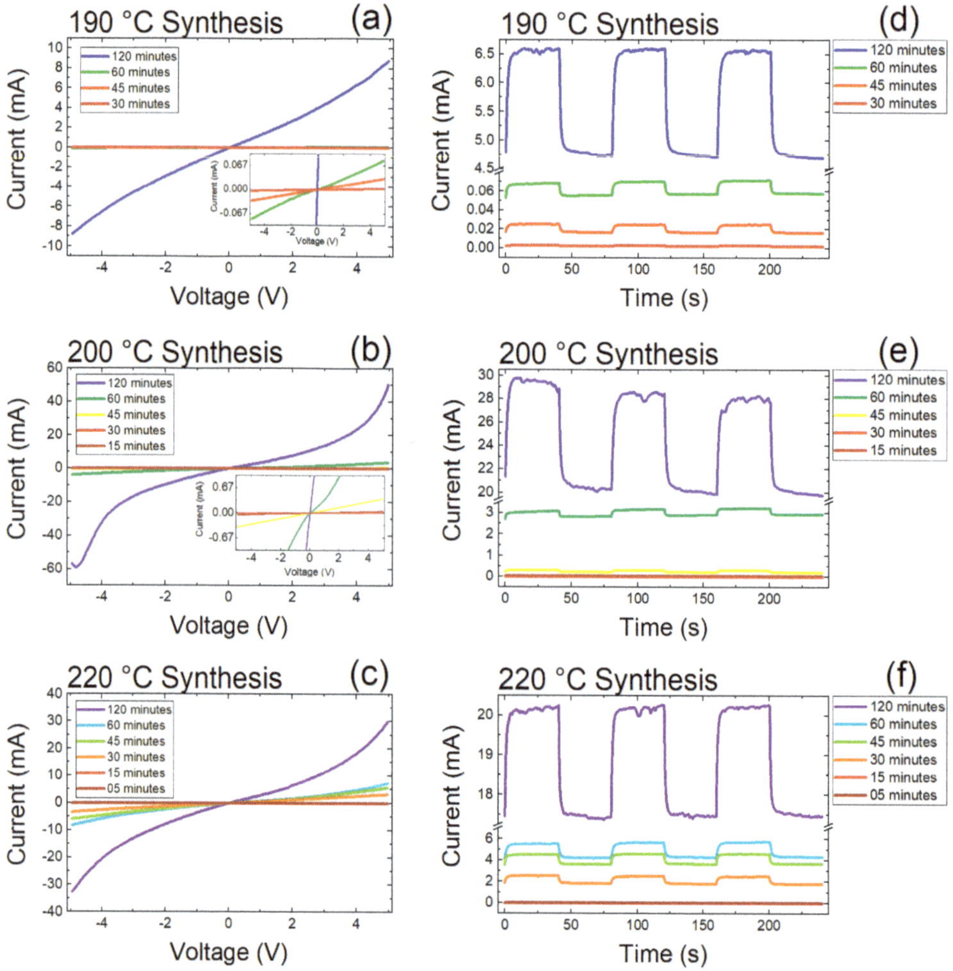

Figure 5. Characterization of infrared photosensors produced with MoS_2 samples grown directly on paper substrates at different times, being: (**a**) IxV of samples synthesized at 190 °C, (**b**) IxV of samples synthesized at 200 °C, (**c**) IxV of samples synthesized at 220 °C, (**d**) Ixt of samples synthesized at 190 °C, (**e**) Ixt of samples synthesized at 200 °C, and (**f**) Ixt of samples synthesized at 220 °C. The Ixt measurements were performed with a bias voltage of 4 V, lighting cycles lasting 40 s and power of 20 mW.

The curves in Figure 5b,c, for longer synthesis times, present a strong increase in the current x bias voltage (with a non-linear behavior) for bias voltages higher than ~2V. This non-linearity in the IxV curve is less intense in the samples synthesized at lower temperature, 190 °C—Figure 5a, and in the samples with shorter synthesis times for the synthesis temperatures of 200 and 220 °C, possibly due to the smaller amount of MoS_2 material. Figure 5d–f show that there is an increase in both the dark current and the photocurrent, a result that goes according to the discussion held for Figure 5a–c. The increase in the dark current with synthesis time can be associated with the presence of the metallic phase and trap states.

To evaluate the infrared photosensors, figures of merit Responsivity (R) and Specific Detectivity (D^*) were used, which can be calculated using the equations $R = (I_{on}\text{-}I_{off})/P$ and $D^* = R/\sqrt{((2eI_{off})/A)}$. Here, I_{on} is the generated photocurrent in the device under illumination, I_{off} is the dark current, P is the incident light power on the effective area of the device, e is the elementary charge and A is the effective surface area. Figure 6 shows that R and D* values increase with increasing of the synthesis time for three synthesis temperatures and for a bias voltage of 4V. The device built with a sample synthesized at 200 °C for 120 min has the highest value for both D^* and R. There is also a big increase of these values for synthesis time longer than 60 min (see yellow curve). The decrease in the R value obtained in the sample synthesized at 220 °C for 120 min, when compared to the sample synthesized at 200 °C for 120 min, may be attributed to the degradation of the cellulose paper.

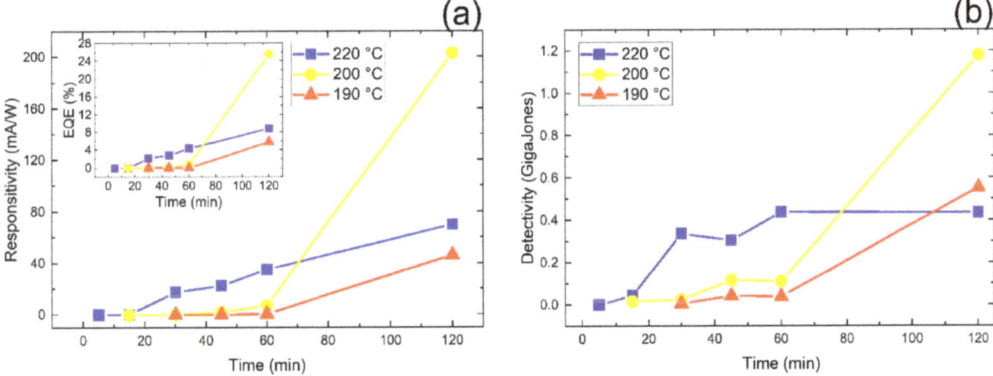

Figure 6. The Responsivity (**a**) and Detectivity (**b**) according to the synthesis time, for samples synthesized with temperatures of 190 °C (red triangles), 200 °C (yellow circles) and 220 °C (blue squares). Responsivity was measured with a laser power of 20 mW. Graph's inset shows the external quantum efficiency (EQE) values as a function of time and synthesis temperatures of the MoS_2 photoactive layer.

Therefore, the experimental data show that the temperature of 200 °C and time of 120 min were the best conditions for the production of near infrared photodetectors based on MoS_2 active layer grown directly on cellulose paper substrates by the microwave-assisted hydrothermal method. This R value is above the value reported in the literature for photodetectors with active MoS_2 layer produced by the hydrothermal method with the same structure. Nahid Chaudhary et al. [55] for example, obtained the maximum R value of 23.8 μA/W for excitation at 635 nm, and Parikshit Sahatiya et al. [28] obtained the R value of 60 μA/W for excitation at 554 nm.

It is important to note that temperatures above 200 °C have generally been used in the literature for synthesis of MoS_2, or heat treatments have been made after the synthesis to increase the fraction of phase 2H over phase 1T. At 200 °C (as in this work), we still have a large proportion of the 1T phase when compared with samples synthesized by the conventional method followed by heat treatment [13,28]. Therefore, we believe that the increase in the 1T/2H interface, caused by the simultaneous and continuous formation of the metallic phase and its conversion to the semiconductor phase, and the consequent increase in the mobility of carriers in the sample, is of great importance to obtain high values of responsivity in MoS_2 photodetectors. The interface between the paper substrate and the MoS_2 layer may be another important characteristic for devices, once it can be a source of trap states [54] and temperature and time synthesis had a greater effect on the photoconductivity. Thus, the quick increase of the photocurrent to bias voltage (>2 V), as observed in Figure 5, can be associated with the detrapping charge from trap states. It is known that photogain in photoconductors increases as carrier lifetime and decreases as transit time $G = \tau_{life}/\tau_{transit}$ [22].

Figure 6 shows the external quantum efficiency (EQE) of the devices. EQE was obtained by using $EQE(\%) = Rhc/(\lambda e)x100$. Here, R is the responsivity, *h* is the Planck's constant, *c* is the speed of light in vacuum, *e* is the elementary charge, and λ is the wavelength of the excitation light. As can be seen, the EQE increases as the time synthesis increases, similar to R and D*. The device built from MoS_2, synthesized at 200 °C for 120 min, presents an EQE value of 26%.

The MoS_2 photodetector has a relatively high R value, compared to other similar devices based on MoS_2 synthesized by the hydrothermal method. The rise and the decay response time has high values (τ_{rise} = 3.7 s e τ_{decay} = 4.7 s) which may be associated with the high concentration of the metallic phase in MoS_2 and trap states in the MoS_2 and/or paper/MoS_2 interface. In comparison, the HfO_2-gated single-layer MoS_2 phototransistors, with high responsivity due to the photogate effect, present a slow response of 0.6–9.0 s. In this case, the slow response time and the high responsivity was induced by the trap states in the MoS_2 or MoS_2/SiO_2 interface [54] Jin et al. [56] deposited Ag NPs on monolayer MoS_2 phototransistor and obtained a very high R value and slow response time of 18.7 s, showing other strategy to induce photoconductive effect. The improvement was attributed to the localized surface plasmon resonance of Ag NPs with the increase of light absorption and carrier injection from Ag NPs to MoS_2 under illumination. Thus, we believe that, in our device, the surface plasmon resonance in the Ag NPs electrode is light excited and generates hot electrons via nonradiative decay plasmonic resonance, since the Ag NPs show a plasmonic resonance signal in the near-infrared region, as shown in Figure S3 in the supplementary file. The energy of the hot electrons may be greater than the small Schottky barrier between Ag NPs and MoS_2, and it can be injected into the MoS_2 multilayer generating a photocurrent [57–59]. Additionally, the effect of the MoS_2 phases (1T and 2H) on the electrical performances of hybrid PDs was compared by Wang et al. [60] and results indicated that the metallic 1T phase exhibited a high *R* value. However, these devices showed low on/off ratios (<2) and slow response times (0.75 s) due to their metallic conducting nature similar to that of the graphene. On the other hand, the semiconducting 2H phase demonstrated lower *R* value and fast response time (<25 ms). Thus, we believe that our PDs built with MoS_2 synthesized at 200 °C for 120 min presented higher metallic phases than other devices built with materials synthesized for other temperatures and times, and that the hot electrons from Ag NPs could be contributing to the photocurrent improvement. Therefore, the contact between Ag NPs and the MoS_2 multilayer must convert from a small Schottky junction to an Ohmic-like contact when the device is illuminated, as observed in Au nanorods/Perovskite photoconductor [56,61]. The device works as a light illuminated photoconductor, explaining the slow response time and high R compared to a similar device based on a MoS_2 synthesized by the conventional hydrothermal method [28,55].

Figure 7 shows the dependence of the photocurrent (I_{on}-I_{off}) as a function of the light power on the photosensor synthesized at 200 °C for 120 min.

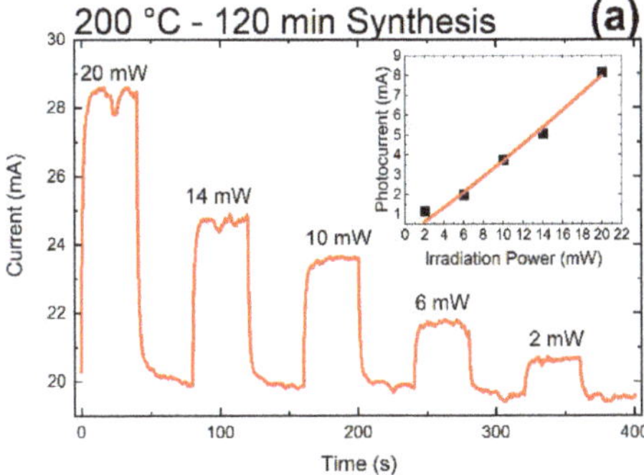

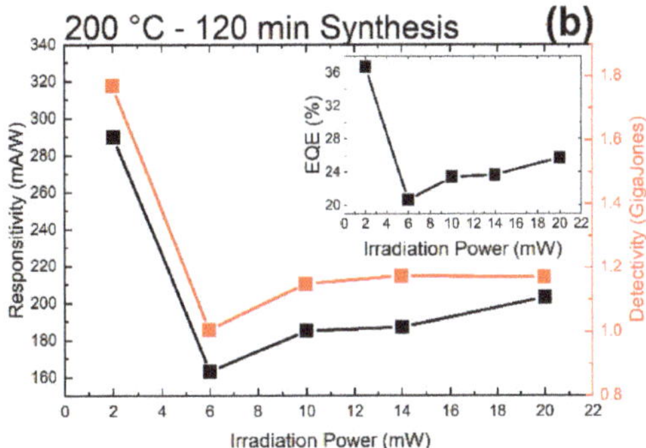

Figure 7. (**a**) Current dependence due to the decrease in the power of the light source (laser 980 nm, ON, OFF) on the sensor. The inset shows the relationship between the generated photocurrent and the irradiation power of the light source on the sensor for the synthesized sample at 200 °C for 120 min. The bias voltage used here was 4 V. (**b**) Responsitivity, Specific Detectivity, and EQE (inset) under different irradiation powers of the device regarding the synthesized sample with a temperature of 200 °C and a time of 120 min.

Figure 7a shows a decrease in photocurrent behavior with decreasing light power intensity. This is an expected result, since with the decrease in the incident light power, there is a decrease in the number of photons reaching the sensor and, consequently, there is a decrease in the number of carriers promoted from the valence band to the conducting band of the MoS_2 nanostructures. With the decrease in the number of carriers in the conduction band, there is a decrease in the generated photocurrent (I_{on}-I_{off}). An analysis of the relationship between the generated photocurrent and the incident light power on the photodetector is showed in the inset of Figure 7a, where a fitting with power law is used ($I = AP^{\theta}$). Here A is a proportionality constant and θ is the exponent that determines the relationship between the photogenerated current and the light source power. From the adjustment, a linear relationship is observed ($\theta = 1.1 \pm 0.1$). Similar behavior of the photocurrent with light intensity was observed for the single-layer MoS_2 phototransitor [54]. Figure 7b shows the relationship between responsivity/detectivity and the light source power used. Note that the highest responsivity and detectivity values (290 mA/W and 1.8×10^9 Jones) occur for the power of 2 mW. In the range of 6 to 2 mW, the responsivity and detectivity values increase rapidly with decreasing power (these results indicate that for lower radiation intensity, in the μW range, the photodetector can present very high responsivity values—these results are not showed here). This behavior has been associated with the increase in the recombination probability as carriers photoexcitation increases [59]. EQE measurements (inset of Figure 7b) obtained at various excitation powers show that the EQE gradually increases as the power decreases, similar to the results for the graphene–MoS_2–graphene devices (responsivity of ≈ 0.2 A W^{-1} and EQE = 55%) [62].

4. Conclusions

MoS_2 nanostructures were grown directly on cellulose paper substrates using the microwave-assisted hydrothermal synthesis method with different temperatures and synthesis times. Variations were observed in the amount of MoS_2 grown on top of cellulose fibers as a function of the synthesis parameters. The analysis also showed the distribution of the metallic and semiconductor phases of the MoS_2 nanostructures throughout the sample, generating metal/semiconductor interfaces. Near-infrared sensors were produced from the synthesized samples, where the screen-printing technique was used to deposit interdigital Ag NPs contacts in the samples. The results of the electrical characterizations showed an improvement in the performance of the sensor with the increase in nanostructures synthesis time, for the three temperatures of synthesis used. Such results can be explained by the increase in the amount of MoS_2, according to the model created for the sensors operation. Based on the data analysis, the photodetector produced from the sample synthesized at 200 °C and 120 min presented a high responsivity value (290 mA/W), a specific detectivity value (1.8 × 10^9 Jones) and an external quantum efficiency of 37%, with response times τ_{rise} = 3.7 s and τ_{decay} = 4.7 s. The sensors produced from the samples synthesized at a temperature of 220 °C also showed high values of responsivity, however, the high temperature causes substrate degradation, making the device fragile and the application difficult. The improvement in the responsivity value of the photodetector can be associated with the photoconductive effect due to three probably factors: (i) the high concentration of the metallic 1T phase in MoS_2 layers; (ii) trap states in the MoS_2 and/or paper/MoS_2 interface; and (iii) hot electron injection from surface plasmon resonance in the Ag NPs electrodes to MoS_2 nanomaterial.

An analysis of the photocurrent dependence as a function of the incident radiation power on the sensor showed a linear dependence of the device's photocurrent in relation to the power of the light source. The responsivity showed a great increase with the reduction of the irradiation power, at the value of 290 mA/W for the power of 2 mW. This result must be associated with the increase in the probability of recombination with the increase in the photoexcited carriers. This work provides a functional example, as well as a promising strategy, to improve the performance of 1T/2H MoS_2-based photodetectors. Further studies on the surface state passivation for device optimization, for example, will be conducted to improve photocurrent in MoS_2 photodetector synthesized by the microwave-assisted hydrothermal synthesis method.

Supplementary Materials: The following are available online at https://www.mdpi.com/2076-3417/11/3/1234/s1, Figure S1: SEM images of the MoS2 nanostructures grown directly on cellulose paper substrates with different temperature and time of synthesis, Figure S2: Micro-Raman mapping and detachment of phases 1T (blue) and 2H (green) in samples of MoS2 nanostructures grown directly on paper substrates with different temperature and time of synthesis, Figure S3: Absorbance spectrum of silver nanoparticles ink in an isopropyl alcohol solution.

Author Contributions: Conceptualization, N.J.A.C., C.G., P.B., L.P., E.F., R.M., E.L. and S.A.L.; Data curation, N.J.A.C., C.G., M.J.d.O., D.N., P.B., L.P., E.L. and S.A.L.; Formal analysis, N.J.A.C., C.G., M.J.d.O., D.N., P.B., L.P., E.L. and S.A.L.; Investigation, N.J.A.C., C.G., D.N., R.M. and S.A.L.; Methodology, N.J.A.C., C.G., M.J.d.O., D.N. and S.A.L.; Project administration, P.B., L.P., E.F., R.M., E.L. and S.A.L.; Resources, D.N., P.B., L.P., E.F., R.M. and S.A.L.; Software, N.J.A.C., C.G., M.J.d.O. and D.N.; Supervision, C.G., D.N., P.B., L.P., E.F., R.M., E.L. and S.A.L.; Validation, N.J.A.C.; Visualization, N.J.A.C. and S.A.L.; Writing—original draft, N.J.A.C., M.J.d.O., D.N., L.P., R.M., E.L. and S.A.L.; Writing—review & editing, N.J.A.C., L.P., R.M., E.L. and S.A.L. All authors have read and agreed to the published version of the manuscript.

Funding: This study was financed in part by the Coordenação de Aperfeiçoamento de Pessoal de Nível Superior—Brasil (CAPES)—Finance Code 001.

Institutional Review Board Statement: Not applicable.

Informed Consent Statement: Not applicable.

Data Availability Statement: The data presented in this study are available on request from the corresponding author.

Acknowledgments: The authors gratefully acknowledge the financial support from the following Brazilian agencies and programs CNPq, Capes, and Fundação Araucária. This research was also funded by FEDER funds through the COMPETE 2020 Programme and National Funds through FCT—Portuguese Foundation for Science and Technology under project number POCI-01-0145-FEDER-007688, Reference UID/CTM/50025. The authors would like to acknowledge the European Commission under project NewFun (ERC-StG-2014 GA 640598) and BET-EU (H2020-TWINN-2015, GA 692373). The authors would like to thank the Multiuser Laboratory of Universidade Tecnológica Federal do Paraná—Londrina Campus—for the performed analyses.

Conflicts of Interest: The authors declare no conflict of interest.

References

1. Torrisi, F.; Carey, T. Graphene, Related Two-Dimensional Crystals and Hybrid Systems for Printed and Wearable Electronics. *Nano Today* **2018**, *23*, 73–96. [CrossRef]
2. Bhatt, K.; Kumar, S.; Tripathi, C.C. High-Performance Ultra-Low Leakage Current Graphene-Based Screen-Printed Field-Effect Transistor on Paper Substrate. *Pramana J. Phys.* **2020**, *94*, 2–5. [CrossRef]
3. Song, H.; Liu, J.; Lu, H.; Chen, C.; Ba, L. High Sensitive Gas Sensor Based on Vertical Graphene Field Effect Transistor. *Nanotechnology* **2020**, *31*, 165503. [CrossRef] [PubMed]
4. Xiong, J.; Di, J.; Zhu, W.; Li, H. Hexagonal Boron Nitride Adsorbent: Synthesis, Performance Tailoring and Applications. *J. Energy Chem.* **2020**, *40*, 99–111. [CrossRef]
5. Lin, H.; Qiu, W.; Liu, J.; Yu, L.; Gao, S.; Yao, H.; Chen, Y.; Shi, J. Silicene: Wet-Chemical Exfoliation Synthesis and Biodegradable Tumor Nanomedicine. *Adv. Mater.* **2019**, *31*, 1–12. [CrossRef] [PubMed]
6. Kiraly, B.; Liu, X.; Wang, L.; Zhang, Z.; Mannix, A.J.; Fisher, B.L.; Yakobson, B.I.; Hersam, M.C.; Guisinger, N.P. Borophene Synthesis on Au(111). *ACS Nano* **2019**, *13*, 3816–3822. [CrossRef]
7. Li, B.; Lai, C.; Zeng, G.; Huang, D.; Qin, L.; Zhang, M.; Cheng, M.; Liu, X.; Yi, H.; Zhou, C.; et al. Black Phosphorus, a Rising Star 2D Nanomaterial in the Post-Graphene Era: Synthesis, Properties, Modifications, and Photocatalysis Applications. *Small* **2019**, *15*, 1–30. [CrossRef]
8. Choi, W.; Choudhary, N.; Han, G.H.; Park, J.; Akinwande, D.; Lee, Y.H. Recent Development of Two-Dimensional Transition Metal Dichalcogenides and Their Applications. *Mater. Today* **2017**, *20*, 116–130. [CrossRef]
9. Sriram, P.; Manikandan, A.; Chuang, F.C.; Chueh, Y.L. Hybridizing Plasmonic Materials with 2D-Transition Metal Dichalcogenides toward Functional Applications. *Small* **2020**. [CrossRef]
10. Wang, F.; Zhang, Y.; Gao, Y.; Luo, P.; Su, J.; Han, W.; Liu, K.; Li, H.; Zhai, T. 2D Metal Chalcogenides for IR Photodetection. *Small* **2019**, *15*. [CrossRef]
11. Wypych, F.; Schoellhorn, R. 1T-MoS2, A New Metallic Modification of Molybdenum Disulfide. *ChemInform* **1992**, *24*, 1386–1388. [CrossRef]
12. Xu, D.; Zhu, Y.; Liu, J.; Li, Y.; Peng, W.; Zhang, G.; Zhang, F.; Fan, X. Microwave-Assisted 1T to 2H Phase Reversion of MoS2 in Solution: A Fast Route to Processable Dispersions of 2H-MoS2 Nanosheets and Nanocomposites. *Nanotechnology* **2016**, *27*, 1–7. [CrossRef]
13. Eda, G.; Yamaguchi, H.; Voiry, D.; Fujita, T.; Chen, M.; Chhowalla, M. Photoluminescence from Chemically Exfoliated MoS 2. *Nano Lett.* **2011**, *11*, 5111–5116. [CrossRef]
14. Munkhbayar, G.; Palleschi, S.; Perrozzi, F.; Nardone, M.; Davaasambuu, J.; Ottaviano, L. A Study of Exfoliated Molybdenum Disulfide (MoS2) Based on Raman and Photoluminescence Spectroscopy. *Solid State Phenom.* **2018**, *271*, 40–46. [CrossRef]
15. Neville, R.A.; Evans, B.L. The Band Edge Excitons in 2H-MoS2. *Phys. Status Solidi* **1976**, *73*, 597–606. [CrossRef]
16. Mak, K.F.; Lee, C.; Hone, J.; Shan, J.; Heinz, T.F. Atomically Thin MoS2: A New Direct-Gap Semiconductor. *Phys. Rev. Lett.* **2010**, *105*, 2–5. [CrossRef]
17. Kumar, R.; Zheng, W.; Liu, X.; Zhang, J.; Kumar, M. MoS2-Based Nanomaterials for Room-Temperature Gas Sensors. *Adv. Mater. Technol.* **2020**, 1–28. [CrossRef]
18. Joyner, J.; Oliveira, E.F.; Yamaguchi, H.; Kato, K.; Vinod, S.; Galvao, D.S.; Salpekar, D.; Roy, S.; Martinez, U.; Tiwary, C.S.; et al. Graphene Supported MoS2 Structures with High Defect Density for an Efficient HER Electrocatalysts. *ACS Appl. Mater. Interfaces* **2020**. [CrossRef]
19. Singh, E.; Singh, P.; Kim, K.S.; Yeom, G.Y.; Nalwa, H.S. Flexible Molybdenum Disulfide (MoS 2) Atomic Layers for Wearable Electronics and Optoelectronics. *ACS Appl. Mater. Interfaces* **2019**, *11*, 11061–11105. [CrossRef]
20. Zhang, T.; Feng, Y.; Zhang, J.; He, C.; Itis, D.; Song, J. Ultrahigh-Rate Sodium-ion Battery Anode Enable by Vertically Aligned (1T-2H MoS2)/CoS2 Hetero-nanosheets. *Mater. Today Nano* **2020**. [CrossRef]
21. Liao, F.; Deng, J.; Chen, X.; Wang, Y.; Zhang, X.; Liu, J.; Zhu, H.; Chen, L.; Sun, Q.; Hu, W.; et al. A Dual-Gate MoS2 Photodetector Based on Interface Coupling Effect. *Small* **2020**, *16*, 1–7. [CrossRef]
22. Huo, N.; Konstantatos, G. Recent Progress and Future Prospects of 2D-Based Photodetectors. *Adv. Mater.* **2018**, *30*. [CrossRef]
23. Varrla, E., Backes, C.; Paton, K.R.; Harvey, A.; Gholamvand, Z.; Cauley, J.; Colemanm, J.N. Large-scale production of size-controlled MoS2 nanosheets by shear exfoliation. *Chem. Mater.* **2015**, *27*, 1129–1139. [CrossRef]
24. Ottaviano, L.; Palleschi, S.; Perozzi, F.; D'Olimpio, G.; Priante, F.; Donarelli, M.; Benassi, P.; Nardone, M.; Gonchigsuren, M.; Gombosuren, M.; et al. Mechanical Exfoliation and Layer Number Identification of MoS2 Revisited. *2D Mater.* **2017**, *4*, 045013. [CrossRef]
25. Liu, H.F.; Wong, S.L.; Chi, D.Z. CVD Growth of MoS2-Based Two-Dimensional Materials. *Chem. Vap. Depos.* **2015**, *21*, 241–259. [CrossRef]
26. Tang, G.; Sun, J.; Wei, C.; Wu, K.; Ji, X.; Liu, S.; Tang, H.; Li, C. Synthesis and Characterization of Flowerlike MoS 2 Nanostructures through CTAB-Assisted Hydrothermal Process. *Mater. Lett.* **2012**, *86*, 9–12. [CrossRef]
27. Lee, C.M.; Park, G.C.; Lee, S.M.; Choi, J.H.; Jeong, S.H.; Seo, T.Y.; Jung, S.B.; Lim, J.H.; Joo, J. Effects of Precursor Concentration on Morphology of MoS2 Nanosheets by Hydrothermal Synthesis. *J. Nanosci. Nanotechnol.* **2016**, *16*, 11548–11551. [CrossRef]
28. Sahatiya, P.; Jones, S.S.; Badhulika, S. Direct, Large Area Growth of Few-Layered {MoS}2 Nanostructures on Various Flexible Substrates: Growth Kinetics and Its Effect on Photodetection Studies. *Flex. Print. Electron.* **2018**, *3*, 15002. [CrossRef]
29. Pimentel, A.; Nunes, D.; Duarte, P.; Rodrigues, J.; Costa, F.M.; Monteiro, T.; Martins, R.; Fortunato, E. Synthesis of Long ZnO Nanorods under Microwave Irradiation or Conventional Heating. *J. Phys. Chem. C* **2014**, *118*, 14629–14639. [CrossRef]
30. Gao, M.R.; Chan, M.K.Y.; Sun, Y. Edge-Terminated Molybdenum Disulfide with a 9.4-Å Interlayer Spacing for Electrochemical Hydrogen Production. *Nat. Commun.* **2015**, *6*, 1–8. [CrossRef]

31. Bilecka, I.; Niederberger, M. Microwave Chemistry for Inorganic Nanomaterials Synthesis. *Nanoscale* **2010**, *2*, 1358–1374. [CrossRef] [PubMed]
32. Miao, H.; Hu, X.; Sun, Q.; Hao, Y.; Wu, H.; Zhang, D.; Bai, J.; Liu, E.; Fan, J.; Hou, X. Hydrothermal Synthesis of MoS2 Nanosheets Films: Microstructure and Formation Mechanism Research. *Mater. Lett.* **2016**, *166*, 121–124. [CrossRef]
33. Gomathi, P.T.; Sahatiya, P.; Badhulika, S. Large-Area, Flexible Broadband Photodetector Based on ZnS–MoS2 Hybrid on Paper Substrate. *Adv. Funct. Mater.* **2017**, *27*, 1–9. [CrossRef]
34. Sahatiya, P.; Kadu, A.; Gupta, H.; Gomathi, T.P.; Badhulika, S. Flexible, Disposable Cellulose-Paper-Based MoS2/Cu2S Hybrid for Wireless Environmental Monitoring and Multifunctional Sensing of Chemical Stimuli. *ACS Appl. Mater. Interfaces* **2018**, *10*, 9048–9059. [CrossRef]
35. Sahatiya, P.; Badhulika, S. Wireless, Smart, Human Motion Monitoring Using Solution Processed Fabrication of Graphene–MoS2 Transistors on Paper. *Adv. Electron. Mater.* **2018**, *4*, 1–9. [CrossRef]
36. Grey, P.; Gaspar, D.; Cunha, I.; Barras, R.; Carvalho, J.T.; Ribas, J.R.; Fortunato, E.; Martins, R.; Pereira, L. Handwritten Oxide Electronics on Paper. *Adv. Mater. Technol.* **2017**, *2*, 2–8. [CrossRef]
37. Martins, R.; Nathan, A.; Barros, R.; Pereira, L.; Barquinha, P.; Correia, N.; Costa, R.; Ahnood, A.; Ferreira, I.; Fortunato, E. Complementary Metal Oxide Semiconductor Technology with and on Paper. *Adv. Mater.* **2011**, *23*, 4491–4496. [CrossRef]
38. Martins, R.; Ferreira, I.; Fortunato, E. Electronics with and on Paper. *Phys. Status Solidi Rapid Res. Lett.* **2011**, *5*, 332–335. [CrossRef]
39. Barras, R.; Cunha, I.; Gaspar, D.; Fortunato, E.; Martins, R.; Pereira, L. Printable Cellulose-Based Electroconductive Composites for Sensing Elements in Paper Electronics. *Flex. Print. Electron.* **2017**, *2*. [CrossRef]
40. Gaspar, C.; Olkkonen, J.; Passoja, S.; Smolander, M. Paper as Active Layer in Inkjet-Printed Capacitive Humidity Sensors. *Sensors* **2017**, *17*, 1464. [CrossRef]
41. Águas, H.; Mateus, T.; Vicente, A.; Gaspar, D.; Mendes, M.J.; Schmidt, W.A.; Pereira, L.; Fortunato, E.; Martins, R. Thin Film Silicon Photovoltaic Cells on Paper for Flexible Indoor Applications. *Adv. Funct. Mater.* **2015**, *25*, 3592–3598. [CrossRef]
42. Cunha, I.; Barras, R.; Grey, P.; Gaspar, D.; Fortunato, E.; Martins, R.; Pereira, L. Reusable Cellulose-Based Hydrogel Sticker Film Applied as Gate Dielectric in Paper Electrolyte-Gated Transistors. *Adv. Funct. Mater.* **2017**, *27*. [CrossRef]
43. Purandare, S.; Gomez, E.F.; Steckl, A.J. High Brightness Phosphorescent Organic Light Emitting Diodes on Transparent and Flexible Cellulose Films. *Nanotechnology* **2014**, *25*. [CrossRef] [PubMed]
44. Nunes, D.; Freire, T.; Barranger, A.; Vieira, J.; Matias, M.; Pereira, S.; Pimentel, A.; Cordeiro, N.J.A.; Fortunato, E.; Martins, R. TiO2 Nanostructured Films for Electrochromic Paper Based-Devices. *Appl. Sci.* **2020**, *10*, 1200. [CrossRef]
45. Zhou, J.; Guo, M.; Wang, L.; Ding, Y.; Zhang, Z.; Tang, Y.; Liu, C.; Luo, S. 1T-MoS2 Nanosheets Confined among TiO2 Nanotube Arrays for High Performance Supercapacitor. *Chem. Eng. J.* **2019**, *366*, 163–171. [CrossRef]
46. Geng, X.; Zhang, Y.; Han, Y.; Li, J.; Yang, L.; Benamara, M.; Chen, L.; Zhu, H. Two-Dimensional Water-Coupled Metallic MoS2 with Nanochannels for Ultrafast Supercapacitors. *Nano Lett.* **2017**, *17*, 1825–1832. [CrossRef]
47. Wang, Z.; Chen, T.; Chen, W.; Chang, K.; Ma, L.; Huang, G.; Lee, J.Y. CTAB-assisted synthesis of single-layer MoS2–graphene composites as anode materials of Li-ion batteries. *J. Mater. Chem. A.* **2013**, *1*, 2202–2210. [CrossRef]
48. Jiang, H.; Ren, D.; Wang, H.; Hu, Y.; Guo, S.; Yuan, H.; Li, C. 2D monolayer MoS2-carbon interoverlapped superstructure: Engineering ideal atomic interface for lithium ion storage. *Adv. Mater.* **2015**, *24*, 3687–3695. [CrossRef]
49. Liu, Q.; Li, X.; He, Q.; Khalil, A.; Liu, D.; Xiang, T.; Song, L. Gram-scale aqueous synthesis of stable few-layered 1T-MoS2: Applications for visible-light-driven photocatalytic hydrogen evolution. *Small* **2015**, *11*, 5556–5564. [CrossRef]
50. Shao, J.; Qu, Q.; Wan, Z.; Gao, T.; Zuo, Z.; Zheng, H. From dispersed microspheres to interconnected nanospheres: Carbon-sandwiched monolayered MoS2 as high-performance anode of Li-ion batteries. *ACS Appl. Mater. Interfaces* **2015**, *7*, 22927–22934.
51. Wu, M.; Zhan, J.; Wu, K.; Li, Z.; Wang, L.; Geng, B.; Pan, D. Metallic 1T MoS 2 nanosheet arrays verticaly grown on activated carbon fiber cloth for enhanced Li-ion storage performance. *J. Mater. Chem. A* **2017**, *5*, 14061–14069. [CrossRef]
52. Lei, Z.; Zhan, J.; Tang, L.; Zhang, Y.; Wang, Y. Recent development of metallic (1T) phase of molybdenum disulfide for energy conversion and strage. *Adv. Energy Mater.* **2018**, *8*. [CrossRef]
53. Mohan, M.; Timung, R.; Deshavath, N.N.; Banerjee, T.; Goud, V.V.; Dasu, V.V. Optimization and Hydrolysis of Cellulose under Subcritical Water Treatment for the Production of Total Reducing Sugars. *RSC Adv.* **2015**, *5*, 103265–103275. [CrossRef]
54. Li, M.; Wang, D.; Li, J.; Pan, Z.; Ma, H.; Jiang, Y.; Tian, Z. Facile hydrothermal synthesis of MoS 22 nano-sheets with controllable structures and enhanced catalytic performance for anthracene hydrogeneration. *RSC Adv.* **2016**, *6*, 71534–71542. [CrossRef]
55. Chaudhary, N.; Khanuja, M.; Islam, S.S. Hydrothermal Synthesis of MoS2 Nanosheets for Multiple Wavelength Optical Sensing Applications. *Sens. Actuators A Phys.* **2018**, *277*, 190–198. [CrossRef]
56. Jing, W.; Ding, N.; Li, L.; Jiang, F.; Xiong, X.; Liu, N.; Zhai, T.; Gao, Y. Ag Nanoparticles Modified Large Area Monolayer MoS_2 Phototransistors with High Responsivity. *Opt. Express* **2017**, *25*. [CrossRef]
57. Wang, W.; Klots, A.; Prasai, D.; Yang, Y.; Bolotin, K.I.; Valentine, J. Hot Electron-Based Near-Infrared Photodetection Using Bilayer MoS 2. *Nano Lett.* **2015**, *15*, 7440–7444. [CrossRef]
58. Hong, T.; Chamlagain, B.; Hu, S.; Weiss, S.M.; Zhou, Z.; Xu, Y.Q. Plasmonic Hot Electron Induced Photocurrent Response at MoS 2 –Metal Junctions. *ACS Nano* **2015**, *9*, 5357–5363. [CrossRef]
59. Liu, B.; Gutha, R.R.; Kattel, B.; Alamri, M.; Gong, M.; Sadeghi, S.M.; Chan, W.L.; Wu, J.Z. Using Silver Nanoparticles-Embedded Silica Metafilms as Substrates to Enhance the Performance of Perovskite Photodetectors. *ACS Appl. Mater. Interfaces* **2019**, *11*, 32301–32309. [CrossRef]

60. Wang, Y.; Fullon, R.; Acerce, M.; Petoukhoff, C.E.; Yang, J.; Chen, C.; Du, S.; Lai, S.K.; Lau, S.P.; Voiry, D.; et al. Solution-Processed MoS2/Organolead Trihalide Perovskite Photodetectors. *Adv. Mater.* **2017**, *29*. [CrossRef]
61. Wang, H.; Lim, J.W.; Quan, L.N.; Chung, K.; Jang, Y.J.; Ma, Y.; Kim, D.H. Perovskite–Gold Nanorod Hybrid Photodetector with High Responsivity and Low Driving Voltage. *Adv. Opt. Mater.* **2018**, *6*. [CrossRef]
62. Yu, W.J.; Liu, Y.; Zhou, H.; Yin, A.; Li, Z.; Huang, Y.; Duan, X. Highly Efficient Gate-Tunable Photocurrent Generation in Vertical Heterostructures of Layered Materials. *Nat. Nanotechnol.* **2013**, *8*, 952–958. [CrossRef] [PubMed]

Article

Weekly and Longitudinal Element Variability in Hair Samples of Subjects Non-Occupationally Exposed

Pasquale Avino [1,2], Monica Lammardo [3], Andrea Petrucci [4,*] and Alberto Rosada [4]

1 Department of Agriculture, Environment and Food Sciences (DiAAA), University of Molise, Via F. De Sanctis, 86100 Campobasso, Italy; avino@unimol.it
2 Institute of Ecotoxicology & Environmental Sciences, Kolkata 700156, India
3 FSN-FISS-RNR, ENEA, R.C.-Casaccia, Via Anguillarese 301, 00123 Rome, Italy; monica.lammardo@enea.it
4 ENEA, R.C.-Casaccia, Via Anguillarese 301, 00123 Rome, Italy; albertorosada@libero.it
* Correspondence: andrea.petrucci@enea.it; Tel.: +39-06-3048-6538

Featured Application: Knowledge of weekly and longitudinal variability of metals in the hair of non-professionally exposed people is of particular importance for understanding the levels/effects of chemicals on workers.

Abstract: Hair is an ideal tissue for tracing the human health conditions. It can be cut easily and painlessly, and the relative clinical results can give an indication of mineral status and toxic metal accumulation following long-term or even acute exposure. Different authors have found outdoor pollution phenomena, such as the levels, significantly alter metal and metalloid hair contents. This paper investigates the element concentration variability in hair samples collected from a not-exposed teenager, neither environmentally nor professionally. The sampling was carried out for one week, and the samples were collected from different locations on the scalp. A nuclear analytical methodology, i.e., the Instrumental Neutron Activation Analysis, is used for determining about 30 elements. Some differences have been found among the samplings as well as between the proximal and distal sections. A deep comparison with other similar studies worldwide present in the literature has been performed for evidencing the relationships and the differences due to different ethnical origins, lifestyles, diets, and climates among the different young populations.

Keywords: hairs; variability; week; longitudinal; element; metals; INAA; occupational exposure; unexposed subject

Citation: Avino, P.; Lammardo, M.; Petrucci, A.; Rosada, A. Weekly and Longitudinal Element Variability in Hair Samples of Subjects Non-Occupationally Exposed. *Appl. Sci.* **2021**, *11*, 1236. https://doi.org/10.3390/app11031236

Academic Editor: Samuel B. Adeloju
Received: 27 December 2020
Accepted: 25 January 2021
Published: 29 January 2021

1. Introduction

During these last few decades, the human biomonitoring through biological fluids (blood, urine) or tissues (hair, nails) has been largely used for the assessment of health effects following an occupational or environmental exposure [1–9], particularly for the absorbed element content. Basically, the researchers have focused their attention on identifying baseline element values in different population samples, living in different areas characterized neither by air/water/soil contamination nor by exposure to chemicals. Different countries conducted large-scale surveys to assess the exposure profile of different populations and better understand serious environmental public health problems [10–14] and to establish the baseline ranges of trace elements in their populations. Other examples are the analysis of trace elements in the U.S population by the U.S. National Health and Nutrition Examination Survey (NHANES) [15], in Canada [16] and in European countries [17–21]. This topic is considered relevant both to obtain finger-print data related to a certain area [22] and, in forensic studies, to the provenance of subjects in a specific site [23–26]. Hairs are also important because they can be used to understand the element variation in case of prolonged intoxication [27–29]. Metals are integrated into hairs during their growth: in this sense the element composition of hairs is the signature of the living area and the lifestyle of each person [30,31]. Although there is in literature a great deal of papers regarding the determination of the elemental content in hairs [32–35], to the authors' knowledge, no paper is focused on the element variability during a week. On the other hand, a few studies concern the longitudinal element distribution along human hairs: Yukawa et al. [36] reported the variation of the trace element concentration in long human hairs, showing the profiles according to the distance from the scalp; Kempson and coauthors [37,38] discussed how the exogenous contamination does not influence the levels of some elements (for instance, Zn); on the contrary, Kempson and Skinner [39] report that some other elements, such as Al, are subjected to being accumulated during pollution events and such as Ca, which was demonstrated to be sensitive to endogenous and exogenous contributions [40]; Park et al. discussed the longitudinal association between toenail zinc levels and the incidence of diabetes among American young adults [41]. Finally, Maurice et al. [42] reported a forensic case of poisoning by thallium: the authors developed an analytical method in order to determine the Tl profile all along the hair. The present authors agree with the consideration reported in a recent paper focused on the variation of longitudinal concentration of trace elements in elephant and giraffe hairs [43]. In here, it is highlighted that an important role for this concentration is played by the animals' behavior traits, which suggests that these traits have to be considered also in the study of human hair. Such studies are necessary for having knowledge of endogenous and exogenous roles of hair elements.

Starting from these considerations, this paper would like to investigate the element concentration variability in hair samples collected from a not-exposed teenager, neither environmentally nor professionally. The sampling was carried out for one week, and the samples were collected from different locations on the scalp. Each hair sample was divided into two parts in order to study the section closest (proximal) to the scalp and the most distant (distal) from it. The analyses were performed by Instrumental Neutron Activation Analysis (INAA), and about 30 elements were determined. The use of such analytical technique allowed minimizing the pre-treatment of the samples and hence the relative positive/negative artifacts and performing a multi-elemental determination simultaneously [44–46].

2. Materials and Methods

2.1. Sample Collection and Storage

Hair samples were taken over the course of a week from a female young person (10 years old) chosen among the primary school students. The samples were taken in the right, central, and left nape area on alternate days, with strands of about 10 cm (Figure 1a). All samples were taken by cutting the hair in the chosen area as close as possible to the root (Figure 1b). The cut was made by means of stainless-steel scissors, with zero release of elements [12], in order to avoid any possible contamination caused by the friction between the blades of the scissors and the surface of the hair. The hair is light brown in color, and it is not brittle. In the period preceding the sampling, the subject did not wash her hair, did not use any type of treatment, and did not bathe in the sea. The subject is a non-smoker and was not subjected to second-hand smoke; she lives in a medium-small suburban center (9000 inhabitants) of Central Italy, where there are no industrial settlements, whereas farms that employ biological agriculture are far from the site, and the vehicle traffic is limited except on Saturday morning for the weekly food market.

Immediately after collection, the wisp of hair was placed in high-purity Kartell nuclear grade containers. In laboratory, the containers with the wisp of hair were immediately placed in a silica gel dryer and kept in the dark in an environment with a temperature between 20 and 27 °C, until the time of pre-treatment and subsequent weighing.

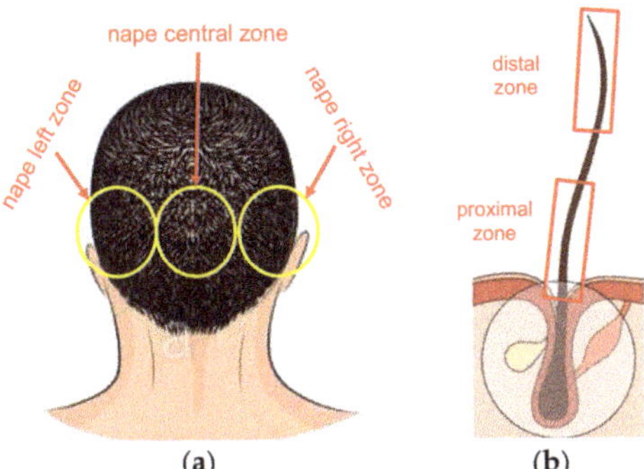

Figure 1. Scheme of the sampling area (**a**) and representation of the proximal and distal hair zones analyzed (**b**).

2.2. Hair Pre-Treatment

The sample pre-treatment was performed by means of a standard procedure, already adopted in previous samplings [12], following a protocol suggested by the International Atomic Energy Agency (IAEA) [47]: washing the samples in 25 mL of ultrapure acetone for 10 min and subsequently rinsing them with ultrapure deionized H_2O (resistivity 18 M$\Omega \times$ cm), repeated three times.

After drying in an oven at 40 °C for 15 min, all the samples were placed back in the desiccator, brought back to room temperature, and weighed in containers of nuclear-grade polythene for neutron irradiation. The weighing was carried out by means of the AE 160 analytical balance (Mettler-Toledo GmbH, Greifensee, Switzerland) having a sensitivity of $\pm$0.1 mg.

2.3. Neutron Irradiation and Gamma Measurements

The neutron irradiation of the samples was carried out in the rotating rack of the TRIGA Mark II research nuclear reactor in the ENEA Casaccia Research Centre. The irradiation time was 24 h, and the neutron flux was $\Phi = 2.34 \times 10^{12}$ n $\times$ cm^{-2} $\times$ s^{-1} (corresponding to fluence F = 1.997 $\times$ 10^{17} n $\times$ cm^{-2}). These parameters were decided in order to detect those elements that, once activated, produce radionuclides whose half-life was between 3 h and several years [48,49].

After the irradiation in the rotating rack, two sets of measurements were carried out: the first one was after 20 h of decay and lasted between 40 min and 1.5 h; the second one was after 15 days of decay and lasted between 20 and 72 h. The position of the samples in the two sets of measurements was vertical at a distance of 4 cm from the detector for the first set and vertical but in contact with the detector for the second set (Figure 2).

The measurements were carried out by an HPGe coaxial Canberra detector with a resolution (FWHM) of 1.88 keV at 1332.5 keV, a relative efficiency of 42.1, and a Peak-to-Compton ratio of 65.8:1. The system has a Canberra multichannel analyzer with 8192 channels.

The energy and efficiency calibrations for the different counting geometry were carried out respectively with a source of ^{137}Cs and ^{60}Co and with a source of ^{152}Eu, whose activities were certified by the Centre Communitaire de Référence (CEA).

Primary standards and secondary reference materials (SRMs) were used: as primary standards, single-element solutions at a concentration of 1000 µg mL^{-1}; as SRM material with similar composition to the investigated matrix, the BCR CRM 397 (human hair), just used in a previous study [50], as well as the NIST1515 (Apple Leaves) for an overall checking.

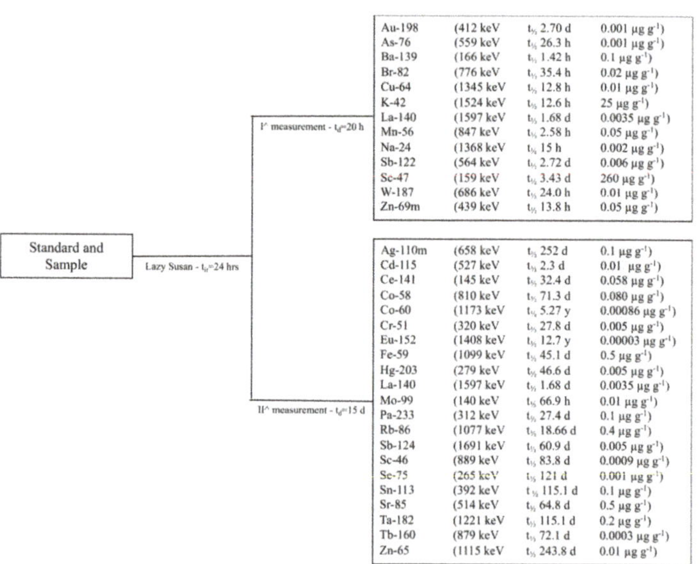

Figure 2. Instrumental Neutron Activation Analysis (INAA) master scheme (t_{ir} irradiation time; t_d decay time; $t_{\frac{1}{2}}$ half-life time). In brackets for each product, nuclides are reported as well as some nuclear analytical chemistry parameters, i.e., peak energy (keV), half-life time (m minute, h hour, d day, y year), and limit of detection (LOD).

3. Results and Discussion

3.1. Quality Control and Quality Assurance (QC/QA)

The environmental studies regarding the correlation between pollutants and biological tissues, such as heavy elements in human hair, need very sensitive and accurate analytical techniques in order to determine contaminants at trace and ultra-trace levels [51]. Among the different possibilities (e.g., spectroscopy, electrochemistry, etc.), the nuclear methods are still the main available techniques to address such stringent requirements. Their high accuracy and precision and the very low limits of detection (LODs) allow investigating a matrix deeply [52]. Furthermore, the possibility of avoiding chemical–physical treatments or performing radiochemical separations is of fundamental importance for achieving these results. Even if INAA is considered a primary analytical technique—i.e., it is possible to analyze the sample just knowing all the nuclear parameters (such as all the nuclear cross-sections of each radionuclide and the nuclear reactor data)—the easier way to analyze a sample is by comparing its activity with that of a standard irradiated in the same conditions. The comparison between the data analysis obtained by INAA and the certified values is the first step for assessing the quality assurance and the quality control (QA/QC). For this aim, standard reference materials (SRMs) such as BCR CRM 397 (Human Hair) and NIST 1515 (Apple Leaves) were chosen according to the matrix similarity and biological origin. Table 1 reports the differences (Δ expressed as %) between our data and the certified values of both the SRMs for each element.

It should be noted that the Human Hair SRM shows a Δ below 6.5%: Se and Zn show high precision and accuracy as well as the indicative or informative element values (i.e., As, Co, Cr, Cu, Fe, Hg, and Mn). The comparison profile for the Apple Leaves is slightly different: among the certified elements, good Δ (below 16%) are achieved for Ba, Ca, Fe, K, Mn, Na, Rb, Sr, and Zn, as also reported in previous paper [53], whereas Hg and Ni show quite relevant differences such as −48.4% and −21.9%, respectively. For Ni, the reason is due to the poor INAA sensitivity. Mercury shows two different results: for Human Hair SRM, the difference between measurements is 5.7%, whereas for Apple Leaves, the SRM is −48.4%. According to our evaluation, the cause has to do with the different Hg levels in the two SRMs. In the first SRM, Hg is at 12.3 μg g^{-1}, whereas the authors find 13.0 μg g^{-1} with a coefficient of variation (CV%, defined as the ratio between standard deviation and mean value × 100) of 6.1, a good value according to the references [49,54,55]. In the second SRM, the certified Hg value is 0.0432 μg g^{-1} (or 43.2 ng g^{-1}), whereas the authors find 0.0223 μg g^{-1} with a CV% of 45.7, which is an unsatisfactory value according to the ref [54]. The second SRM shows an Hg content almost 300 times lower than the first SRM: this occurrence justifies the high Δ between ours and the certified data in the second SRM.

Table 1. Analytical standard comparison (mean ± s.d.; μg g^{-1}) between our data and certified values: BCR CRM 397 (human hair) and NIST-SRM 1515 (Apple Leaves).

Elem.	Nucl. [1]	BCR CRM 397			NIST-SRM 1515		
		Found	Certified	Δ	Found	Certified	Δ
As	^{76}As	0.29 ± 0.02	(0.31 ± 0.02) [a]	−6.5	n.d.	-	-
Ba	^{131}Ba	n.d.	-	-	51.0 ± 8.0	48.8 ± 2.3	4.3
Br	^{82}Br	n.d.	-	-	3.1 ± 0.7	(1.8) [b]	41.9
Ca	^{47}Ca	n.d.	(1560 ± 40) [b]	-	15267 ± 676	15250 ± 100	0.3
Ce	^{141}Ce	n.d.	-	-	3.6 ± 0.3	(3) [b]	20.0
Co	^{60}Co	0.56 ± 0.02	(0.55 ± 0.03) [b]	1.8	0.10 ± 0.01	(0.09) [b]	10.0
Cr	^{51}Cr	93 ± 9	(91 ± 33) [b]	2.2	0.4 ± 0.1	(0.3) [b]	33.3
Cu	^{64}Cu	113 ± 7	(110 ± 5) [a]	2.7	n.d.	5.69 ± 0.13	-
Fe	^{59}Fe	575 ± 4	(580 ± 10) [b]	−0.9	72 ± 1	82.7 ± 2.6	−12.9
Hg	^{203}Hg	13.0 ± 0.8	12.3 ± 0.5	5.7	0.0223 ± 0.0102	0.0432 ± 0.0023	−48.4
K	^{42}K	n.d.	-	-	16035 ± 717	16080 ± 210	−0.6
La	^{140}La	n.d.	-	-	20.4 ± 0.5	(20) [b]	2.0
Mn	^{56}Mn	10.9 ± 0.6	(11.2 ± 0.3) [b]	−2.7	54.4 ± 1.6	54.1 ± 1.1	0.6
Na	^{24}Na	n.d.	-	-	28.3 ± 0.7	24.4 ± 2.1	16.0
Ni	^{58}Co	n.d.	(46.0 ± 1.4) [a]	-	0.731 ± 0.080	0.936 ± 0.094	−21.9
Rb	^{86}Rb	n.d.	-	-	9.4 ± 0.9	10.2 ± 1.6	−7.8
Sc	^{46}Sc	n.d.	-	-	0.028 ± 0.001	(0.03) [b]	−6.7
Se	^{75}Se	2.1 ± 0.1	2.00 ± 0.08	5.0	n.d.	-	-
Sm	^{153}Sm	n.d.	-	-	2.9 ± 0.1	(3) [b]	3.3
Sr	^{85}Sr	n.d.	-	-	28.0 ± 3.0	25.1 ± 1.1	11.6
Zn	^{65}Zn	196 ± 4	199 ± 5	−1.5	12.50 ± 0.11	12.45 ± 0.43	0.4

[1] Product nuclide; [a] indicative values expressed as μg g^{-1}; [b] informative values expressed as μg g^{-1}; n.d.: not detected; Δ: difference (%) between our mean values calculated and certified one as $\frac{(ourvalue - certifiedvalue)}{certifiedvalue} \times 100$. The standard deviation is calculated among five replicas.

3.2. Element Content in Human Hair Samples

Before performing the analysis, the authors worried about the effects of cleaning and cutting hair before sampling. The problem of hair cleaning is largely discussed as well as the effect of washing using different procedures, as reported in the literature [56–62]. The IAEA recommends a cleaning procedure for hair [47]. Frequent head washing does not affect (the significance limit greater than 0.05) the content of some elements (Br, Co, Cu, Mn, Se, Zn) or of some pollutants (Ni, Hg); on the other hand, As does get washed out but with no such great amount (lower by 1.7 times) [39,47,63]. The hairs were cut by metallic scissors: before that, the oxides were carefully removed. Any effect of the friction between blades and hair shaft were limited in order to prevent possible sample contamination.

Table 2 shows the level of 24 elements measured in the samples investigated along with the minimum and maximum values and the CV%. First, two important elements, i.e., As and Hg, considered hazardous elements for the human health and of exogenous origin, are below the corresponding LODs: this is a preliminary confirmation that the subject is not exposed to sources of As and Hg. Other considerations could be drawn about essential elements, i.e., Cr, Cu, Fe, Mn, Se, and Zn: their variability, expressed as CV%, is below 40% except for Cr and Cu, which have 66.2% and 68.8%, respectively.

Table 2. Element levels and single daily concentrations (μg g^{-1}) determined in all the investigated independent samples (five independent measurements).

Element	Concentration	Concentration (Mean ± s.d. [1])		
	x (Min–Max; CV%)	1st Day	2nd Day	3rd Day
Ag	0.740 (0.34–1.2; 36.6)	0.630 ± 0.354	0.855 ± 0.417	0.735 ± 0.134
As	<LOD	<LOD	<LOD	<LOD
Au	0.022 (0.008–0.37; 63.6)	<LOD	0.010 ± 0.03	0.028 ± 0.013
Ba	1.44 (1.20–1.69; 15.9)	1.20 ± 0.18	1.58 ± 0.31	1.50 ± 0.27
Br	26.5 (22.9–32.1; 13.2)	25.0 ± 3.0	24.1 ± 0.6	30.4 ± 2.3
Ca	2010 (1363–2601; 25.4)	1982 ± 875	2264 ± 468	1784 ± 303
Co	0.041 (0.031–0.066; 31.7)	0.032 ± 0.002	0.038 ± 0.006	0.054 ± 0.017
Cr	0.086 (0.002–0.180; 66.2)	0.086 ± 0.006	0.081 ± 0.007	0.092 ± 0.128
Cu	15.1 (2.30–27.8; 68.8)	19.9 ± 9.0	21.2 ± 9.4	4.17 ± 2.70
Eu	0.001(0.001; N/A)	<LOD	<LOD	0.001
Fe	17.7 (13.5–28.4; 33.3)	14.1 ± 0.1	14.3 ± 1.1	24.6 ± 5.4
Hg	<LOD	<LOD	<LOD	<LOD
K	249 (178–364; 26.5)	195 ± 23	228 ± 8	323 ± 58
La	0.957 (0.288–1.54; 61.8)	0.915 ± 0.886	0.805 ± 0.641	1.35 ± 0.12
Mn	3.75 (1.91–5.22; 33.0)	3.10 ± 1.70	3.20	4.66 ± 0.69
Na	109 (88.2–141; 16.5)	95.1 ± 9.8	109 ± 6	122 ± 27
Rb	0.269 (0.105–0.540; 60.9)	0.117 ± 0.017	0.319 ± 0.070	0.370 ± 0.241
Sb	0.075 (0.045–0.120; 37.3)	0.055 ± 0.013	0.076 ± 0.029	0.096 ± 0.034
Sc	0.007 (0.002–0.011; 71.4)	0.007 ± 0.007	<LOD	0.007 ± 0.006
Se	0.413 (0.035–0.050; 13.3)	0.390 ± 0.057	0.455 ± 0.064	0.395 ± 0.049
Tb	0.020 (0.020; N/A)	<LOD	<LOD	0.020
Th	0.009 (0.000–0.036; 200)	0.00033	0.00043	0.018 ± 0.025
W	0.019 (0.012–0.043; 63.1)	0.012 ± 0.001	0.015 ± 0.003	0.031 ± 0.017
Zn	52.3 (66.3–92.5; 37.2)	77.1 ± 12.0	89.7 ± 3.9	72.1 ± 8.2

[1] s.d. standard deviation.

Br, Ca, K, and Na, the last three considered labile elements (because they are strongly influenced by washing [64]) show a low CV% below 27%, confirming that the hair was not washed. Silver and gold are present at low concentrations, 740 and 22 ng g^{-1}, respectively: their presence could suggest a previous use of shampoos containing nanoparticles of these two elements into the composition for antimicrobial activity [65,66]. Other elements such as La, Rb, Sc, Th, and W can be considered of environmental origin: their CVs% are high, above 60%, especially Th, up to 200%. Finally, Ba, Br, and Sb show low CVs%: their presence and levels could be due to anthropogenic pollution and particularly to airborne particulate matter, as just evidenced by authors in previous papers [46,49].

Table 3 shows the concentration trend (the number of samples per nuclide is too low—only three scalp locations and two longitudinal positions—to obtain any actual reasonable trend that could be considered real and not obtained by chance) in the areas of the nape where the sampling was carried out (left, center, and right) as well as the longitudinal variation of the concentrations along the hair.

If the trends of Br, Ca, K, and Na from Tables 2 and 3 are taken into account, similar concentrations are noted for Br both during the days and in correspondence with the sampling zones and in the longitudinal variation: mean value 26.5 ± 3.46 μg g^{-1}; 1st day 25.0 ± 3.0 μg g^{-1}; 2nd day 24.1 ± 0.6 μg g^{-1}; 3rd day 30.4 ± 2.3 μg g^{-1} (Table 2); left zone 27.1 ± 2.1 μg g^{-1}; central 23.7 ± 0.45 μg g^{-1}; right 28.8 ± 1.7 μg g^{-1}; hair proximal section 26.5 ± 2.6 μg g^{-1}; distal 26.5 ± 4.9 μg g^{-1}. The same stability in the concentration trend is evident for Na, Ca, and K. These four elements, i.e., Br, Ca, K, and Na, are ubiquitous in the environment; besides, Ca, K, and Na are among the fundamental components of the tissues and biological fluids in the human body, and K and Na are also present in the body secretions (exudate, etc.). In the samples, Br, K, and Na show an increasing concentration trend between the first and third day (Table 2). Similarly, the same trend is shown, but in a much more marked way, by elements of environmental origin such as La, + 47.5%

increase in concentration on the third day compared to the first; Rb, + 216%; Th, + 98%; and W + 158%. The authors would like to underline that a fundamental requirement of this study was to have unwashed hairs, in order to understand the natural element levels in human hair. Confirming this statement (i.e., unwashed hair), a similar trend is found for all the elements mentioned above, with the exception of bromine which remains constant, whereas greater increases are found for the elements of environmental origin.

Table 3. Element concentrations (μg g^{-1}) in the nape different areas and along the hair (proximal and distal zones), (five independent measurements).

	Element Concentration (Mean ± s.d. [1])				
	Scalp Location			Longitudinal Variability	
	Left Zone	Central Zone	Right Zone	Proximal	Distal
Ag	0.882 ± 0.250	1.15 ± 0.295	0.830 ± 0.095	0.953 ± 0.172	0.527 ± 0.133
Au	<LOD	<LOD	0.037 ± 0.009	0.037	0.015 ± 0.006
Ba	<LOD	<LOD	1.30 ± 0.19	1.30	1.49 ± 0.26
Br	27.1 ± 2.1	23.7 ± 0.5	28.8 ± 1.7	26.5 ± 2.6	26.5 ± 4.9
Ca	1363 ± 619	1933 ± 331	1998 ± 214	1765 ± 349	2255 ± 594
Co	0.033 ± 0.001	0.034 ± 0.004	0.042 ± 0.012	0.036 ± 0.005	0.046 ± 0.018
Cr	0.081 ± 0.004	0.086 ± 0.005	0.092 ± 0.090	0.056 ± 0.047	0.116 ± 0.058
Cu	26.3 ± 6.4	27.8 ± 1.8	6.12 ± 1.91	20.1 ± 12.1	10.1 ± 6.8
Eu	<LOD	<LOD	0.00070 ± 0.00035	0.01	<LOD
Fe	14.2 ± 0.1	13.5 ± 0.8	20.8 ± 3.8	16.2 ± 4.0	19.2 ± 8.0
K	211 ± 16	234 ± 6	282 ± 41	242 ± 36	255 ± 97
La	0.288 ± 0.626	0.351 ± 0.453	1.35 ± 0.67	0.663 ± 0.595	1.40 ± 0.20
Mn	4.30 ± 1.21	3.20 ± 1.61	4.20 ± 0.48	3.90 ± 0.61	3.52 ± 2.30
Na	102 ± 7	105 ± 4	103 ± 19	103 ± 1	114 ± 26
Rb	0.129 ± 0.011	0.270 ± 0.049	0.200 ± 0.170	0.199 ± 0.070	0.338 ± 0.219
Sb	0.064 ± 0.009	0.096 ± 0.014	0.072 ± 0.019	0.077 ± 0.017	0.073 ± 0.040
Sc	0.012 ± 0.005	<LOD	0.007 ± 0.004	0.008 ± 0.007	0.007 ± 0.006
Se	0.352 ± 0.004	0.414 ± 0.045	0.430 ± 0.035	0.397 ± 0.042	0.430 ± 0.070
Tb	<LOD	<LOD	0.020 ± 0.010	0.020	<LOD
Th	<LOD	<LOD	0.001 ± 0.017	0.001	0.012 ± 0.020
W	0.012 ± 0.000	0.013 ± 0.002	0.019 ± 0.012	0.015 ± 0.04	0.024 ± 0.016
Zn	68.6 ± 4.1	87.0 ± 2.7	66.3 ± 5.8	74.0 ± 11.3	85.3 ± 7.3

[1] s.d. standard deviation.

A similar trend is shown by the daily trend (the number of samples per nuclide is too low—only 3 days—to obtain any actual reasonable trend that could be considered real and not obtained by chance) of some elements considered essential, such as Cr, Fe, and Mn; for the last two, the increase is significant: + 74.5% and + 50.3% respectively. Cr and Fe increase their level in the distal area with respect to the proximal one, and a similar trend is also shown by Se and Zn, which, instead, have a daily trend with a maximum on the second day.

However, there are elements that maintain a constant concentration both with regard to the daily trend (Table 2) and in the nape sampling areas and in the distal and proximal ones (Table 3), as highlighted by the coefficient of variation (CV%) calculated taking into account all the concentrations obtained. These elements are the ones present either in the body's tissues or in the biochemical systems, such as Ca, K, Na, and Br (CV% 15.2%, 16.6%, 7.76%, and 8.64%, respectively), and those considered essential such as Mn, Se, and Zn (15.7%, 7.73%, and 11.4%, respectively) and those considered pollutants of anthropogenic origin: Ba, Sb, and Co (10.7%, 18.7%, and 19.2%, respectively). For the last ones, it is possible to hypothesize a hair contamination deriving from their levels in the environment, whereas the level stability of the essential elements is due to regular biochemical systems/reactions. This occurrence allows defining basal elements concentration levels in the hair analysis both in good health conditions and in "anomalies" caused by occupational and/or environmental exposure.

On the other hand, the concentration of elements of environmental origin (i.e., La, Rb, Sc, Th, and W) is less stable: their CVs vary between 38.4%, tungsten, and 126.9%, thorium. The hypothesis that can be advanced is that their presence in the environment, and therefore their levels in the hair, is connected to the variation of local climatic conditions: wind direction, wind speed, temperature, humidity, etc.

A trend similar to that of the elements of environmental origin is also shown by two chemicals considered essential, i.e., Cr and Cu, and, albeit to a lesser extent, by Fe (CV% 45%, 53.2%, and 23.5%, respectively). For these three elements, the hypothesis of a double origin could be put forward; i.e., there is an overlapping between the element basal levels (i.e., the concentration levels naturally present in the human body) and the concentrations deriving from environmental pollution. This hypothesis can be confirmed because these three elements (i.e., Cr, Cu, and Fe) significantly increase their concentrations in the distal area of the hair (i.e., the section more in contact with the environment) compared to the proximal area (Table 3), even if their average concentration level agrees with hair data reported for Rome and Italy (Table 4). Finally, it is interesting to note that some elements of environmental origin but coming from anthropogenic pollution and considered particularly toxic such as As, Hg, and Ni were not revealed by INAA, i.e., they show levels below the LOD (Figure 2). Therefore, it can be assumed that As, Hg, and Ni are < 1 ng g^{-1}, < 5 ng g^{-1}, and < 80 ng g^{-1}, respectively. Very low levels of Eu and Tb, considered Rare Earth Elements (REEs) of environmental origin, were measured just in one hair sample at concentrations of 1 and 20 ng g^{-1} respectively (LOD for Eu 0.03 and for Tb 0.3 ng g^{-1}).

Table 4. Concentration (μg g^{-1}) comparison between these data and those found in other Italian and international studies.

	Our Data	Literature Values				
		Rome [50]			Italy [67]	International [68–76]
		Female	Male + Female	Light Brown		
Ag	0.740 ± 0.271	0.23 ± 0.42	0.40 ± 1.58	0.29 ± 0.30	0.83 ± 1.96	0.005–9.9
Au	0.022 ± 0.014	0.06 ± 0.08	0.09 ± 0.16	0.10 ± 0.12	0.036 ± 0.038	0.0005–591
Ba	1.44 ± 0.23	0.33 ± 0.94	0.57 ± 1.50	0.87 ± 1.96		0.12–29
Br	26.5 ± 3.5	9.63 ± 15.2	8.66 ± 13.8	5.54 ± 4.92	12.9 ± 10.1	0.15–490
Ca	2010 ± 512	812 ± 580	1120 ± 1060	1776 ± 1776	750 ± 1150	7–10887
Co	0.041 ± 0.013	0.17 ± 0.26	0.17 ± 0.25	0.12 ± 0.17	0.145 ± 0.133	0.002–15
Cr	0.086 ± 0.057	0.15 ± 0.10	0.17 ± 0.15	0.17 ± 0.09	0.234 ± 0.155	0.026–65.3
Cu	15.1 ± 10.4	9.56 ± 4.78	10.5 ± 7.11	11.6 ± 8.91		3.68–72.5
Eu	0.001					-
Fe	17.7 ± 5.9	16.8 ± 26.4	16.5 ± 23.9	17.5 ± 19.3	13.5 ± 6.2	3–2400
K	249 ± 66	335 ± 272	314 ± 274	258 ± 280	940 ± 1110	0.94–2370
La	0.957 ± 0.592	0.03 ± 0.04	0.03 ± 0.03	0.03 ± 0.03	0.038 ± 0.031	<0.003–37
Mn	3.75 ± 1.24	0.40 ± 0.33	0.42 ± 0.32	0.41 ± 0.24		0.03–81.5
Na	109 ± 18	647 ± 545	581 ± 521	476 ± 426	1180 ± 1260	0.04–3500
Rb	0.269 ± 0.164				3.20 ± 5.06	0.01–0.14
Sb	0.075 ± 0.028	0.06 ± 0.06	0.06 ± 0.05	0.07 ± 0.05	0.047 ± 0.034	0.007–38
Sc	0.007 ± 0.005	0.0010 ± 0.0003	0.0010 ± 0.0002	0.0011 ± 0.0008	0.0023 ± 0.0025	0.0004–550
Se	0.413 ± 0.055	0.96 ± 1.55	0.99 ± 1.56	1.17 ± 1.62	0.64 ± 0.28	0.002–66
Tb	0.020					-
Th	0.009 ± 0.018				0.0011 ± 0.0014	-
W	0.019 ± 0.012					-
Zn	52.3 ± 19.5	178 ± 46	189 ± 82	240 ± 144	139.9 ± 65.1	<1–1770

3.3. Comparison with Studies on Adult and Teenager Population

Table 4 shows a comparison between these data and similar levels found in a previous study performed in Rome [50].

First, the presence of elements such as Ag, Au, and Co can be due to cosmetics and/or personal hygiene products. In particular, Ag shows concentrations comparable with the data of the Rome group (0.740 $\pm$ 0.271 $\mu g\ g^{-1}$ versus 0.40 $\pm$ 1.58 $\mu g\ g^{-1}$) [50] and the Italian population (0.83 $\pm$ 1.96 $\mu g\ g^{-1}$) [67] as well as Au (0.022 $\pm$ 0.014 $\mu g\ g^{-1}$ vs. 0.036 $\pm$ 0.038 $\mu g\ g^{-1}$ for the Italian population) [50], whereas Co shows decidedly lower levels than the Italian population (0.041 $\pm$ 0.013 $\mu g\ g^{-1}$ vs. 0.145 $\pm$ 0.133 $\mu g\ g^{-1}$), but they fall within the international data (0.002–15 $\mu g\ g^{-1}$) [68–76].

Levels of Ca, K, and Na are in line with the data of the Rome and Italy groups. In particular, Ca and K show similar levels to the light brown hair in the Rome group [50] (Ca: 2010 $\pm$ 512 $\mu g\ g^{-1}$ vs. 1776 $\pm$ 1776 $\mu g\ g^{-1}$; K: 249 $\pm$ 66 $\mu g\ g^{-1}$ vs. 258 $\pm$ 280 $\mu g\ g^{-1}$) with a much lower standard deviation. Similar considerations can be also put forward for K, which are compatible with the relevant standard deviation of both Rome and Italian data. Br is instead at concentration levels equal to more than double the concentration data of all the Rome and Italian samples, even though it is slightly above the variability of the Italian data (26.5 $\pm$ 3.5 $\mu g\ g^{-1}$ vs. 12.9 $\pm$ 10.1 $\mu g\ g^{-1}$) but it falls within the international data range (0.15–490 $\mu g\ g^{-1}$).

Among the essential elements, Cr, Cu, Fe, Se, and Zn show good agreement with the data of Rome and Italy groups, whereas Co presents lower levels than those of Rome and Italy groups and Mn higher than the ones of the Rome samples.

As for the elements of environmental origin, the levels of Rb, Sc, and Th are in full agreement with the data reported in Table 4, whereas W is absent in all the other comparison samples, and La is at significantly higher levels. For La, as well as for Mn and Br, all the concentration data are higher than the comparison values [50]: 0.957 $\pm$ 0.592 $\mu g\ g^{-1}$ vs. 0.038 $\pm$ 0.031 $\mu g\ g^{-1}$ in the Italian reference group, i.e., 25 times higher; Mn 3.75 $\pm$ 1.24 $\mu g\ g^{-1}$ vs. 0.42 $\pm$ 0.32 $\mu g\ g^{-1}$ in the entire series of the Rome samples, 9 times higher; Br 26.5 $\pm$ 3.5 $\mu g\ g^{-1}$ versus 12.9 $\pm$ 10.12 $\mu g\ g^{-1}$ in the Italian group, 2 times higher. This anomaly can be explained by environmental contamination, since these three elements are present in the urban atmosphere, especially in airborne and dust depositions [77,78]. In previous papers [79,80], the authors report data on the elements in PM_{10} and $PM_{2.5}$ airborne sampled in downtown Rome along with the trends during the last three decades: both La and Mn and Br show significant levels (particularly, La 170 $\pm$ 101 pg m^{-3} in PM_{10}; Mn 60 $\pm$ 44 ng m^{-3}, Br 17.1 $\pm$ 13.9 ng m^{-3} in $PM_{2.5}$).

Table 5 shows a large comparison among our data and data collected from teenagers (boys and girls) worldwide [81–91]. The table also reports some reference data from Korean [92], Canadian [93], and American [94] teenagers hair values. First of all, it could be seen that our data are broader and more accurate than those determined in other studies, considering the high performance of the analytical method used, i.e., INAA. This occurrence allows drawing only few significant considerations on the content of some metals: Ca (from 2 to 10 times) and Mn (up to 10 times; ours and some other data are similar, an exception is present in the data collected from girl teenagers in Rome, i.e., 69.2) are higher than those reported in the other studies [81–91] and in the reference values contained in [92–94], whereas Cr and As are always less than those reported in the literature. Finally, Ag, Cu, Fe, Se, and Zn show levels close to those reported worldwide.

For a better correlation and significance of this comparison, the authors have reported the correlations in Table 6 and the relative plots in Figure 3 between our data and those determined in girl teenagers hair worldwide [81–91]. As it can be seen, there is high correlation (above 0.9) with the data found, especially in West Europe (different Italian locations, Spain), and low correlation (ranging between 0.1 and 0.6) with those determined in Artic area (Arkhangelsk) and Korea, for this latter also for boys. This occurrence can be interpreted considering the different ethnical origins, lifestyles, diets, and climates (including the presence of different and massive anthropogenic and/or natural sources) among the different young populations.

Table 5. Comparison of element levels (μg g^{-1}) found in hair samples of teenagers worldwide.

(a)																			
	Our	Pakistan [81]		Russian Federation [82]				Italy [83]		Korean [84]		Brazil [85]		Italy [86]		East Aral Sea [87]			
				North Eur.		Arkhangelsk		Rome						Palermo		Kazalinsk		Zhanakorgan	
	G [1]	B	G	B	G	B	G	B	G	B	G	B	G	B	G	B	G	B	G
Ag	0.74			0.25	0.25	0.17	0.16												
As	<LOD [2]							0.08	0.2	8.99	8.56	0.004	0.01	0	0	0.66	0.61	0.78	0.58
Au	0.022																		
Ba	1.44			0.5	0.5	2.45	2.19			0.33	0.32			1.06	1.5				
Br	26.5															7.37	10.38	7.56	5.79
Ca	2010	567	577.4	1125	850	478.58	1496.76	347	630	198.31	227.75					806	904	887	1170
Co	0.041	1.925	0.881	0.5	0.6	0.2	0.19	0.48	0.19	0.01	0.01	0.008	0.007	0.26	0.16	0.73	1.03	1.41	0.96
Cr	0.086	1.128	2.952	1	1	0.73	0.62	3.24	1.73	0.48	0.45			0.21	0.08	0.52	0.43	0.52	0.7
Cu	15.1	8.06	22.12	11.5	11	10.61	9.42	9.2	16.7	15.09	15.96			18.88	24.44	12.4	12.8	25.8	20.8
Eu	0.001																		
Fe	17.7	35.21	77.4	31	32.5	57.4	27.75	38.5	14.5	12.4	12.86					47.5	43.8	56	44.8
Hg	0.005									0.49	0.51	0.22	0.09			0.87	1.51	0.88	0.88
K	249			530	800	1410	93.04			32.71	35.59					387	714	838	819
La	0.957																		
Mn	3.75	2.172	4.651	1.75	1.75	2.86	2.2	0.28	69.2	0.3	0.27	0.35	0.29	0.34	0.22	3	3.59	2.76	2.97
Na	109									27.8	26.42					619	782	401	492
Rb	0.269													0.03	0.01				
Sb	0.075											0.005	0.005	0.04	0.02				
Sc	0.007																		
Se	0.413			1.9	1.9	0.85	0.88	0.51	0.48	0.75	0.74	0.12	0.12	0.59	0.41	0.91	0.85	0.69	0.85
Sr	0.5			2.75	2.75	1.93	8.12	1.02	2.32					4.05	7.56	10.12	11.7	8.34	14.34
Tb	0.02																		
Th	0.009																		
W	0.019																		
Zn	52.3	117.4	225.1	187.5	187.5	155.59	184.76	153	162	72.53	67.25			137.12	197.2	199.6	175	197.6	193.7

Table 5. *Cont.*

(b)

	Italy [88]		Spain [89]		Poland [90]		Italy [91]						Reference Values		
	Rome		Madrid				Sicily (Urban)		Sicily (Rural)		Sicily (Ind.)		Korean	Canadian	American
	B	G	B	G	B	G	B	G	B	G	B	G	[92]	[93]	[94]
Ag			0.194	0.335											
As	0.096	0.09	0.07	0.07			0.0003	0.0003	0.06	0.03	0.06	0.03	0.18–0.24	<1	0.10–0.20
Au															
Ba			0.3	0.6			0.91	1.49	0.21	1.44	0.53	1.39	0.4–0.6	<0.6	0.00–2.60
Br															
Ca	447.9	374.6			186	252							270–430	190–738	220–970
Co	0.954	0.052	0.0112	0.02183			0.07	0.08	0.17	0.2	0.12	0.16	0.01–0.01	0.01–0.05	0.01–0.03
Cr	0.559	0.427	0.4	0.4			0.17	0.06	0.24	0.28	0.14	0.11	0.80–0.90	0.23–0.80	0.20–0.80
Cu	21.74	10.5	16.2	40.3	10	11	18.16	20.83	14.81	18.93	10.18	12.85	13–16	3.9–23.3	9–39
Eu															
Fe	19.48	12.8	18.2	18.1	9	9							11–13	11–24	5–16
Hg													0.5–0.7	<1.5	0.09–0.18
K													770–1490	5–40	20–240
La															
Mn	0.426	0.27	0.34	0.398			0.22	0.31	0.32	0.41	0.28	0.37	0.23–0.32	0.20–0.80	0.10–1.30
Na													340–650	18–85	40–360
Rb							0.03	0.01	0.1	0.11	0.14	0.12			
Sb							0.03	0.01	0.05	0.05	0.03	0.04			
Sc															
Se	0.838	0.45	0.4	0.4			0.44	0.39	0.72	0.71	0.39	0.32	0.80–0.90	0.50–2.00	0.30–1.80
Sr			0.57	1.9			3.43	7.87	0.8	3.53	5.31	24.35			
Tb															
Th															
W															
Zn	159.1	144.5	101	90.6	138	163	132.41	197.21	199.16	257.51	168.7	173.06	50–60	180–220	100–210

[1] B: boys, G: girls; [2] LOD: limit of detection.

Table 6. Linear regression and coefficient of determination (R^2) between our data and each dataset for different girl teenagers hair determined worldwide [81–91].

Country	Sex	Linear Regression	R^2
Pakistan			
	girls	y = 4.3399 x − 8.841	0.9596
	boys	y = 2.2490 x − 5.702	0.9486
Russian Federation			
North Europe	girls	y = 3.2442 x − 6.0669	0.9962
	boys	y = 2.1688 x + 2.6307	0.9770
Arkhangelsk	girls	y = 0.4252 x + 16.754	0.2921
	boys	y = 5.6558 x − 26.022	0.9887
Italy			
Rome (2000)	girls	y = 2.6415 x − 8.3798	0.9097
	boys	y = 2.8262 x − 5.304	0.9436
Korean			
	girls	y = 0.1558 x + 8.2496	0.3186
	boys	y = 0.1489 x + 8.7558	0.2621
Brazil			
	girls	y = 0.0685 x + 0.0381	0.8539
	boys	y = 0.0768 x + 0.0630	0.6464
Italy			
Palermo	girls	y = 3.6786 x − 3.6842	0.9690
	boys	y = 2.5653 x − 2.4715	0.9739
East Aral Sea			
Kazalinsk	girls	y = 3.4555 x + 9.0975	0.7891
	boys	y = 2.1209 x + 21.810	0.6089
Zhanakorgan	girls	y = 3.4786 x − 4.0669	0.9684
	boys	y = 3.4353 x − 6.7967	0.9882
Italy			
Rome (1992)	girls	y = 2.7462 x + 2.2731	0.7651
	boys	y = 2.9344 x − 7.3893	0.9395
Spain			
Madrid	girls	y = 1.7397 x − 0.6410	0.9506
	boys	y = 1.8624 x − 3.0717	0.9646
Poland			
	girls	y = 4.2421 x − 59.335	0.9945
	boys	y = 3.5639 x − 48.761	0.9952
Italy			
Sicily—urban	girls	y = 3.6786 x − 3.6842	0.9690
	boys	y = 2.4794 x − 2.5041	0.9741
Sicily—rural	girls	y = 4.7754 x − 6.3693	0.9532
	boys	y = 3.7000 x + 5.1878	0.9541
Sicily—industrial	girls	y = 3.1540 x − 1.8459	0.9286
	boys	y = 3.1144 x − 4.0461	0.9461

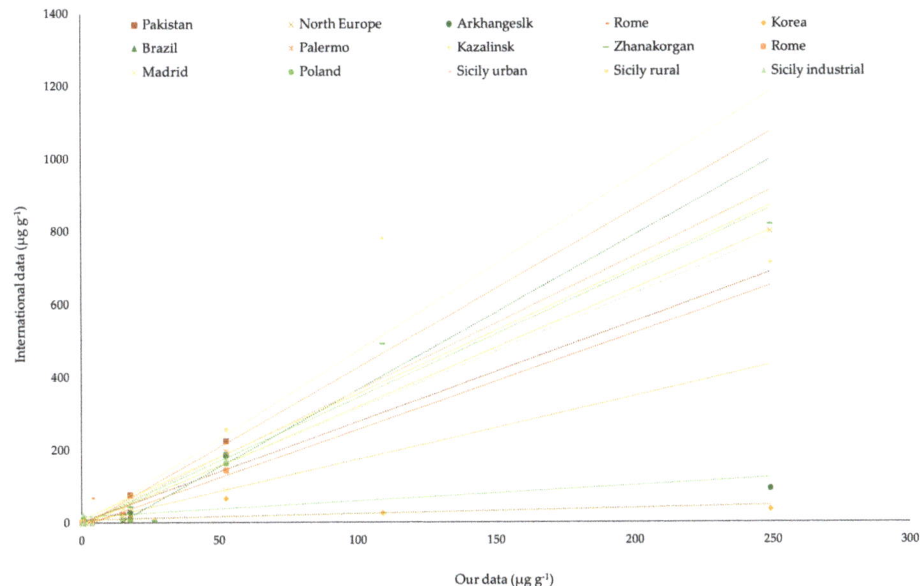

Figure 3. Correlation plots among our data and similar data determined in girl teenagers' hair worldwide [81–91].

4. Conclusions

The data determined in this study are a first tentative study of the element profile both in different hair sections and in different days of a week, also in relationship to the anthropogenic and/or natural sources present in an area, specifically in a suburban area in Central Italy. The element content in the hairs of a girl is compared with studies reporting the relative levels in hairs collected from girl teenagers worldwide: the diverse correlations found with West Europe and East Europe/Central Asia values highlight differences depending on habits, lifestyles, and nutritional diets.

Author Contributions: Conceptualization, A.R. and P.A.; methodology, M.L. and A.R.; software, A.P. and M.L.; validation, A.P. and A.R.; formal analysis, M.L. and A.R.; investigation, P.A.; resources, A.P. and M.L.; data curation, P.A.; writing—original draft preparation, A.R. and P.A.; writing—review and editing, A.P., A.R. and P.A.; visualization, A.R.; supervision, A.P. and P.A. All authors have read and agreed to the published version of the manuscript.

Funding: This research received no external funding.

Institutional Review Board Statement: The study was conducted according to the guidelines of the Declaration of Helsinki, and approved by the Institutional Review Board of ENEA (approval June 2006).

Informed Consent Statement: Informed consent was obtained from all subjects involved in the study.

Data Availability Statement: Data is available upon request by contacting the corresponding author.

Conflicts of Interest: The authors declare no conflict of interest.

References

1. Gault, A.G.; Rowland, H.A.; Charnock, J.M.; Wogelius, R.A.; Gomez-Morilla, I.; Vong, S.; Lang, M.; Samreth, S.; Sampsom, M.L.; Polya, D.A. Arsenic in hair and nails of individuals exposed to Arsenic-rich groundwaters in Kandal province, Cambodia. *Sci. Total Environ.* **2007**, *393*, 168–176. [CrossRef] [PubMed]
2. Rodrigues, J.L.; Batista, B.L.; Nunes, J.A.; Passos, C.J.; Barbosa, F., Jr. Evaluation of the use of human hair for biomonitoring the deficiency of essential and exposure to toxic elements. *Sci. Total Environ.* **2008**, *405*, 370–376. [CrossRef] [PubMed]

3. Protano, C.; Avino, P.; Manigrasso, M.; Vivaldi, V.; Perna, F.; Valeriani, F.; Vitali, M. Environmental electronic vape exposure from four different generations of electronic cigarettes: Airborne particulate matter levels. *Int. J. Environ. Res. Public Health* **2018**, *15*, 2172. [CrossRef] [PubMed]
4. Buononato, E.V.; De Luca, D.; Galeandro, I.C.; Congedo, M.L.; Cavone, D.; Intranuovo, G.; Guastadisegno, C.M.; Corrado, V.; Ferri, G.M. Assessment of environmental and occupational exposure to heavy metals in Taranto and other provinces of Southern Italy by means of scalp hair analysis. *Environ. Monit. Assess.* **2016**, *188*, 337.
5. Sazakli, E.; Leotsinidis, M. Hair biomonitoring and health status of a general population exposed to nickel. *J. Trace Elem. Med. Biol.* **2017**, *43*, 161–168.
6. Jursa, T.; Stein, C.R.; Smith, D.R. Determinants of hair manganese, lead, cadmium and arsenic levels in environmentally exposed children. *Toxics* **2018**, *6*, 19. [CrossRef]
7. Vinnikov, D.; Semizhon, S.; Rybina, T.; Zaitsev, V.; Pleshkova, A.; Rybina, A. Occupational exposure to metals and other elements in the tractor production. *PLoS ONE* **2018**, *13*, e0208932. [CrossRef] [PubMed]
8. Nouioui, M.A.; Araoud, M.; Milliand, M.L.; Bessueille-Barbier, F.; Amira, D.; Ayouni-Derouiche, L.; Hedhili, A. Evaluation of the status and the relationship between essential and toxic elements in the hair of occupationally exposed workers. *Environ. Monit. Assess.* **2018**, *190*, 731. [CrossRef]
9. Cabral Pinto, M.M.S.; Marinho-Reis, P.; Almeida, A.; Pinto, E.; Neves, O.; Inácio, M.; Gerardo, B.; Freitas, S.; Simões, M.R.; Dinis, P.A.; et al. Links between cognitive status and trace element levels in hair for an environmentally exposed population: A case study in the surroundings of the Estarreja industrial area. *Int. J. Environ. Res. Public Health* **2019**, *16*, 4560. [CrossRef]
10. Iyengar, V. Reference values for trace element concentrations in whole blood, serum, hair, liver, milk, and urine specimens from human subjects. In *Trace Elements in Man and Animals*; Hurley, L.S., Keen, C.L., Lönnerdal, B., Rucker, R.B., Eds.; Springer: Boston, MA, USA, 1988; Volume 6, pp. 535–537.
11. Batista, B.L.; Rodrigues, J.L.; Tormen, L.; Curtius, A.J.; Barbosa, F., Jr. Reference concentrations for trace elements in urine for the Brazilian population based on q-ICP-MS with a simple dilute-and-shoot procedure. *J. Braz. Chem. Soc.* **2009**, *20*, 1406–1413. [CrossRef]
12. Avino, P.; Capannesi, G.; Renzi, L.; Rosada, A. Instrumental neutron activation analysis and statistical approach for determining baseline values of essential and toxic elements in hairs of high school students. *Ecotox. Environ. Saf.* **2013**, *92*, 206–214. [CrossRef] [PubMed]
13. Rocha, G.H.O.; Steinbach, C.; Munhoz, J.R.; Madia, M.A.O.; Faria, J.K.; Hoeltgebaum, D.; Barbosa Jr., F.; Batista, B.L.; Souza, V.C.O.; Nerilo, S.B.; et al. Trace metal levels in serum and urine of a population in southern Brazil. *J. Trace Elem. Med. Biol.* **2016**, *35*, 61–65. [CrossRef] [PubMed]
14. Santos, M.D.; Soares, F.M.C.; Baisch, M.P.R.; Baisch, M.A.L.; da Silva, R.F.M., Jr. Biomonitoring of trace elements in urine samples of children from a coal-mining region. *Chemosphere* **2018**, *197*, 622–626. [CrossRef] [PubMed]
15. Watson, C.V.; Lewin, M.; Ragin-Wilson, A.; Jones, R.; Jarret, J.M.; Wallon, K.; Ward, C.; Hilliard, N.; Irvin-Barnwell, E. Characterization of trace elements exposure in pregnant women in the United States, NHANES 1999–2016. *Environ. Res.* **2020**, *183*, 109208. [CrossRef] [PubMed]
16. Saravanabhavan, G.; Werry, K.; Walker, M.; Haines, D.; Malowany, M.; Khoury, C. Human biomonitoring reference values for metals and trace elements in blood and urine derived from the Canadian Health Measures Survey 2007–2013. *Int. J. Hyg. Environ. Health* **2017**, *220*, 189–200. [CrossRef] [PubMed]
17. Minoia, C.; Sabbioni, E.; Apostoli, P.; Pietra, R.; Pozzoli, L.; Gallorini, M.; Nicolaou, G.; Alessio, L.; Capodaglio, E. Trace element reference values in tissues from inhabitants of the European community I. A study of 46 elements in urine, blood and serum of Italian subjects. *Sci. Total Environ.* **1990**, *95*, 89–105. [CrossRef]
18. White, M.A.; Sabbioni, E. Trace element reference values in tissues from inhabitants of the European Union. X. A study of 13 elements in blood and urine of a United Kingdom population. *Sci. Total Environ.* **1998**, *216*, 253–270. [CrossRef]
19. Heitland, P.; Köster, H.D. Biomonitoring of 37 trace elements in blood samples from inhabitants of northern Germany by ICP-MS. *J. Trace Elem. Med. Biol.* **2006**, *20*, 253–262. [CrossRef]
20. Roca, M.; Sánchez, A.; Pérez, R.; Pardo, O.; Yusa, V. Biomonitoring of 20 elements in urine of children: Levels and predictors of exposure. *Chemosphere* **2016**, *144*, 1698–1705. [CrossRef]
21. Aprea, M.C.; Apostoli, P.; Bettinelli, M.; Lovreglio, P.; Negri, S.; Perbellini, L.; Perico, A.; Ricossa, M.C.; Salamon, F.; Scapellato, M.L.; et al. Urinary levels of metal elements in the non-smoking general population in Italy: SIVR study 2012–2015. *Toxicol. Lett.* **2018**, *298*, 177–185. [CrossRef]
22. Font, L.; van der Peijl, G.; van Wetten, I.; Vroon, P.; van der Wagt, B.; Davies, G. Strontium and lead isotope ratios in human hair: Investigating a potential tool for determining recent human geographical movements. *J. Anal. At. Spectrom.* **2012**, *27*, 719–732. [CrossRef]
23. Foo, S.C.; Khoo, N.Y.; Heng, A.; Chua, L.H.; Chia, S.E.; Ong, C.N.; Ngim, C.H.; Jeyaratnam, J. Metals in hair as biological indices for exposure. *Int. Arch. Occup. Environ. Health* **1993**, *65*, 83–86. [CrossRef] [PubMed]
24. Daniel III, C.R.; Piraccini, B.M.; Tosti, A. The nail and hair in forensic science. *J. Am. Acad. Dermatol.* **2004**, *50*, 258–261. [CrossRef]
25. Kučera, J.; Kameník, J.; Havránek, V. Hair elemental analysis for forensic science using nuclear and related analytical methods. *Forensic Chem.* **2018**, *7*, 65–74. [CrossRef]

26. Avino, P.; Mercaldo, F.; Nardone, V.; Notardonato, I.; Santone, A. Machine learning to identify gender via hair elements. *Proc. Int. Jt. Conf. Neural Netw.* **2019**, *2019*, 8851914.
27. D'Urso, F.; Salomone, A.; Seganti, F.; Marco Vincenti, M. Identification of exposure to toxic metals by means of segmental hair analysis: A case report of alleged chromium intoxication. *Forensic Toxicol.* **2017**, *35*, 195–200. [CrossRef]
28. Ash, R.D.; He, M. Details of a thallium poisoning case revealed by single hair analysis using laser ablation inductively coupled plasma mass spectrometry. *Forensic Sci. Int.* **2018**, *292*, 224–231.
29. Katz, S.A. On the use of hair analysis for assessing arsenic intoxication. *Int. J. Environ. Res. Public Health* **2019**, *16*, 977. [CrossRef]
30. Patra, R.C.; Swarup, D.; Naresh, R.; Kumar, P.; Nandi, D.; Shekhar, P.; Roy, S.; Ali, S.L. Tail hair as an indicator of environmental exposure of cows to lead and cadmium in different industrial areas. *Ecotox. Environ. Saf.* **2007**, *66*, 127–131. [CrossRef]
31. Valderrama, J.T.; Beach, E.F.; Yeend, I.; Sharma, M.; Van Dun, B.; Dillon, H. Effects of lifetime noise exposure on the middle-age human auditory brainstem response, tinnitus and speech-in-noise intelligibility. *Hear. Res.* **2018**, *365*, 36–48. [CrossRef]
32. Sky-Peck, H.H. Distribution of trace elements in human hair. *Clin. Physiol. Biochem.* **1990**, *8*, 70–80. [PubMed]
33. Kosanovic, M.; Jokanovic, M. Quantitative analysis of toxic and essential elements in human hair. Clinical validity of results. *Environ. Monit. Assess.* **2011**, *174*, 635–643. [CrossRef] [PubMed]
34. Pozebon, D.; Scheffler, G.L.; Dressler, V.L. Elemental hair analysis: A review of procedures and applications. *Anal. Chim. Acta* **2017**, *992*, 1–23. [CrossRef] [PubMed]
35. Liang, G.; Pan, L.; Liu, X. Assessment of typical heavy metals in human hair of different age groups and foodstuffs in Beijing, China. *Int. J. Environ. Res. Public Health* **2017**, *14*, 914.
36. Yukawa, M.; Suzuki-Yasumoto, M.; Tanaka, S. The variation of trace element concentration in human hair: The trace element profile in human long hair by sectional analysis using neutron activation analysis. *Sci. Total Environ.* **1984**, *38*, 41–54. [CrossRef]
37. Kempson, I.M.; Skinner, W.M. ToF-SIMS analysis of elemental distributions in human hair. *Sci. Total Environ.* **2005**, *338*, 213–227. [CrossRef]
38. Kempson, I.M.; Skinner, W.M.; Kirkbridge, K.P. Advanced analysis of metal distributions in human hair. *Environ. Sci. Technol.* **2006**, *40*, 3423–3428.
39. Kempson, I.M.; Skinner, W.M. A comparison of washing methods for hair mineral analysis: Internal versus external effects. *Biol. Trace Elem. Res.* **2012**, *150*, 10–14. [CrossRef]
40. Kempson, I.M.; Skinner, W.M.; Martin, R.R. Changes in the metal content of human hair during diagenesis from 500 years, exposure to glacial and aqueous environments. *Archaeometry* **2010**, *52*, 450–466.
41. Park, J.; Xun, P.; Li, J.; Morris, S.J.; Jacobs, D.R.; Liu, K.; He, K. Longitudinal association between toenail zinc levels and the incidence of diabetes among American young adults: The CARDIA trace element study. *Sci. Rep.* **2016**, *6*, 23155. [CrossRef]
42. Maurice, J.F.; Wibetoe, G.; Endre Sjåstad, K.-E. Longitudinal distribution of thallium in human scalp hair determined by isotope dilution electrothermal vaporization inductively coupled plasma mass spectrometry. *J. Anal. At. Spectrom.* **2002**, *17*, 485–490. [CrossRef]
43. Hu, L.; Fernandez, D.P.; Cerling, T.E. Longitudinal and transverse variation of trace element concentrations in elephant and giraffe hair: Implication for endogenous and exogenous contributions. *Environ. Monit. Assess.* **2018**, *190*, 644. [CrossRef] [PubMed]
44. Avino, P.; Carconi, P.L.; Lepore, L.; Moauro, A. Nutritional and environmental properties of algal products used in healthy diet by INAA and ICP-AES. *J. Radioanal. Nucl. Chem.* **2000**, *244*, 247–252. [CrossRef]
45. Avino, P.; Capannesi, G.; Rosada, A. Ultra-trace nutritional and toxicological elements in Rome and Florence drinking waters determined by Instrumental Neutron Activation Analysis. *Microchem. J.* **2011**, *97*, 144–153. [CrossRef]
46. Avino, P.; Capannesi, G.; Manigrasso, M.; Sabbioni, E.; Rosada, A. Element assessment in whole blood, serum and urine of three Italian healthy sub-populations by INAA. *Microchem. J.* **2011**, *99*, 548–555.
47. International Atomic Energy Agency (IAEA). *Activation Analysis of Hair as an Indicator of Contamination of Man by Environmental Trace Element Pollutants (IAEA-RL-50)*; Ryabukhin, Y.S., Ed.; International Atomic Energy Agency, Agency's Laboratories, Analytical Quality Control Services: Seibersdorf, Austria, 1976; p. 135. Available online: http://www.iaea.org/programmes/aqcs/pdf/a_rl_050.pdf (accessed on 27 November 2020).
48. Campanella, L.; Crescentini, G.; Avino, P.; Moauro, A. Determination of macrominerals and trace elements in the alga Spirulina platensis. *Analusis* **1998**, *26*, 210–214. [CrossRef]
49. Avino, P.; Capannesi, G.; Rosada, A. Characterization and distribution of mineral content in fine and coarse airborne particle fractions by neutron activation analysis. *Toxicol. Environ. Chem.* **2006**, *88*, 633–647. [CrossRef]
50. Avino, P.; Capannesi, G.; Renzi, L.; Rosada, A. Physiological parameters affecting the hair element content of young Italian population. *J. Radioanal. Nuc. Chem.* **2011**, *306*, 737–743. [CrossRef]
51. Bass, D.A.; Hickok, D.; Quig, D.; Urek, K. Trace element analysis in hair: Factors determining accuracy, precision, and reliability. *Altern. Med. Rev.* **2001**, *6*, 472–481.
52. Capannesi, G.; Rosada, A.; Avino, P. Elemental characterization of impurities at trace and ultra-trace levels in metallurgical lead samples by INAA. *Microchem. J.* **2009**, *93*, 188–194.
53. Avino, P.; Capannesi, G.; Rosada, A.; Manigrasso, M.; Sabbioni, E. Multivariate analysis applied to some elements in human fluids and whole bloods of hemodialysis patients determined by INAA. *J. Radioanal. Nucl. Chem.* **2013**, *298*, 1957–1968. [CrossRef]
54. Djingova, R.; Kuleff, I. Chapter 5 Instrumental techniques for trace analysis. In *Trace Metals in the Environment*; Markert, B., Friese, K., Eds.; Elsevier: Amsterdam, The Netherlands, 2000; Volume 4, pp. 137–185.

55. Smodiš, B.; Bleise, A. IAEA quality control study on determining trace elements in biological matrices for air pollution research. *J. Radioanal. Nucl. Chem.* **2007**, *271*, 269–274. [CrossRef]
56. Leblanc, A.; Dumas, P.; Lefebvre, L. Trace element content of commercial shampoos: Impact on trace element levels in hair. *Sci. Total Environ.* **1999**, *229*, 121–124. [CrossRef]
57. Chyla, M.A.; Zyrnicki, W. Determination of metal concentrations in animal hair by the ICP method: Comparison of various washing procedures. *Biol. Trace Elem. Res.* **2000**, *75*, 187–194. [CrossRef]
58. Morton, J.; Carolan, V.A.; Gardiner, P.H.E. Removal of exogenously bound elements from human hair by various washing procedures and determination by inductively coupled plasma mass spectrometry. *Anal. Chim. Acta* **2002**, *455*, 23–34. [CrossRef]
59. Razagui, I.B.-A. A comparative evaluation of three washing procedures for minimizing exogenous trace element contamination in fetal scalp hair of various obstetric outcomes. *Biol. Trace Elem. Res.* **2008**, *123*, 47–57. [CrossRef] [PubMed]
60. Li, P.; Feng, X.; Qiu, G.; Wan, Q. Hair can be a good biomarker of occupational exposure to mercury vapor: Simulated experiments and field data analysis. *Sci. Total Environ.* **2011**, *409*, 4484–4488. [CrossRef]
61. Mantinieks, D.; Wright, P.; Di Rago, M.; Gerostamoulos, D. A systematic investigation of forensic hair decontamination procedures and their limitations. *Drug Test. Anal.* **2019**, *11*, 1542–1555. [CrossRef]
62. Nielsen, M.K.K.; Johansen, S.S. Internal quality control samples for hair testing. *J. Pharm. Biomed. Anal.* **2020**, *188*, 113459. [CrossRef]
63. Sen, J.; Das Chaudhuri, A.B. Brief communication: Choice of washing method of hair samples for trace element analysis in environmental studies. *Am. J. Phys. Anthropol.* **2001**, *115*, 289–291.
64. Sera, K.; Futatsugawa, S.; Murao, S. Quantitative analysis of untreated hair samples for monitoring human exposure to heavy metals. *Nucl. Instrum. Methods Phys. Res. B* **2002**, *189*, 174–179. [CrossRef]
65. Qiu, K.; Durham, P.G.; Anselmo, A.C. Inorganic nanoparticles and the microbiome. *Nano Res.* **2018**, *11*, 4936–4954.
66. Pulit-Prociak, J.; Chwastowski, J.; Siudek, M.; Banach, M. Incorporation of metallic nanoparticles into cosmetic preparations and assessment of their physicochemical and utility properties. *J. Surfactants Deterg.* **2018**, *21*, 575–591. [CrossRef]
67. Wolfsperger, M.; Hauser, G.; Gößler, W.; Schlagenhaufen, C. Heavy metals in human hair samples from Austria and Italy: Influence of sex and smoking habits. *Sci. Total Environ.* **1994**, *156*, 235–242. [CrossRef]
68. Chojnacka, K.; Michalak, I.; Zielińska, A.; Górecka, H.; Górecki, H. Inter-relationship between elements in human hair: The effect of gender. *Ecotoxicol. Environ. Saf.* **2010**, *73*, 2022–2028. [PubMed]
69. Rodushkin, I.; Axelsson, M.D. Application of double focusing sector field ICP-MS for multielemental characterization of human hair and nails. Part II. A study of the inhabitants of northern Sweden. *Sci. Total Environ.* **2000**, *262*, 21–36. [CrossRef]
70. Sturaro, A.; Parvoli, G.; Doretti, L.; Allegri, G.; Costa, C. The influence of color, age, and sex on the content of zinc, copper, nickel, manganese, and lead in human hair. *Biol. Trace Elem. Res.* **1994**, *40*, 1–8. [CrossRef]
71. Iyengar, G.V. Reevaluation of the trace element content in Reference Man. *Radiat. Phys. Chem.* **1998**, *51*, 545–560. [CrossRef]
72. Tsai, Y.-Y.; Wang, C.-T.; Chang, W.-T.; Wang, R.-T.; Huang, C.-W. Concentrations of potassium, sodium, magnesium, calcium, copper, zinc, manganese and iron in black and gray hairs in Taiwan. *J. Health Sci.* **2000**, *46*, 46–48.
73. Zaichick, V. Medical elementology as a new scientific discipline. *J. Radioanal. Nucl. Chem.* **2006**, *269*, 303–309. [CrossRef]
74. Chojnacka, K.; Górecka, H.; Górecki, H. The effect of age, sex, smoking habit and hair color on the composition of hair. *Environ. Toxicol. Pharmacol.* **2006**, *22*, 52–57. [CrossRef] [PubMed]
75. Mohammed, N.K.; Mizera, J.; Spyrou, N.M. Elemental contents in hair of children from Zanzibar in Tanzania as bio-indicator of their nutritional status. *J. Radioanal. Nucl. Chem.* **2008**, *276*, 125–128. [CrossRef]
76. Mohammed, N.K.; Spyrou, N.M. Determination of trace elements in hair from Tanzanian children: Effect of dietary factors. *J. Radioanal. Nucl. Chem.* **2008**, *278*, 455–458. [CrossRef]
77. Manigrasso, M.; Vernale, C.; Avino, P. Traffic aerosol lobar doses deposited in the human respiratory system. *Environ. Sci. Pollut. Res.* **2017**, *24*, 13866–13873. [CrossRef]
78. Stabile, L.; Buonanno, G.; Avino, P.; Fuoco, F.C. Dimensional and chemical characterization of airborne particles in schools: Respiratory effects in children. *Aerosol Air Qual. Res.* **2013**, *13*, 887–900. [CrossRef]
79. Avino, P.; Capannesi, G.; Rosada, A. Heavy metal determination in atmospheric particulate matter by Instrumental Neutron Activation Analysis. *Microchem. J.* **2008**, *88*, 97–106. [CrossRef]
80. Avino, P.; Capannesi, G.; Rosada, A. Source identification of inorganic airborne particle fraction (PM_{10}) at ultratrace levels by means of INAA short irradiation. *Environ. Sci. Pollut. Res.* **2014**, *21*, 4527–4538. [CrossRef]
81. Khalique, A.; Ahmad, S.; Anjum, T.; Jaffar, M.; Shah, M.H.; Shaheen, N.; Tariq, S.R.; Manzoor, S. A comparative study based on gender and age dependence of selected metals in scalp hair. *Environ. Monit. Assess.* **2005**, *104*, 45–57. [CrossRef]
82. Soroko, S.I.; Maximova, I.A.; Protasova, O.V. Age and gender features of the content of macro and trace elements in the organisms of children from the European North. *Hum. Physiol.* **2014**, *40*, 603–612. [CrossRef]
83. Senofonte, O.; Violante, N.; Caroli, S. Assessment of reference values for elements in human hair of urban schoolboys. *J. Trace Elem. Med. Biol.* **2000**, *14*, 6–13. [CrossRef]
84. Park, H.-P.; Shin, K.-O.; Kim, J.-S. Assessment of reference values for hair minerals of Korean preschool children. *Biol. Trace Elem. Res.* **2007**, *116*, 119–130. [CrossRef] [PubMed]
85. Carneiro, M.F.H.; Moresco, M.B.; Chagas, G.R.; de Oliveira Souza, V.C.; Rhoden, C.R.; Barbosa, F., Jr. Assessment of trace elements in scalp hair of a young urban population in Brazil. *Biol. Trace Elem. Res.* **2011**, *143*, 815–824. [CrossRef] [PubMed]

86. Dongarrà, G.; Lombardo, M.; Tamburo, E.; Varrica, D.; Cibella, F.; Cuttitta, G. Concentration and reference interval of trace elements in human hair from students living in Palermo, Sicily (Italy). *Environ. Toxicol. Pharmacol.* **2011**, *32*, 27–34. [CrossRef] [PubMed]
87. Chiba, M.; Sera, K.; Hashizume, M.; Shimoda, T.; Sasaki, S.; Kunii, O.; Inaba, Y. Element concentrations in hair of children living in environmentally degraded districts of the East Aral Sea region. *J. Radioanal. Nucl. Chem.* **2004**, *259*, 149–152. [CrossRef]
88. Caroli, S.; Senofonte, O.; Violante, N.; Fornarelli, L.; Powar, A. Assessment of reference values for elements in hair of urban normal subjects. *Microchem. J.* **1992**, *46*, 174–183. [CrossRef]
89. Llorente Ballesteros, M.T.; Navarro Serrano, I.; Izquierdo Álvarez, S. Reference levels of trace elements in hair samples from children and adolescents in Madrid, Spain. *J. Trace Elem. Med. Biol.* **2017**, *43*, 113–120. [CrossRef]
90. Długaszek, M.; Skrzeczanowski, W. Relationships between element contents in Polish children's and adolescents' hair. *Biol. Trace Elem. Res.* **2017**, *180*, 6–14. [CrossRef]
91. Dongarrà, G.; Varrica, D.; Tamburo, E.; D'Andrea, D. Trace elements in scalp hair of children living in differing environmental contexts in Sicily (Italy). *Environ. Toxicol. Pharmacol.* **2012**, *34*, 160–169. [CrossRef]
92. Kim, G.N.; Song, H.J. Hair mineral analysis of normal Korean children. *Korean J. Dermatol.* **2002**, *40*, 1518–1526.
93. Cvitković, A.; Capak, K.; Jurasović, J.; Barišin, A.; Ivić-Hofman, I.; Poljak, V.; Valjetić, M. Metal concentration study in a population living in the vicinity of an oil refinery. In *WIT Transactions on Ecology and The Environment*; Almorza, D., Longhurst, J.W.S., Brebbia, C.A., Barnes, J., Eds.; WIT Press: Ashurst Lodge, UK, 2017; Volume 211, pp. 255–261.
94. Druyan, M.E.; Bass, D.; Puchyr, R.; Urek, K.; Quig, D.; Harmon, E.; Marquardt, W. Determination of reference ranges for elements in human scalp hair. *Biol. Trace Elem. Res.* **1998**, *62*, 183–197. [CrossRef]

Article

Stabilization of Municipal Solid Waste Fly Ash, Obtained by Co-Combustion with Sewage Sludge, Mixed with Bottom Ash Derived by the Same Plant

Ahmad Assi [1], Fabjola Bilo [1], Alessandra Zanoletti [1], Laura Borgese [1], Laura Eleonora Depero [1], Mario Nenci [2] and Elza Bontempi [1,*]

[1] INSTM and Chemistry for Technologies Laboratory, Department of Mechanical and Industrial Engineering, University of Brescia, via Branze, 38, 25123 Brescia, Italy; a.assi@unibs.it (A.A.); fabjola.bilo@unibs.it (F.B.); alessandra.zanoletti@unibs.it (A.Z.); laura.borgese@unibs.it (L.B.); laura.depero@unibs.it (L.E.D.)

[2] A2A Ambiente S.p.A., 25124 Brescia, Italy; mario.nenci@a2a.eu

* Correspondence: elza.bontempi@unibs.it

Received: 6 August 2020; Accepted: 1 September 2020; Published: 2 September 2020

Abstract: This study presents an innovative stabilization method of fly ash derived from co-combustion of municipal solid waste and sewage sludge. Bottom ash, obtained from the same process, is used as a stabilizing agent. The stabilization method involved the use of two other components—flue gas desulfurization residues and coal fly ash. Leaching tests were performed on stabilized samples, aged in a laboratory at different times. The results reveal the reduction of the concentrations of heavy metals, particularly Zn and Pb about two orders of magnitude lower with respect to fly ash. The immobilization of heavy metals on the solid material mainly depends on three factors—the amount of used ash, the concentrations of Zn and Pb in as-received fly ash and the pH of the solution of the final materials. The inert powder, obtained after the stabilization, is a new eco-material, that is promising to be used as filler in new sustainable composite materials.

Keywords: sewage sludge disposal; municipal solid waste; co-combustion; fly ash; bottom ash; heavy metal stabilization

---

1. Introduction

In recent years, the diffuse practice of sewage sludge (SS) land spreading has generated several concerns, due to the presence of potential contaminants (such as pathogenic agents, toxic inorganic substances, and microorganisms) [1,2], with the consequence of emerging alternatives for SS treatments strategies, such as mono- and co-combustion. For example, co-combustion has been realized with municipal solid waste incineration (MSWI). The co-combustion has several advantages, the main related to the possibility to use the already existing plants (such as MSW incineration plants), avoiding investments costs, due to the need to construct new incinerators expressly devoted to mono-combustion. Moreover, the wastes generated from co-combustion of MSW and SS, i.e., fly ash (FA), and bottom ash (BA), must be properly managed to avoid landfilling and/or pollution, generated by unsuitable treating strategies. MSWI-FA is generally considered the most problematic incineration waste, due to the presence of leachable heavy metals [3]. Several technologies for MSWI-FA stabilization have been already proposed [4–8] with the aim to promote its reuse [9–11]. Even though some recently proposed MSWI-FA treatments were defined as zero-waste technologies [12], they often require MSWI-FA pre-treatments, which need the use of some additional processes and raw materials.

In a very recently published paper, we have demonstrated that MSWI-BA can be used to stabilize MSWI-FA [13]. Indeed, after metal separation, BA is generally recovered for use in the building industry. On the contrary, FA is generally destined to landfill. However, the recently proposed strategy allows the use of a waste (BA) to stabilize another waste (FA). It is important to highlight that the proposed procedure has several advantages—it employs wastes produced at the same location, strongly suggesting the possibility to directly apply the new technology on the incinerator plant sites; in addition, it avoids the transport of wastes in different locations and the landfilling of MSWI-FA. The idea to combine different wastes to take advantage of their valuable components to reduce the contained pollutants is not new [8,14,15]. In particular, the suggestion to use wastes and by-products to minimize energy, materials, and emissions for remediation was recently defined as the Azure Chemistry approach [16]. In the present work, the results recently proposed, that involve the combination of Sewage-MSWI BA and Sewage-MSWI FA to obtain a new safe eco-material, are considered to investigate if the process can be applied also to Sewage-MSWI FA derived from co-combustion. The obtained inert can be defined as eco-materials due to the low energies and emissions required to perform the proposed stabilization [14]. In particular, the aim of this work is the investigation if different amounts of SS in the co-combustion can influence the FA stabilization procedure that must be used to reduce the leachable pollutants.

2. Experimental

2.1. Materials

In this study, FA and BA from the co-incineration of MSW and SS were recovered from the incineration plant located in Brescia (Northern Italy). In this plant, Sewage-MSWI-FA is collected with APC (air-pollution-control) residues, originating from cleaning the flue gases before emission to air. These residues consist of fine particulates, that are generally defined FA.

The co-incineration of MSW and SS was implemented in this plant since 2017. The plant normally processes an SS concentration of 7% on the flow rate of MSW. Because the aim of the present experiment is to study the effects of different SS flow rates on the co-incineration residue stabilization, the plant worked for one day, with the three separate combustion lines at different SS concentrations—0, 2, and 4 tons/h. The amount of MSW incinerated on each line was 31 tons/h.

The schema representing the operation of the first, second and third combustion lines (also reporting the SS concentrations) is shown in Figure 1.

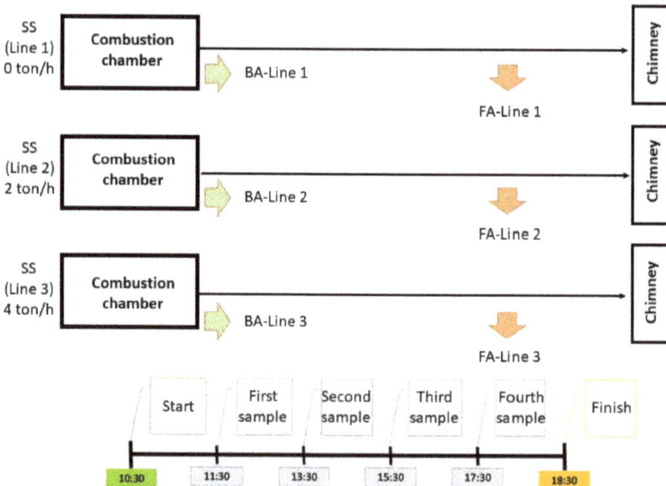

Figure 1. Schema of the co-incineration experiment made on the industrial plant, where the three different combustion lines are represented.

During the experimentation day, Sewage-MSWI FA and Sewage-MSWI BA were collected from the three separated lines at predetermined time intervals of two hours (see Figure 1). At the end of the day, the ashes collected at different interval times were adequately mixed to obtain a homogeneous and representative sample of Sewage-MSWI FA and Sewage-MSWI BA for each line. In total, 6 ash samples (2 samples for each line) were recovered (see Figure 1).

The samples obtained and considered in this study can be divided into two categories—three samples of Sewage-MSWI BA obtained from the bottom of the combustion chamber and three samples of Sewage-MSWI FA obtained from the bag filters.

After sampling, Sewage-MSWI BA was manually sorted—metals and particles with diameter higher than 2 cm were separated. Then, due to the Sewage-MSWI BA moisture, a thermal treatment at 100 °C for about 2 h was made. After drying, Sewage-MSWI BA was grounded with Mixer Mill Retsh (MM400) and sieved until 106 μm.

Other as-received ashes are used with the aim to stabilize FA. In particular, CFA (coal combustion ash) is a by-product in thermal coal power plants. This ash is removed by a dust-collection system from the combustion gases before they are emitted into the atmosphere [17].

Flue gas desulphurization (FGD) residues are produced during the removal of sulphur oxides from coal-burning power plants. In this process, insoluble calcium sulphite and calcium sulphate solids are formed because absorbed SO_2 reacts with lime in scrubbing liquor [18]. They also are a by-product of the coal combustion system. FGD residues contain high amounts of S and Ca [5]. Calcium hydroxide is very important to promote the carbonation reactions.

Both CFA and FGD residues were collected from Brescia pulverized coal thermal power plant.

2.2. Stabilization Procedure

The as-received Sewage-MSWI FA samples were stabilized following a very recent proposed technology based on the use of MSWI-BA deriving from the corresponding combustion line to stabilize FA [4]—for each FA, the procedure (defined procedure a) involved the mix of about 130 g of FA with of 20 g of BA, about 30 g of CFA, and about 40 g of FGD [4,13] (see Table 1). Another procedure, with the same amount of ashes reported for procedure a)), but that did not involve the BA addition, was also realized (defined procedure b)). Procedure b) was selected to have a comparative procedure suitable to allow us to evaluate the BA role in the stabilization. In this case, all samples exhibited a relative weight percentage, already used for similar stabilization technology [5], as follows—65% FA, 15% CFA and 20% FGD. For both stabilization procedures, all powders were carefully mixed before adding approx. 200 mL of milliQ water (Millipore DirectQ-5 TM, Millipore S.A.S., 67120, Molsheim, France); then they were additionally mixed for 20 min. The obtained eco-materials were aged in laboratory for 3 months at room temperature.

Table 1. Sample descriptions.

Line	Procedure	Mass (g)				V (mL)
		Sewage-MSWI FA	**CFA**	**FGD**	**Sewage-MSWI BA**	**H_2O**
1	a)	130.2	31	40.8	20.1	200
	b)	130.6	30.4	40.7	-	200
2	a)	130	30.8	40	20	200
	b)	130.1	30.3	40.4	-	200
3	a)	130.2	30.1	40.8	20	200
	b)	130.1	30	40.3	-	200
	b)	130.1	30	40.2	-	200

Sewage-MSWI FA: sewage and municipal solid waste incineration-fly ash; CFA: Coal Fly Ash; FGD: Flue Gas Desulphurization; Sewage-MSWI BA: Bottom Ash.

2.3. Leaching Test and TXRF Analysis

Leaching tests were carried out according to CEN EN 12457-2 regulation (EN 13055-1:2002-CN/TC 154-CEN-CEN, 2002) on all as-received powders and stabilized samples (in this case the test was made each month, after the stabilization). The procedure for the leaching test, reported by [5,19,20], consists of mixing at room temperature approximately 20 g of each sample with 200 mL of milliQ water (1:10) by means of an agitator for 2 h [20]. pH measurement by a pH-meter (Metrohm, model 827 Lab, Origgio, Italy) is conducted immediately after the leaching test on the samples filtered by 0.45 μm pore membranes.

After filtration, elemental chemical analysis is realized by Total-reflection X-ray Fluorescence (TXRF) using a S2 Picofox system from Bruker (Bruker AXS Microanalysis GmbH, Berlin, Germany) equipped with Mo tube operating at 50 kV and 750 μA and a Silicon Drift Detector (SDD). For this aim, 0.010 g of a Ga solution with a concentration of 100 mg/L is used as internal standard (Ga-ICP Standard Soluyion, Fluka, Sigma Aldrich, Saint Louis, MO, USA). It is added to the leachate solutions and homogenized by a mean of vortex shaker at 2500 rpm for 1 min. Three replicates are always prepared for each sample by adding a droplet of 10 μL of sample. By the use of a dedicated instrumental software based on mono-element profiles, the spectra are deconvolution to evaluate the peak areas. The TXRF lower detection limits (LOD) evaluated with similar experimental conditions are reported [21]. Chemical analysis of soluble elements, with atomic numbers less than 19, such Na, cannot be made by TXRF due to their low fluorescence yield [22]. Furthermore, Si cannot be evaluated, because the sample holder for TXRF analysis is made of quartz.

3. Results and Discussion

In order to understand the effects of the addition of different SS amounts to the three combustion lines, all starting ashes (FA and BA) and the obtained stabilized eco-materials were analyzed. Elemental chemical analysis using TXRF was performed to estimate the amount of leachable metals that can be found in the ashes. As reported in the experimental section, as-received FAs were analyzed as well; BA samples were pre-treated before the analysis with a reduction of the grains dimension to 106 μm. Table 2 reports the results of the leaching tests made on these ashes (BA and FA). It shows the concentration of soluble elements derived from the three combustion lines, as resulted from the TXRF analysis. Furthermore, data about BA leaching are shown.

The leaching data of the BA and FA derived from different lines are in agreement with results reported in literature, also considering the alkaline pH of the solutions [23–25]. The major leachable elements in FA are Cl, Ca, K, S and Br. Relevant quantities of other elements such as Zn, Pb and Rb are also found. The main soluble elements found in the BA solutions are Ca, Cl, K and S. The presence of a high amount of Ca in FA in comparison to BA is expected, due to the addition of lime as a stabilizing agent. Indeed, during the incineration process, acid gases are produced (like HCl and SO_2) and lime is normally added to absorb these gases. A higher amount of Pb and Zn are present in FA, in comparison to BA, due to the fact that these metals are moderately volatile elements, then they are generally found in ashes collected at the chimney after a combustion process carried out at a maximum temperature about 1000 °C [15]. Comparing the leaching data of different combustion lines, it is very interesting to notice that in FA, the concentration of leachable Zn is substantially irrespective of the amount of co-incinerated SS. On the contrary, the amount of soluble Pb increases with the amount of SS, probably due to the presence of this metal in sewage sludge [26]. Concerning P, an element that is well-known to be abundant in SS [27], its concentration appears to decrease in FA with the increase of co-incinerated SS. On the contrary, in BA the leachable P shows an inverse behavior. These results are substantially due to the presence of this element in soluble phases (leaching tests only allow to detect soluble elements). Moreover, it means that the increase of SS amount in co-incineration seems to increase the P leachability in BA (see Table 2). Similar consideration can be extended to Cu, for BA.

Table 2. Results of the Total-reflection X-ray Fluorescence (TXRF) analysis and pH values of leachate solutions of Sewage-MSWI FA and Sewage-MSWI BA derived from different combustion lines.

Samples	FA-Line 1	FA-Line 2	FA-Line 3	BA-Line 1	BA-Line 2	BA-Line 3
pH	**11.89**	**11.92**	**11.9**	**12.09**	**12.2**	**12.18**
Element	**(mg/L)**	**(mg/L)**	**(mg/L)**	**(mg/L)**	**(mg/L)**	**(mg/L)**
P *	40.3 ± 16.0	25.4 ± 8.4	21.9 ± 5.7	5.4 ± 2.5	9.3 ± 2.6	12.7 ± 5.3
S *	273 ± 48	267 ± 63	217 ± 61	71 ± 3.6	131 ± 21	81 ± 21
Cl	8890 ± 193	5969 ± 954	6545 ± 958	62 ± 36	318 ± 185	408 ± 81
K	1087 ± 126	711 ± 280	595 ± 161	111 ± 17	81 ± 21	105 ± 32
Ca	6387 ± 1044	4930 ± 913	4677 ± 706	651 ± 68	979 ± 175	995 ± 258
Cr	<LOD	<LOD	<LOD	<LOD	<LOD	<LOD
Mn	<LOD	<LOD	<LOD	<LOD	<LOD	<LOD
Fe	0.4 ± 0.2	0.2 ± 0.1	0.2 ± 0.2	0.2 ± 0.1	0.07 ± 0.03	0.09 ± 0.04
Cu	0.23 ± 0.05	0.09 ± 0.03	0.33 ± 0.05	4.5 ± 0.5	5.6 ± 0.7	6.1 ± 0.7
Zn	11.7 ± 1.4	9.0 ± 0.4	10.1 ± 1.4	0.3 ± 0.2	0.7 ± 0.2	0.61 ± 0.03
Br	211 ± 26	183 ± 19	236 ± 21	1.7 ± 0.3	1.9 ± 0.2	1.8 ± 1.0
Rb	7.3 ± 1.5	7.1 ± 0.9	7.6 ± 0.8	0.4 ± 0.2	0.2 ± 0.0	0.25 ± 0.02
Sr	16.0 ± 4.3	13.8 ± 0.4	27.8 ± 3.9	3.9 ± 0.4	5.5 ± 0.6	4.7 ± 0.2
Pb	92.0 ± 15.6	94.1 ± 6.8	127.1 ± 5.4	0.58 ± 0.04	1.6 ± 0.3	1.6 ± 0.4

Values are expressed as the average ± standard deviation of three TXRF measurements. Relative sensitivities for elements with * were calculated based on a calibration curve. LOD-lower detection limit.

After the stabilization procedure, a significant decrease in heavy metal leachability can be noticed, if samples are analyzed one, two, and three months after the stabilization (see Tables 3 and 4), in comparison to data reported in Table 2. In particular, three months after the stabilization the concentration of Pb is sometimes lower than LOD by TXRF spectrometry, which is 0.002 mg/L, thus demonstrates that the stabilization procedure was effective in reducing the solubility of heavy metals and that the stabilization efficacy increases with time. Moreover, comparing data reported in Tables 3 and 4 (considering the corresponding month of stabilization), it seems that the use of BA is not so fundamental in reducing the heavy metal mobility. Indeed, eco-materials obtained applying the procedures a) and b) (the procedure b) is made without the addition of BA) sometimes show comparable Pb concentrations in their leaching solutions. Another difference concerns the Cu concentration, that is higher in samples treated by BA addition. Even if this metal appears to be stabilized after aging, it is evident that this origin can be attributed to BA due to its higher concentrations in the BA than FA (see Table 2).

In a very recent paper, the stabilization mechanism was proposed and discussed—it was shown that dissolved amorphous silica and alumina (derived from BA) in the presence of calcium ions (and in a highly alkaline environment) promote a pozzolanic reaction with FA, with the formation of cementitious compounds such as C–S–H and calcium aluminate hydrates (C-A-H). Furthermore, carbonation reactions occurred, due to the calcium hydroxide that is present in the used wastes (for example in FGD residues) [28,29].

In particular, considering the pH of all raw FA and final stabilized eco-materials, it means that all FAs have a starting pH about 12 (see Table 2). Instead, stabilized materials have a pH of about 11 (see Table 3) and 8 (see Table 4), depending on the procedure used for stabilization. Indeed, carbonation produces a reduction of pH [30].

Table 3. Results of the TXRF analysis and pH values of samples stabilized following the a) procedure with Sewage-MSWI FA from different lines after (1), (2) and (3) months.

Samples	FA + BA Line 1			FA + BA Line 2			FA + BA Line 3		
Time	**1 M**	**2 M**	**3 M**	**1 M**	**2 M**	**3 M**	**1 M**	**2 M**	**3 M**
pH	**13.5**	**12.1**	**10.5**	**13.5**	**12.0**	**10.8**	**13.8**	**12.1**	**11.2**
Elements	**(mg/L)**	**(mg/L)**	**(mg/L)**	**(mg/L)**	**(mg/L)**	**(mg/L)**	**(mg/L)**	**(mg/L)**	**(mg/L)**
P *	42.8 ± 5.6	57.4 ± 6.9	33.2 ± 10.5	28.9 ± 15.2	59.6 ± 9.0	33.3 ± 13.6	37.2 ± 26.7	58.3 ± 18.2	24.9 ± 6.2
S *	423 ± 168	358 ± 21	436 ± 81	318 ± 53	369 ± 82	366 ± 129	100 ± 30	208 ± 32	251 ± 41
Cl	3164 ± 234	4911 ± 1052	2350 ± 571	2175 ± 885	4735 ± 168	3069 ± 978	4568 ± 93	4797 ± 248	2308 ± 699
K	366 ± 58	593 ± 182	237 ± 34	222 ± 78	638 ± 55	336 ± 79	493 ± 47	525.1 ± 106	174 ± 41
Ca	2222 ± 2008	3569 ± 835	1518 ± 175	1331 ± 232	3339 ± 145	1938 ± 274	3636 ± 197	3676 ± 200	1498 ± 439
Cr	<LOD	0.1 ± 0.002	0.2 ± 0.1	<LOD	0.1 ± 0.1	0.1 ± 0.1	<LOD	<LOD	<LOD
Mn	<LOD	<LOD	<LOD	<LOD	<LOD	<LOD	0.1 ± 0.04	<LOD	<LOD
Fe	0.10 ± 0.01	0.4 ± 0.2	0.2 ± 0.1	0.6 ± 0.3	0.3 ± 0.1	0.4 ± 0.2	0.2 ± 0.1	0.74 ± 0.04	0.2 ± 0.1
Cu	0.07 ± 0.01	0.1 ± 0.01	0.08 ± 0.01	0.1 ± 0.1	0.1 ± 0.1	0.09 ± 0.04	0.3 ± 0.01	0.2 ± 0.01	0.1 ± 0.1
Zn	0.2 ± 0.1	0.4 ± 0.1	0.1 ± 0.0	0.6 ± 0.04	0.6 ± 0.1	0.07 ± 0.04	3.5 ± 1.9	1.4 ± 0.3	0.2 ± 0.1
Br	79.5 ± 9.0	75.1 ± 4.7	88.4 ± 4.2	93.3 ± 8.9	62.0 ± 1.6	76.4 ± 1.7	62.5 ± 6.7	69.9 ± 6.5	81.7 ± 4.3
Rb	3.4 ± 0.7	3.3 ± 0.4	4.5 ± 0.3	4.6 ± 0.8	3.2 ± 0.2	3.8 ± 0.5	2.2 ± 0.3	3.1 ± 0.2	3.7 ± 1.0
Sr	12.5 ± 0.8	9.9 ± 0.5	13.2 ± 1.5	15.0 ± 2.1	10.1 ± 0.4	11.0 ± 1.5	14.2 ± 2.6	16.4 ± 0.7	18.7 ± 2.3
Pb	2.1 ± 0.4	3.0 ± 0.4	<LOD	1.7 ± 0.4	2.6 ± 0.5	<LOD	17.1 ± 10.0	6.5 ± 1.6	0.27 ± 0.05

Values are expressed as the average ± standard deviation of three measurements. Relative sensitivities for elements with * were calculated based on a calibration curve. LOD lower detection limit.

Table 4. Results of the TXRF analysis and pH values of samples stabilized following the b) procedure with Sewage-MSWI FA from different lines after (1), (2) and (3) months.

Samples	FA + BA Line 1			FA + BA Line 2			FA + BA Line 3		
Time	**1 M**	**2 M**	**3 M**	**1 M**	**2 M**	**3 M**	**1 M**	**2 M**	**3 M**
pH	**13.6**	**8.7**	**7.7**	**13.7**	**12.2**	**8.0**	**13.6**	**12.1**	**8.4**
Elements	**(mg/L)**	**(mg/L)**	**(mg/L)**	**(mg/L)**	**(mg/L)**	**(mg/L)**	**(mg/L)**	**(mg/L)**	**(mg/L)**
P *	65.4 ± 23.1	30.2 ± 1.9	28.2 ± 14.6	40.4 ± 17.7	57.7 ± 2.8	27.9 ± 2.8	39.5 ± 1.8	49.2 ± 2.3	32.5 ± 18.9
S *	640 ± 180	422 ± 219	604 ± 244	330 ± 121	363 ± 50	469 ± 177	244 ± 24	378 ± 54	562 ± 172
Cl	3740 ± 484	3083 ± 1181	2185 ± 677	2237 ± 1059	5123 ± 106	1953 ± 835	3677 ± 281	5381 ± 1236	2287 ± 640
K	474 ± 105	311 ± 8	257 ± 83	242 ± 126	636 ± 13	202 ± 136	369 ± 24	606 ± 201	215 ± 61
Ca	3156. ± 474	2190 ± 722	1643 ± 443	1400 ± 925	3810 ± 79	1290 ± 642	2553 ± 416	4248 ± 1396	1587 ± 524
Cr	<LOD	<LOD	<LOD	<LOD	<LOD	<LOD	0.110 ± 0.002	0.2 ± 0.1	<LOD
Mn	<LOD	<LOD	0.13 ± 0.02	<LOD	<LOD	<LOD	0.11 ± 0.04	<LOD	<LOD
Fe	0.2 ± 0.1	0.3 ± 0.1	0.5 ± 0.1	0.5 ± 0.1	0.17 ± 0.01	0.3 ± 0.2	0.3 ± 0.2	<LOD	0.28 ± 0.02
Cu	<LOD	<LOD	<LOD	<LOD	0.1 ± 0.001	<LOD	<LOD	<LOD	<LOD
Zn	0.2 ± 0.1	0.3 ± 0.1	1.0 ± 0.1	0.9 ± 0.2	0.9 ± 0.1	0.3 ± 0.1	1.7 ± 0.5	2.0 ± 1.1	0.07 ± 0.04
Br	64.9 ± 11.4	82.2 ± 13.1	75.3 ± 3.4	102.3 ± 27.6	85.8 ± 7.4	90 ± 32	94.1 ± 7.4	92.1 ± 6.5	107.3 ± 2.3
Rb	2.7 ± 0.2	4.6 ± 1.5	3.4 ± 0.5	4.3 ± 2.1	3.5 ± 0.3	4.5 ± 2.5	3.1 ± 0.2	3.3 ± 0.4	4.5 ± 0.6
Sr	9.7 ± 0.6	10.9 ± 1.8	9.0 ± 1.6	12.9 ± 4.0	10.4 ± 0.3	9.5 ± 4.3	17.4 ± 0.5	15.9 ± 1.4	19.0 ± 1.2
Pb	2.4 ± 0.4	<LOD	<LOD	7.1 ± 1.9	7.6 ± 1.3	<LOD	8.3 ± 2.8	10.7 ± 5.7	<LOD

Values are expressed as the average ± standard deviation of three measurements. Relative sensitivities for elements with * were calculated based on a calibration curve. LOD lower detection limit.

Both pozzolanic and carbonation reactions have been demonstrated to be effective in heavy metal mobilization [31], but the comparison between procedures a) and b) highlights the fundamental contribution of carbonation. As explained in the introduction part, the difference in the amount of SS in the co-combustion with MSW is considered to evaluate if the addition of this waste plays a role in the stabilization mechanism. For this aim, the procedure not involving the use of BA (procedure b) is considered. Moreover, this procedure allows for the reaching of lower pH values of the obtained eco-materials (see Table 4), increasing the efficacy of stabilization.

Figure 2 reports the concentration of Pb and Zn in the leaching solutions of stabilized samples, considered after different times, versus the pH of the solution (for both procedures a) and b)). Zn and Pb are found in a high concentration in the raw FA (ranging from 9–11.7 and 92–127.1 mg/L, respectively) with high pH values (about 12), while in the stabilized samples the concentration of these elements is found to be often two orders of magnitude lower. Figure 2 clearly highlights that an increase of the aging time corresponds to a decrease of pH of the solutions [32], in agreement with the already reported evidence of carbonation. This also corresponds to a decrease of the heavy metal leachability, as already observed and discussed [4,33]. In particular, concerning Pb, three months after stabilization it is sometimes lower than LOD by TXRF spectrometry, as reported in Tables 3 and 4, therefore it cannot be found in Figure 2.

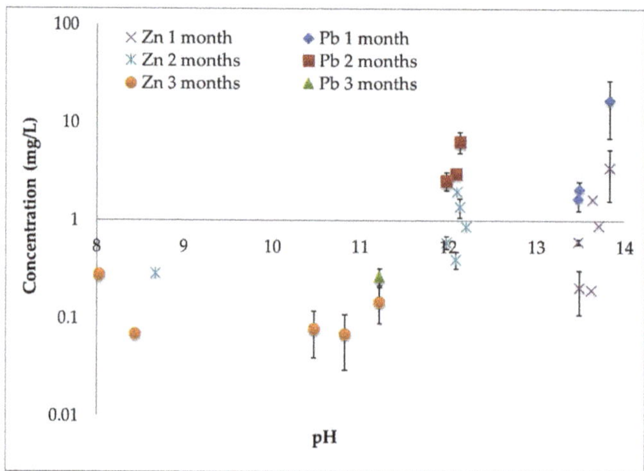

Figure 2. Concentration values of Pb and Zn in the leaching solution of stabilized samples (involving both procedures a) and b)) during the first three months versus the pH.

To better highlight the pH role, the variation of the concentration of Pb and Zn in the leaching solutions of stabilized samples can be considered analyzing data reported in Figure 3. In this Figure, the concentration of Pb and Zn in the solutions of stabilized samples (considering both procedures a) and b)), are plotted versus their values in the leaching solution of raw FA (initial metals concentrations, before the stabilization). Figure 3 allows for the comparison of the results of stabilization, considering that raw FA (corresponding to different combustion lines) had a different amount of leachable Pb and Zn. In particular, it is evident that all samples show a reduction of the Pb and Zn concentration, after the stabilization. Obviously, samples that contained a lower amount of leachable heavy metals in raw FA show a lower concentration of corresponding metals in the solutions of stabilized eco-materials. This means that samples with higher initial concentrations of Pb and Zn need more time (or more stabilizing agents) to reduce the leachable elements concentration.

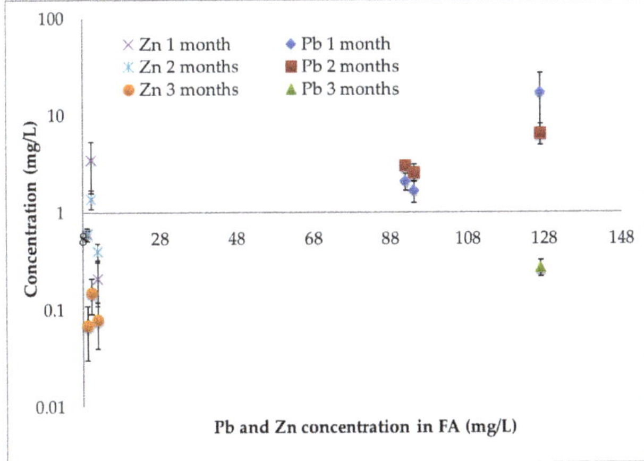

Figure 3. Concentration values of Pb and Zn in the leaching solution of stabilized samples (involving both procedures a) and b)) during the first three months versus the concentration of Pb and Zn in MSWI FA.

The role of BA in the stabilization can be highlighted by analyzing data reported in Figure 4, that show the concentration of Pb and Zn during the first three months evaluated in the leaching solutions of all stabilized samples (considering both procedures a) and b)), versus the ratio between the total amount of ash (FA, CFA, FGD, and BA) and the sum of CFA and BA amount. It means that the samples that contain the higher CFA + BA amount are better stabilized in comparison to those containing a lower amount of these stabilization agents (the concentration of Pb and Zn are lower). Indeed, it is important to highlight that BA contains amorphous silica [4] and CFA contains aluminosilicate glass [13,34]. Then, it is possible to suppose that dissolved amorphous silica in the presence of calcium ions (and in a highly alkaline environment) promotes a pozzolanic reaction with FA. Then, it is possible to conclude that BA plays a fundamental role in reducing the Pb and Zn presence in solution of the produced eco-materials. Obviously, it was also shown that the pH role is fundamental, and it must be controlled and possibly adjusted to obtain the best results in terms of stability of the obtained eco-materials.

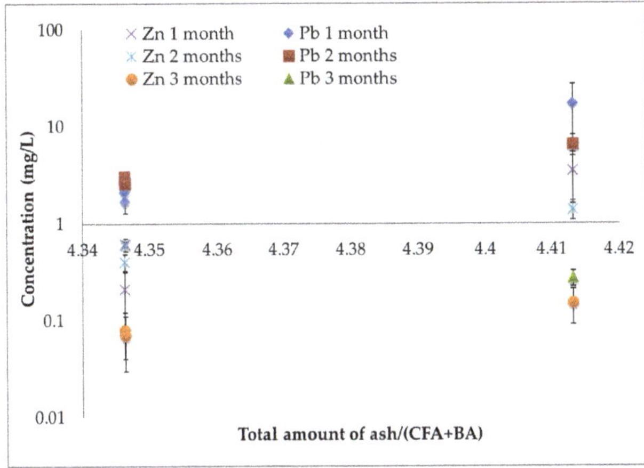

Figure 4. Concentration values of Pb and Zn in the leaching of stabilized samples (involving procedure a)) during the first three months versus the partition of total amount of ash and the sum of CFA and BA quantities.

27. van der Kooij, S.; Van Vliet, B.J.M.; Stomph, T.J.; Sutton, N.B.; Anten, N.P.R.; Hoffland, E. Phosphorus recovered from human excreta: A socio-ecological-technical approach to phosphorus recycling. *Resour. Conserv. Recycl.* **2020**, *157*, 104744. [CrossRef]
28. Bosio, A.; Rodella, N.; Depero, L.E.; Bontempi, E. Rice Husk Ash Based Composites, Obtained by Toxic Fly Ash Inertization, and their Applications as Adsorbents. *Chem. Eng.* **2014**, *37*. [CrossRef]
29. Assi, A.; Federici, S.; Bilo, F.; Zacco, A.; Depero, L.E.; Bontempi, E. Increased sustainability of carbon dioxide mineral sequestration by a technology involving fly ash stabilization. *Materials* **2019**, *12*, 2714. [CrossRef]
30. Ji, L.; Yu, H. Carbon dioxide sequestration by direct mineralization of fly ash. In *Carbon Dioxide Sequestration in Cementitious Construction Materials*; Elsevier: Amsterdam, The Netherlands, 2018; pp. 13–37. ISBN 9780081024447.
31. Leonard, R.J.; Davidson, D.T. Pozzolanic reactivity study of fly ash. *Highw. Res. Board Bull.* **1959**, 1–17.
32. Zhang, Y.; Cetin, B.; Likos, W.J.; Edil, T.B. Impacts of pH on leaching potential of elements from MSW incineration fly ash. *Fuel* **2016**, *184*, 815–825. [CrossRef]
33. Rodella, N.; Pasquali, M.; Zacco, A.; Bilo, F.; Borgese, L.; Bontempi, N.; Tomasoni, G.; Depero, L.E.; Bontempi, E. Beyond waste: New sustainable fillers from fly ashes stabilization, obtained by low cost raw materials. *Heliyon* **2016**, *2*. [CrossRef]
34. Liu, P.; Huang, R.; Tang, Y. Comprehensive Understandings of Rare Earth Element (REE) Speciation in Coal Fly Ashes and Implication for REE Extractability. *Environ. Sci. Technol.* **2019**, *53*, 5369–5377. [CrossRef] [PubMed]
35. Junakova, N.; Junak, J.; Balintova, M. Reservoir sediment as a secondary raw material in concrete production. *Clean Technol. Environ. Policy* **2015**, *17*, 1161–1169. [CrossRef]
36. Assi, A.; Bilo, F.; Zanoletti, A.; Ducoli, S.; Ramorino, G.; Gobetti, A.; Zacco, A.; Federici, S.; Depero, L.E.; Bontempi, E. A Circular Economy Virtuous Example—Use of a Stabilized Waste Material Instead of Calcite to Produce Sustainable Composites. *Appl. Sci.* **2020**, *10*, 754. [CrossRef]
37. Guarienti, M.; Gianoncelli, A.; Bontempi, E.; Moscoso Cardozo, S.; Borgese, L.; Zizioli, D.; Mitola, S.; Depero, L.E.; Presta, M. Biosafe inertization of municipal solid waste incinerator residues by COSMOS technology. *J. Hazard. Mater.* **2014**, *279*, 311–321. [CrossRef] [PubMed]
38. Bilo, F.; Moscoso, S.; Borgese, L.; Delbarba, M.V.; Zacco, A.; Bosio, A.; Federici, S.; Guarienti, M.; Presta, M.; Bontempi, E.; et al. Total reflection X-ray fluorescence spectroscopy to study Pb and Zn accumulation in zebrafish embryos. *X-ray Spectrom.* **2015**, *44*, 124–128. [CrossRef]
39. Guarienti, M.; Cardozo, S.M.; Borgese, L.; Lira, G.R.; Depero, L.E.; Bontempi, E.; Presta, M. COSMOS-rice technology abrogates the biotoxic effects of municipal solid waste incinerator residues. *Environ. Pollut.* **2016**, *214*, 713–721. [CrossRef]

Article

Microalgal Growth in Paper Industry Effluent: Coupling Biomass Production with Nutrients Removal

Bruna Porto [1,2], Ana L. Gonçalves [3,*], Ana F. Esteves [3], Selene M. A. Guelli Ulson de Souza [1], Antônio A. Ulson de Souza [1], Vítor J. P. Vilar [2] and José C. M. Pires [3]

1 Laboratory of Numerical Simulation of Chemical Systems and Mass Transfer (LABSIN-LABMASSA), Federal University of Santa Catarina, Chemical and Food Engineering Department, PO Box 476, 88040-900 Florianópolis/SC, Brazil; b.porto@hotmail.com (B.P.); selene.souza@ufsc.br (S.M.A.G.U.d.S.); antonio.augusto.ulson.souza@gmail.com (A.A.U.d.S.)

2 Laboratory of Separation and Reaction Engineering – Laboratory of Catalysis and Materials (LSRE-LCM), Faculty of Engineering, University of Porto, Rua Dr Roberto Frias, 4200-465 Porto, Portugal; vilar@fe.up.pt

3 LEPABE – Laboratory for Process Engineering, Environment, Biotechnology and Energy, Faculty of Engineering, University of Porto, Rua Dr Roberto Frias, 4200-465 Porto, Portugal; afcesteves@fe.up.pt (A.F.E.); jcpires@fe.up.pt (J.C.M.P.)

* Correspondence: algoncalves@fe.up.pt; Tel.: +351-22-041-3656

Received: 31 March 2020; Accepted: 24 April 2020; Published: 26 April 2020

Abstract: Paper and pulp industries produce effluents with high phosphorus concentrations, which need to be treated before their discharge in watercourses. The use of microalgae for this purpose has attracted the attention of researchers because: (i) microalgae can assimilate phosphorus (one of the main nutrients for their growth); and (ii) growing on effluents can significantly reduce the costs and environmental impact of microalgal biomass production. This study evaluated the growth and ability of *Chlorella vulgaris* to remove the phosphorus from a secondary-treated effluent of a Portuguese paper company. Batch experiments were performed for 11 days using different dilutions of the effluent to evaluate its inhibitory effect on microalgae. Results showed that the non-diluted effluent inhibited microalgal growth, indicating that this bioremediation process is possible after a previous dilution of the effluent. Regarding phosphorus removal, promising results were achieved, especially in the experiments conducted with the most diluted effluent: removal efficiencies obtained in these conditions were (54 ± 1)%. Another interesting finding of this study was microalgal growth in flakes' form (mainly due to the compounds present in the effluent and to the pH values achieved), which can be an important economic advantage for biomass recovery after the remediation step.

Keywords: biomass production; *Chlorella vulgaris*; microalgae; nutrients removal; paper industry effluent; effluent treatment

1. Introduction

Paper and pulp industries require large amounts of water during their manufacturing stages. For example, the production of 1 kg of paper requires 10 to 50 L of water [1]. At the same time, large amounts of effluents (about 2000 $m^3\ d^{-1}$) are generated, presenting as main features [1,2]: (i) high chemical oxygen demand (COD, 1000–13,000 $mg_{O2}\ L^{-1}$); (ii) high total suspended solids contents; (iii) non-biodegradable organic materials; (iv) adsorbable organic halogens (AOX): (v) color; (vi) phenolic compounds; (vii) high total phosphorus contents; and (viii) limiting nitrogen concentrations. Due to the large volumes involved and respective compositions, discharge of these effluents without any treatment can cause several environmental problems [1,3]: (i) colored effluents can affect aesthetics, water transparency and gas solubility in water bodies; (ii) increase in the concentration of toxic compounds, which can affect aquatic flora and fauna; and (iii) eutrophication with consequent decrease of dissolved oxygen concentration and pH oscillations, which can negatively impact aquatic ecosystems. Therefore, treatment of these effluents is necessary before their discharge.

Among the contaminants present in these effluents, phosphorus is of particular concern, as it subsists in the effluents after the secondary treatment step and is one of the main contributors to the eutrophication phenomenon [4]. Currently applied methods to reduce phosphorus concentration in these effluents include physicochemical methods, such as precipitation using aluminum and iron salts. However, these techniques tend to be costly and to produce large amounts of sludge contaminated with the referred chemical compounds, requiring further treatment [5,6]. Therefore, microalgal cultures have appeared as a feasible alternative to conventional physicochemical methods. These microorganisms have shown their ability to effectively remove color, nutrients, such as nitrogen and phosphorus, trace metals and other compounds from the culture medium [7,8].

Microalgae are fast-growing photosynthetic microorganisms that have gained much attention in recent decades, due to their high potential in a wide variety of applications. During photosynthesis, microalgae uptake CO_2 from the atmosphere or flue gas emissions, contributing to the reduction of the atmospheric concentration of this greenhouse gas [9]. These microorganisms also require inorganic sources of nitrogen and phosphorus as macronutrients, enabling the use of microalgal cultures as a tertiary treatment stage (when significant concentrations of these nutrients persist after previous treatment processes) [10]. Finally, microalgal biomass presents a very rich composition in polysaccharides, lipids, proteins, vitamins, and other valuable compounds, which make microalgae a valuable resource for several applications [11,12], such as the production of natural colorants or dyes, bioenergy, and biofertilizers. Also, effluent treatment with microalgae has the following advantages [10]: (i) reduction of nitrogen and phosphorus concentrations to levels below the legislated limits for effluent discharge (EU Directives 1991/271/EEC and 1998/15/EC); (ii) recovery/recycle of these nutrients, which production presents negative environmental impacts; (iii) increase of the oxygen concentration in the treated effluent; (iv) production of biomass that can be integrated into the value chain of the company; and (v) reduction of net carbon dioxide emissions.

Despite the need to search for eco-friendly and cost-effective remediation strategies, only a few studies have reported the treatment of pulp and paper industry effluents using microalgae [1]. Tarlan et al. [7] evaluated the removal of color, AOX, and COD from an effluent resulting from a wood-based pulp and paper Turkish company using a mixed microalgal culture (composed by *Chlorella* and diatoms). Initial composition of this effluent in terms of color, AOX, and COD was: 4018 Pt-Co, 46.3 mg L^{-1} and 1248 mg L^{-1}, respectively. Operating in batch mode and using three different dilutions of this effluent, resulting from the process of pulp production using red pine, the authors reported removal efficiencies of 84%, 80% and 58% for color, AOX, and COD, respectively. Gentili [13] aimed to evaluate the growth of microalgae on mixtures of municipal, dairy, and pulp and paper effluents to achieve the dual goal of nutrients removal and lipids production. The use of mixtures of pulp and paper industry effluents with municipal and dairy ones was to evaluate if these mixtures could (i) promote microalgal growth without previous dilution with freshwater; and (ii) provide the required nutrients for biomass production without the need for nutrients supplementation. Characterization of these mixtures revealed an ammonium–nitrogen (NH_4–N) concentration ranging from 14.75 mg_N L^{-1} to 22.35 mg_N L^{-1}, a nitrate–nitrogen (NO_3–N) concentration between 1.6 mg_N L^{-1} and 10.1 mg_N L^{-1} and a phosphate–phosphorus (PO_4–P) concentration ranging between 1.06 mg_P L^{-1} and 1.25 mg_P L^{-1}. With this study, carried out in laboratory tubes, in batch mode, the authors demonstrated that the microalgae *Scenedesmus* sp., *Scenedesmus dimorphus* and *Selenastrum minutum* were able to achieve nitrogen and phosphorus removal efficiencies of 96%–99% and 91%–99%, respectively. Finally, in the study performed by Usha, et al. [1], a mixed microalgal culture (composed by two *Scenedesmus* species) was grown in different dilutions (0%–95%) of a pulp and paper mill effluent, resulting from an Indian company, with the following composition: (i) 9.932 mg_N L^{-1} of NO_3–N; (ii) 30.25 mg_P L^{-1} of PO_4–P; (iii) 3000.15 mg L^{-1} of COD; and (iv) 2944 mg L^{-1} of biochemical oxygen demand (BOD). The experiments, aimed at evaluating both biomass production and nutrients removal efficiencies, were performed in batch mode, for 28 days, using open ponds as cultivation system (outdoor conditions). Regarding nutrients uptake, the most promising results were obtained for the 40% dilution: (i) 65% for NO_3–N removal; (ii) 81.3% for PO_4–P; and (iii) 75% for COD; and (iv) 82% for BOD.

The main goal of the present study was to evaluate biomass production and phosphorus removal from a secondary-treated effluent of a Portuguese paper company using the microalga *Chlorella vulgaris*. Different dilutions were performed to evaluate possible inhibitory effects of the effluent on microalgal growth and phosphorus uptake ability.

2. Materials and Methods

2.1. Microalgae Strain and Maintenance Medium

The microalga *C. vulgaris* (CCAP 211/11B) was obtained from the Culture Collection of Algae and Protozoa, United Kingdom. The strain was maintained on modified Organization for Economic Co-operation and Development (OECD) culture medium [14], with the following composition (mg L^{-1}): 119 KNO_3; 12 $MgCl_2·2H_2O$; 18 $CaCl_2·2H_2O$; 15 $MgSO_4·7H_2O$; 20 KH_2PO_4; 0.08 $FeCl_3·6H_2O$; 0.1 $Na_2EDTA·2H_2O$; 0.185 H_3BO_3; 0.415 $MnCl_2·4H_2O$; 0.003 $ZnCl_2$; 0.0015 $CoCl_2·6H_2O$; 0.00001 $CuCl_2·2H_2O$; 0.007 $Na_2MoO_4·2H_2O$ and 100 Na_2CO_3.

2.2. Paper Industry Effluent and Culture Conditions

Effluent from a Portuguese paper company, collected after the secondary treatment step, was characterized (Table 1) and employed as a culture medium for microalgal growth. The methodology adopted for effluent characterization was the following: (i) COD and turbidity were determined according to the Standard Methods for the Examination of Water and Wastewater [15] (through the 5220-D and 2130-B tests, respectively); (ii) total dissolved carbon (TDC), dissolved organic carbon (DOC) and inorganic carbon (DIC) were determined using an organic carbon analyzer (TOC-V_{CSN}, Shimadzu); and (iii) chlorides, sulfates, nitrates, nitrites and phosphates were determined through ion chromatography (ICS-2100, Dionex). Due to the low concentration of nitrogen in the effluent, when compared with typical nutritional requirements of microalgae, the effluent was supplemented with $NaNO_3$ to achieve N:P molar ratios ranging between 6:1 and 9:1. Ratios between 5:1 and 30:1 have been considered adequate for several microalgal species [16,17].

Table 1. Physicochemical characterization of the paper industry effluent used in this study.

Parameters	Values	Unit
Turbidity	1.55	NTU[a]
pH	7.02	-
Dissolved organic carbon (DOC)	296	mg L^{-1}
Total dissolved carbon (TDC)	369	mg L^{-1}
Dissolved inorganic carbon (DIC)	72.5	mg L^{-1}
Soluble chemical oxygen demand (COD_S)	323	mg L^{-1}
Chlorides (Cl^-)	671	mg L^{-1}
Sulfates (SO_4^{2-})	808	mg L^{-1}
Phosphate–phosphorus (PO_4–P)	12.3	mg L^{-1}
Nitrate–nitrogen (NO_3–N)	8.73	mg L^{-1}
Nitrite–nitrogen (NO_2–N)	3.42	mg L^{-1}

[a] Nephelometric turbidity unit.

Batch experiments were performed in 1—L borosilicate glass flasks (VWR, Portugal) with a working volume of 950 mL for 11 days. The raw effluent (assay 1) and four different dilutions with freshwater (assays 2–5) were used as the culture medium for microalgal growth, with nitrogen concentrations (corresponding to the sum between nitrate– and nitrite–nitrogen) ranging between

12.7 mg$_N$ L^{-1} and 34.2 mg$_N$ L^{-1} and phosphorus concentrations (phosphate–phosphorus) ranging between 4.01 mg$_P$ L^{-1} and 12.3 mg$_P$ L^{-1}. The medium was inoculated with 250 mL of *C. vulgaris* inoculum to obtain an initial biomass concentration of ~ 68 mg$_{dw}$ L^{-1}. The cultures were continuously exposed to: (i) photosynthetically active radiation between 30–40 μmol m^{-2} s^{-1}, using a 34-W white led panel; and (ii) atmospheric air filtered with 0.45–μm nylon membranes (Specanalitica, Portugal), injected at ~ 90 L h^{-1}, using Trixie AP 180 air pumps (Trixie, Tarp, Germany). The experimental setup is shown in Figure 1. Two independent experiments were performed for each assay.

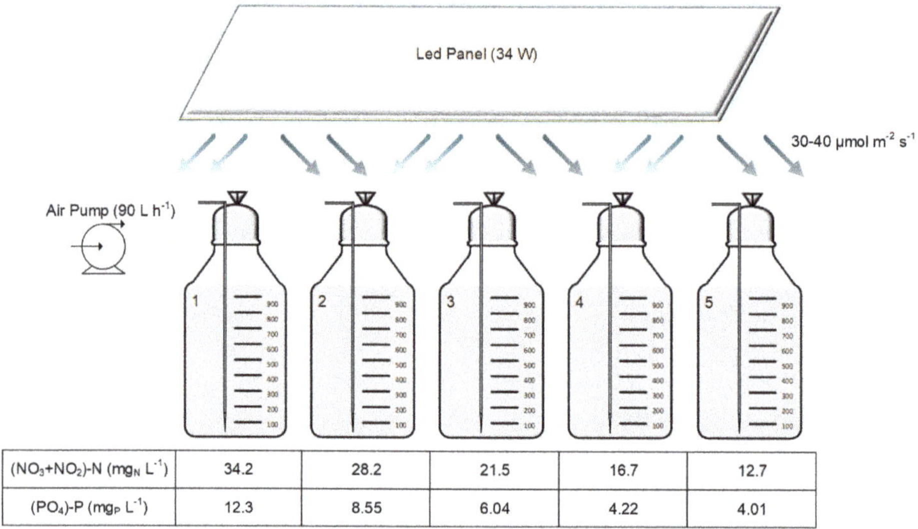

Figure 1. Schematic representation of the experimental setup.

2.3. Microalgal Growth Monitoring and Kinetic Growth Parameters

Operational parameters, such as pH and temperature, were daily monitored using a SympHony SB90M5 pH-meter (VWR, Portugal). Microalgal growth was also daily assessed through optical density measurements at 680 nm (OD$_{680}$) using a UV-6300 PC spectrophotometer (VWR, United States). To eliminate the interference of the effluent color on OD$_{680}$ measurements, the cells were separated from the culture medium by centrifugation (at 4000 rpm, for 10 min), the supernatant was discarded, and the cells were resuspended in an equal volume of distilled water, as described by Hodaifa, et al. [18]. This procedure was repeated twice. The relationship between OD$_{680}$ and biomass concentration (X, mg$_{dw}$ L^{-1}) for *C. vulgaris* was previously established by linear regression, according to Equation (1):

$$\mathrm{X} = 0.0024 \times \mathrm{OD}_{680} + 0.0030\ \mathrm{R}^2 = 0.9999, \tag{1}$$

Biomass concentrations were used to determine the kinetic growth parameters: (i) specific growth rates (μ, d^{-1}); and (ii) maximum and average biomass productivities (P_{max} and P_{aver}, mg$_{dw}$ L^{-1} d^{-1}). The specific growth rates were determined from the first-order kinetic model, according to Equation (2):

$$\frac{\mathrm{dX}}{\mathrm{dt}} = \mu \mathrm{X} \leftrightarrow \mu = \frac{\ln \mathrm{X}_1 - \ln \mathrm{X}_0}{\mathrm{t}_1 - \mathrm{t}_0}, \tag{2}$$

where X_1 and X_0 correspond, respectively, to biomass concentration (mg_{dw} L^{-1}) in the end (t_1, d) and beginning (t_0, d) of the exponential growth phase. Biomass productivities (P, mg_{dw} L^{-1} d^{-1}) were calculated for each pair of consecutive points, through Equation (3):

$$P = \frac{X_{z+1} - X_z}{t_{z+1} - t_z}, \quad (3)$$

where X_z represents the biomass concentration (mg_{dw} L^{-1}) at time t_z (d) and X_{z+1} corresponds to the biomass concentration (mg_{dw} L^{-1}) at time t_{z+1} (d). The maximum productivity was determined from the maximum value obtained from Equation (3). On the other hand, average biomass productivities were determined according to Equation (4):

$$P_{aver} = \frac{X_f - X_i}{t_f - t_i}, \quad (4)$$

where X_f and X_i correspond, respectively, to biomass concentration (mg_{dw} L^{-1}) in the end (t_f, d) and beginning (t_i, d) of the cultivation period.

2.4. Nutrients Removal

Nutrients removal was evaluated in terms of nitrogen and (N) phosphorus (P) present in the culture medium/effluent. Nitrogen was assessed in the forms of nitrate and nitrite ions, whereas phosphorus was monitored through the presence of phosphate ions. From each assay, 5 mL of the microalgal suspension were periodically collected (days 0, 1, 2, 4, 7, 9, and 11). These samples were centrifuged at 4000 rpm, for 10 min, and the supernatants were filtered through 0.45–µm nylon membranes (Specanalitica, Portugal). Nitrate, nitrite and phosphate concentrations were determined in an ion chromatograph (ICS-2100, Dionex) equipped with an anion analytical column (4x 250 mm, AS11-HC) and a self-regeneration suppressor (4 mm, AERS 500). The values obtained in the first and last day of culturing were used to calculate the following removal parameters: (i) removal efficiencies (*%R*, %); (ii) average removal rates (*RR*, mg L^{-1} d^{-1}); and (iii) mass removal (*R*, mg L^{-1}), as shown in Equations (5), (6) and (7), respectively:

$$\%R = \frac{S_f - S_i}{S_i} \times 100, \quad (5)$$

$$RR = \frac{S_f - S_i}{t_f - t_i}, \quad (6)$$

$$R = S_f - S_i, \quad (7)$$

where S_f and S_i correspond to the nitrogen (nitrate + nitrite) or phosphorus (phosphate) concentration (mg L^{-1}) in the end (t_f, d) and beginning (t_i, d) of the cultivation period, respectively.

2.5. Statistical Analysis

Each parameter shown in the present paper was expressed as the mean and standard deviation. The Tukey statistical test was used to investigate if the differences between the different effluent concentrations studied could be considered significant. These statistical tests were performed using Statistica 8.0 (StatSoft Inc., USA) and were carried out at a significance level (p) of 0.05.

3. Results and Discussion

3.1. Microalgal Growth

The *C. vulgaris* growth curves in raw and diluted paper industry effluent are shown in Figure 2. These results evidence the inexistence of an adaptation phase for all assays and an exponential growth phase that lasted approximately four days. In addition, no cell decay was observed during the 11-day batch culture, indicating that the experiments could be extended for a longer period. The increase of biomass concentration during the cultivation period, as well as the lack of an adaptation phase, shows that *C. vulgaris* was able to grow in this effluent. However, biomass concentrations achieved in non-diluted effluent (assay 1) were statistically lower ($p < 0.05$) than those achieved in more diluted effluents from assays 3–5.

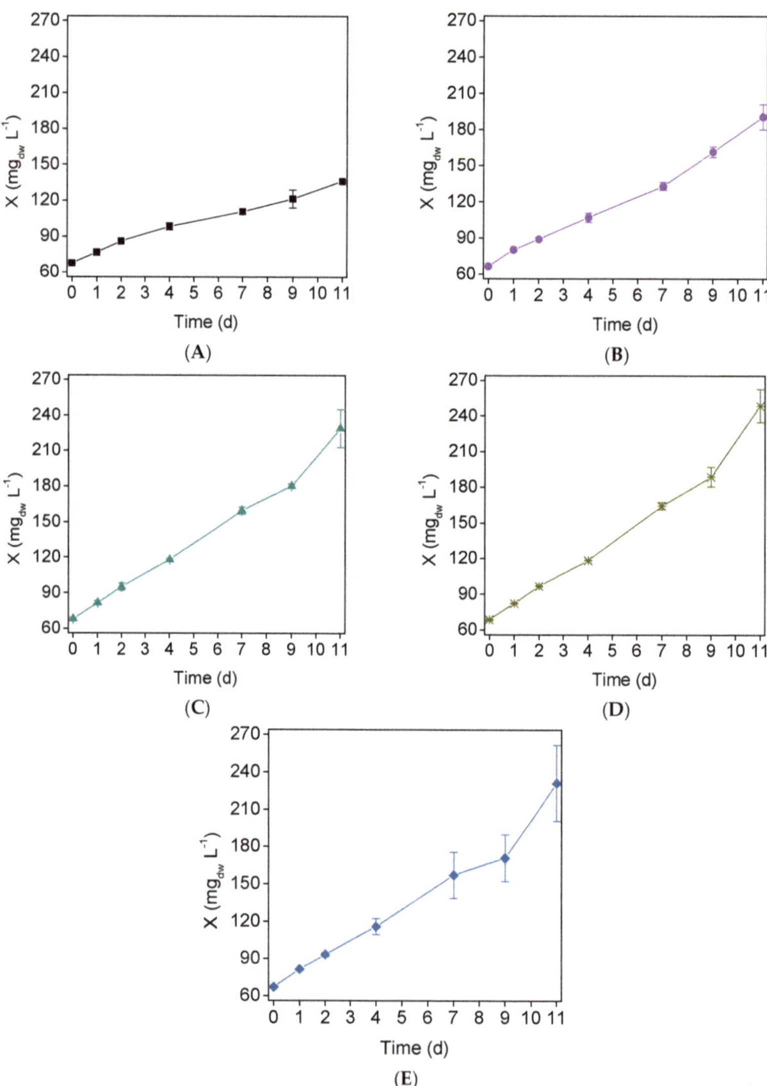

Figure 2. *C. vulgaris* cultures growth curves in raw and diluted secondary-treated paper industry effluent: (**A**) Assay 1—■—; (**B**) Assay 2—●—; (**C**) Assay 3—▲—; (**D**) Assay 4—✳—; and (**E**) Assay 5—◆—. Error bars correspond to the standard deviation of the mean obtained from two independent experiments.

To complement the analysis from growth curves, microalgal growth parameters, such as specific growth rate, maximum biomass concentration, and maximum and average biomass productivities, were determined and presented in Table 2. From these data, it is possible to see a general increase in growth parameters from assay 1 to assay 5, i.e., from the non-diluted effluent to the more diluted one. Regarding specific growth rates, values ranged from (0.093 ± 0.007) d^{-1} to (0.16 ± 0.02) d^{-1} in assays 1 and 5, respectively. The highest values of maximum biomass concentrations were also obtained in more diluted effluents from assays 4 and 5: (249 ± 14) mg_{dw} L^{-1} and (231 ± 31) mg_{dw} L^{-1}, respectively. Similar behavior was observed for both maximum and average biomass productivities. Maximum biomass productivities/average biomass productivities obtained in assays 4 and 5 were (30 ± 3)/(16 ± 1) mg_{dw} L^{-1} d^{-1} and (30 ± 6)/(15 ± 3) mg_{dw} L^{-1} d^{-1}, respectively. In opposition, maximum and average biomass productivities obtained in assay 1 were (9.8 ± 0.2) mg_{dw} L^{-1} d^{-1} and (6.2 ± 0.1) mg_{dw} L^{-1} d^{-1}, respectively.

Table 2. Specific growth rates (μ, in d^{-1}), maximum biomass concentrations (X_{max}, in mg_{dw} L^{-1}), and maximum and average biomass productivities (P_{max} and P_{aver}, in mg_{dw} L^{-1} d^{-1}) determined for *C. vulgaris* grown in raw and diluted secondary-treated paper industry effluent.

Assay	(NO_3+NO_2)–N (mg_N L^{-1})	PO_4–P (mg_P L^{-1})	μ (d^{-1})	X_{max} (mg_{dw} L^{-1})	P_{max} (m_{dw} L^{-1} d^{-1})	P_{aver} (m_{dw} L^{-1} d^{-1})
1	34.2	12.7	0.093 ± 0.007 ^{a}	136 ± 1 ^{a}	9.8 ± 0.2 ^{a}	6.22 ± 0.09 ^{a}
2	28.3	8.55	0.11 ± 0.01 ^{ab}	191 ± 10 ^{ab}	15 ± 2 ^{ab}	11 ± 1 ^{ab}
3	21.5	6.04	0.136 ± 0.004 ^{bc}	229 ± 16 ^{b}	24 ± 7 ^{ab}	15 ± 1 ^{b}
4	16.7	4.22	0.134 ± 0.002 ^{bc}	249 ± 14 ^{b}	30 ± 3 ^{b}	16 ± 1 ^{b}
5	12.7	4.01	0.16 ± 0.02 ^{c}	231± 31 ^{b}	30 ± 6 ^{b}	15 ± 3 ^{b}

Values are presented as the mean ± standard deviation obtained from two independent experiments. Within the same column, mean values sharing at least one common letter (in superscript) are not statistically different ($p > 0.05$).

In contrast to what was observed by Gentili [13], the increment in nitrogen and phosphorus concentration did not contribute to an increase in kinetic growth parameters. Accordingly, these results may indicate inhibitory effects of the effluent on microalgae, which can influence microalgal cultures in different ways [19–21]: (i) the effluent color may act as a barrier to light penetration, thus limiting microalgal access to light and photosynthetic activity; and (ii) paper industry effluents are characterized by the presence of lignin, humic acids, furans, and dioxins and by high levels of aluminum and manganese, which exhibit toxic effects on microalgae.

Most studies regarding the bioremediation of paper industry effluents with microalgae focus on the removal of contaminants and only a few report biomass production yields. Polishchuk, et al. [20] reported that the maximum specific growth rate obtained for *Nannochloropsis oculata* grown in effluents resulting from pulp and paper industry was 0.405 d^{-1}. Tao, et al. [19] revealed that maximum biomass concentrations achieved by *Scenedesmus acuminatus* and *C. vulgaris* grown in paper industry effluents were 291 mg L^{-1} and 822 mg L^{-1}, respectively. Considering the values referred in the literature, microalgal growth parameters obtained in this study were significantly lower, which can be attributed to the inhibitory effects promoted by the effluent used (in assays 1–3) and to the low concentration of some essential nutrients (in more diluted effluents of assays 4 and 5). Another explanation for the low biomass concentrations and productivities achieved may be related to the phenomenon of flakes formation observed within the cultivation period (autoflocculation). Cells' agglomeration can affect the accurate measurement of OD_{680} and, on the other hand, it can reduce light absorption efficiency by cells incorporated within flakes, thus resulting in lower photosynthetic activity. In this study, this phenomenon occurred due to the increase of culture pH (from 7.8 to 8.6) or due to the presence of certain compounds in the effluent, which can induce a change in the surface charge of the cells and affect suspensions' stability [22]. Despite the low microalgal growth rates, the flakes formation enables a cost-effective biomass removal after effluent remediation. The density similar to water and small size of microalgal cells difficult the harvesting process and make this step one of the most expensive within microalgal biomass production processes [22,23]. However, when cells agglomerate, an increase in their density and size is observed, contributing to higher settling rates and allowing biomass recovery using the least expensive harvesting method: sedimentation.

3.2. Nutrients Removal

In this study, nitrogen (in the forms of nitrate and nitrite) and phosphorus (in the form of phosphate) concentrations were monitored within the cultivation time to evaluate the potential of *C. vulgaris* to uptake these nutrients from a paper industry effluent with different concentrations of both nitrogen and phosphorus. Figure 3 shows the variation of nitrogen and phosphorus concentration in each assay. Regarding nitrogen removal (Figure 3A), this element was readily assimilated by *C. vulgaris* in the diluted effluents (assays 2–5). In the raw effluent (corresponding to assay 1), a two-day delay was observed in nitrogen assimilation, which may be related to the adaptation of the microalga to these conditions. Regarding the assimilation patterns observed in assays 2–5, these were approximately linear for assays 2–4, with nitrogen concentration decreasing gradually during the cultivation time. On the other hand, in assay 5, corresponding to the more diluted effluent experiments, nitrogen concentration decreased until the seventh day of culturing and then it was maintained approximately constant. This behavior may be attributed to a decrease in photosynthetic activity, as nitrogen concentration decreased, and explains the lower biomass concentrations achieved in assay 5 when compared to the one obtained in assay 4 (according to Table 2, (231 ± 31) mg_{dw} L^{-1} and (249 ± 14) mg_{dw} L^{-1}, respectively). Also, at the end of the cultivation time, nitrogen concentration remaining in cultures corresponding to assays 4 and 5 was approximately the same ((2.81 ± 0.05) mg_N L^{-1} and (2.6 ± 0.2) mg_N L^{-1}, respectively), indicating a limitation of this nutrient in the last days of assay 5. As for nitrogen concentration, phosphorus concentration also decreased within the cultivation time (Figure 3B), but in a lesser extent, which is related with microalgal nutritional requirements, as given by its typical elemental biochemical composition: $CO_{0.48}H_{1.83}N_{0.11}P_{0.01}$ [24]. The reduction observed in nitrogen and phosphorus concentration in the studied effluent (raw or diluted) shows that *C. vulgaris* can promote an efficient uptake of both nutrients. However, except for nitrogen concentration in assay 5, total depletion of these nutrients did not occur after the 11 days of culturing, reiterating what was stated in relation to cell growth, that the cultures could be extended for an increased period to further improve nutrients removal efficiencies. Another similarity with the microalgal growth parameters already described is the higher variations in nitrogen and phosphorus concentrations observed in the experiments where the effluent was previously diluted (assays 2–5), which indicate that these conditions were more favorable for *C. vulgaris* photosynthetic activity.

Nitrogen and phosphorus removal parameters are presented in Figures 4 and 5, respectively. As with microalgal growth parameters, a general increase in nutrients removal efficiencies was observed from assay 1 to 5, with values ranging from (24 ± 10)% to (80 ± 4)% for nitrogen (Figure 4A) and from (13.0 ± 0.9)% to (54 ± 1)% for phosphorus (Figure 5A). However, Figure 4A shows that there was no statistical difference ($p > 0.05$) in nitrogen removal efficiency between assays 4 and 5, which can be explained by the low concentration achieved in the assay 5 (the one corresponding to the most diluted effluent) that might have been limiting for microalgal growth. In fact, according to Table 2, maximum biomass concentration achieved in assay 4 was higher than that in assay 5, indicating that the highest dilution applied in this study may have contributed to nitrogen limitation to *C. vulgaris*, with effects on their growth and nutrients removal parameters. Regarding nitrogen removal rates (Figure 4B) and mass removal (Figure 4C), the highest values were determined in assays 3 and 4 and no statistical differences were observed ($p > 0.05$): (i) average removal rates were (1.31 ± 0.07) mg_N L^{-1} d^{-1} and (1.26 ± 0.08) mg_N L^{-1} d^{-1}, respectively; and (ii) mass removal values were (14.4 ± 0.8) mg_N L^{-1} and (13.9 ± 0.9) mg_N L^{-1}, respectively. These results are in accordance with maximum biomass concentration achieved and indicate higher photosynthetic activity of *C. vulgaris* in these intermediate conditions. A different behavior was observed for phosphorus. In this case, average removal rates (Figure 5B) and mass removal values (Figure 5C) determined for assays 1 to 4 were not statistically different ($p > 0.05$), but values determined for assay 5 were statistically higher ($p < 0.05$), reaching an average removal rate of (0.20 ± 0.01) $mg_P\ L^{-1}\ d^{-1}$ and a mass removal of (2.2 ± 0.1) $mg_P\ L^{-1}$.

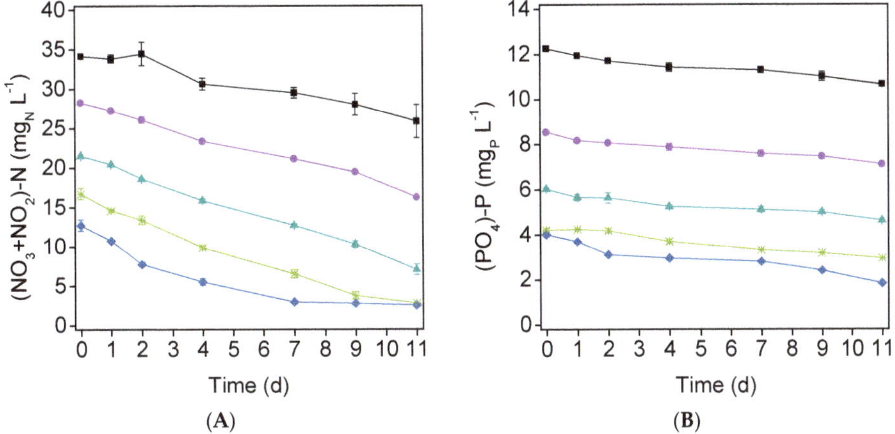

Figure 3. Temporal variation of (**A**) nitrogen (nitrate + nitrite) and (**B**) phosphorus (phosphate) concentration determined in *C. vulgaris* cultures grown in raw and diluted secondary-treated paper industry effluent (Assays: 1—■—, 2—●—, 3—▲—, 4—✳— and 5—◆—). Error bars correspond to the standard deviation of the mean obtained from two independent experiments.

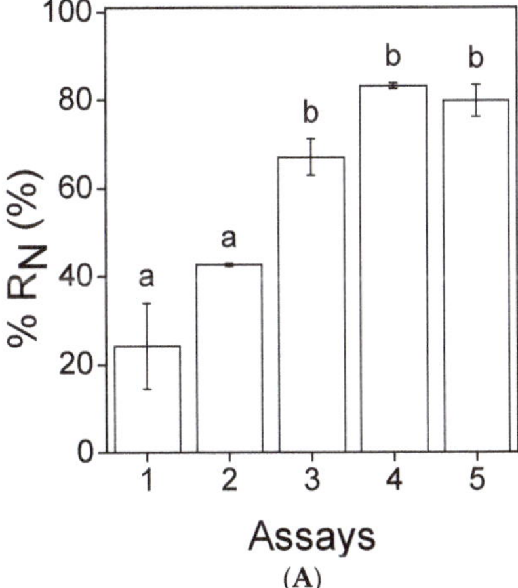

Figure 4. *Cont.*

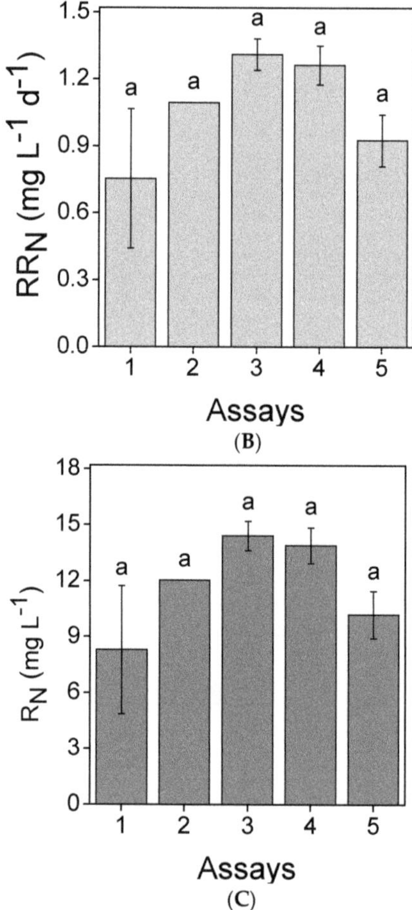

Figure 4. Nitrogen (nitrate + nitrite) removal parameters obtained by *C. vulgaris* cultures grown in raw and diluted secondary-treated paper industry effluent (assays 1–5): (**A**) removal efficiency (%R_N); (**B**) average removal rate (RR_N); and (**C**) mass removal (R_N). Error bars correspond to the standard deviation of the mean obtained from two independent experiments. Mean values sharing at least one common letter (shown above the bars) are not statistically different ($p > 0.05$).

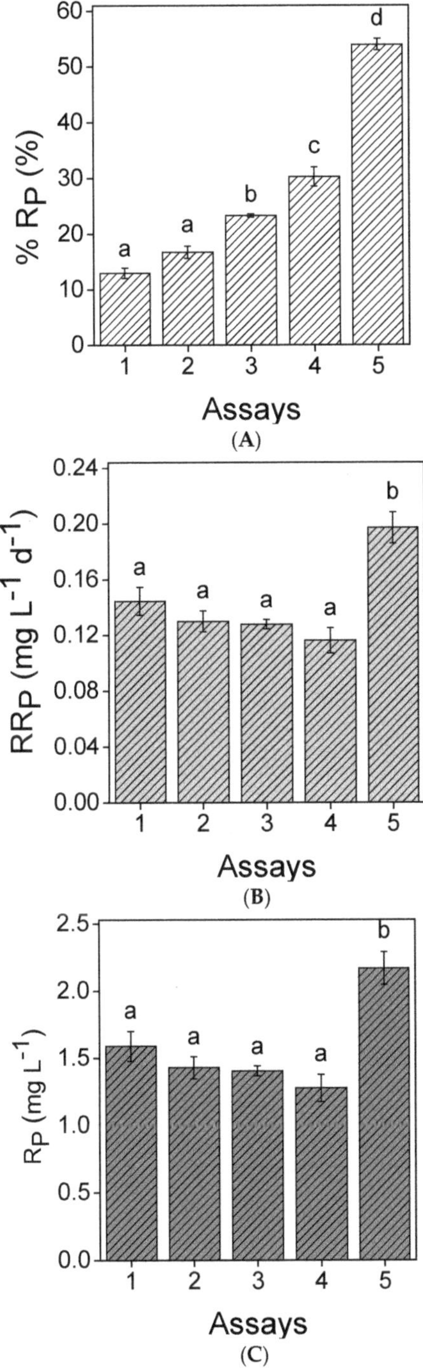

Figure 5. Phosphorus (phosphate) removal parameters obtained by *C. vulgaris* cultures grown in raw and diluted secondary-treated paper industry effluent (assays 1–5): (**A**) removal efficiency (%R_P); (**B**) average removal rate (RR_P); and (**C**) mass removal (R_P). Error bars correspond to the standard deviation of the mean obtained from two independent experiments. Mean values sharing at least one common letter (shown above the bars) are not statistically different ($p > 0.05$).

Nutrients removal from paper industry effluents has already been reported in the literature. Table 3 highlights nitrogen and phosphorus removal efficiencies and removal rates obtained in these studies. According to these data, removal efficiencies reported by Tao, et al. [19] and Gentili [2] are significantly higher than those obtained in this study, whereas values reported by Usha, et al. [1] were closer to those obtained in the present study, especially in assays 3–5. The lower removal efficiencies obtained in this study when compared with those reported by Tao, et al. [19], may be associated with the higher N:P molar ratio used in the reference study, which was ~ 66:1. On the other hand, the higher removal efficiencies reported by Gentili [13] may be associated with the use of other effluents to achieve the dual role of providing the required nutrients for microalgal growth while contributing to a reduction in the toxicity of the paper industry effluent. Another explanation for the increased efficiencies obtained in these studies is the nitrogen source used. As in the present study, Usha, et al. [1] cultivated microalgae in an effluent with nitrate–nitrogen as the main nitrogen source. On the other hand, Tao, et al. [19] tested an effluent with ammonium as the main nitrogen source (digestate obtained from the treatment of a pulp and paper industry effluent) and Gentili [13] evaluated this treatment with both nitrogen forms present. According to several studies, although nitrate–nitrogen is the most thermodynamically stable form (and the most commonly found in aquatic environments), ammonia is directly assimilated and converted into proteins by microalgae, while nitrate must be reduced to nitrite and then to ammonia before being assimilated by microalgal cells [25]. However, for an adequate comparison of nutrients removal performance, it is important to determine the average removal rate, as this parameter takes into account initial nutrients concentrations and cultivation/treatment time. Comparing average removal rates obtained in the present study and in the reference studies, values in the same order of magnitude were obtained, except in what concerns ammonium–nitrogen removal in the studies performed by Tao, et al. [19] and Gentili [2]. In these cases, the higher removal rates obtained may be associated with the higher ability of microalgae to assimilate ammonium–nitrogen than nitrate–nitrogen. Considering values of average RR, it is possible to conclude that promising results were obtained in this study. Moreover, differences found in experimental conditions used in this study and in the studies reported in the literature demonstrate that these results can be significantly enhanced. Besides increasing N:P molar ratio and providing an ammonium–nitrogen source, the increase of light supply should also be considered, as values reported in the literature correspond to cultures grown under light intensities of 130–800 μmol m^{-2} s^{-1}, whereas results reported in the present study were obtained with light intensities of 30–40 μmol m^{-2} s^{-1}.

In summary, the results obtained in this study for both nitrogen and phosphorus removal evidence that the remediation of paper industry effluents using microalgae is possible, provided that it is properly diluted to avoid inhibitory effects related to the presence of strong color or high concentrations of toxic compounds, typically associated with effluents resulting from this industrial sector [19,20]. Considering the results obtained for nitrogen removal, the dilution of the effluent to the concentrations present in assays 3 and 4 is the most adequate. In these conditions, nitrogen concentrations were significantly reduced, reaching (7.1 ± 0.7) mg_N L^{-1} and (2.81 ± 0.05) mg_N L^{-1}, respectively (which corresponds to the highest average removal rates: (1.31 ± 0.07) mg_N L^{-1} d^{-1} and (1.26 ± 0.09) mg_N L^{-1} d^{-1}, respectively). Regarding phosphorus removal, the highest removal rate was obtained for the conditions tested in assay 5: (0.20 ± 0.01) mg_P L^{-1} d^{-1}.

Table 3. Comparison between nutrients removal efficiencies (%R, in %) and average removal rates (RR, in mg L^{-1} d^{-1}) obtained in this study and other studies reporting microalgal growth in effluents resulting from pulp and paper industries.

Effluent	Microalgae	Culture Time (d)	Element / Form	Initial Concentration (mg L^{-1})	%R (%)	RR (mg L^{-1} d^{-1})	Ref.
Paper	*C. vulgaris*	11	(NO_3+NO_2)-N	34.2	24	0.75	This study
			PO_4-P	12.3	13	0.14	
			(NO_3+NO_2)-N	28.2	43	1.1	
			PO_4-P	8.55	17	0.13	
			(NO_3+NO_2)-N	21.5	67	1.3	
			PO_4-P	6.04	23	0.13	
			(NO_3+NO_2)-N	16.7	83	1.3	
			PO_4-P	4.22	30	0.12	
			(NO_3+NO_2)-N	12.7	80	0.93	
			PO_4-P	4.01	54	0.20	
Pulp and paper mill	*C. vulgaris*	14	NH_4-N	240	99	17	[19]
			PO_4-P	8.00	97	0.55	
Pulp and paper with dairy sludge and municipal	*Scenedesmus* sp. *Scenedesmus dimorphus* *Selenastrum minutum*	6	NH_4-N	22.4	99	3.7	[13]
			NO_3-N	1.06	27–53	0.048–0.094	
			PO_4-P	10.1	96–98	1.6–1.7	
			NH_4-N	14.8	96–98	2.3–2.4	
			NO_3-N	1.08	41–46	0.074–0.083	
			PO_4-P	1.60	96–97	0.25–0.26	
			NH_4-N	21.0	99	3.5	
			NO_3-N	1.25	27–43	0.056–0.090	
			PO_4-P	2.99	90–94	0.45–0.47	
Paper mill	*Scenedesmus* sp.	28	NO_3-N	2.24	65	0.052	[1]
			PO_4-P	9.86	71	0.25	

Despite the promising nitrogen and phosphorus removal rates, the results obtained in this study demonstrated that the cultures were limited by nitrogen, as nitrogen and phosphorus were assimilated by *C. vulgaris* at a N:P molar ratio ranging from 10:1 to 24:1. Considering these results and the N:P molar ratios used in this study (between 6:1 and 9:1), nutrients uptake could be enhanced by increasing nitrogen supply. Another alternative to achieve an adequate N:P molar ratio and reduce the toxicity of this effluent would be to dilute it with other effluents, as proposed in other studies [13]. Finally, the remediation process could be further improved by modulating microalgal cultivation conditions. According to Gonçalves, et al. [25], light conditions, temperature, and pH are also important parameters that can influence microalgal growth and, hence, the efficiency of the bioremediation process. Thus, from the prospecting of this work, other studies evaluating these parameters should be carried out to further improve nitrogen and phosphorus uptake from paper industry effluents.

4. Conclusions

This study showed the feasibility of using *C. vulgaris* for the bioremediation of a paper industry effluent fortified with a nitrogen source, targeting phosphorus removal. *C. vulgaris* was able to grow in all studied effluent conditions (in non-diluted and diluted ones). However, it was possible to conclude that growing on non-diluted effluent resulted in lower biomass productivities, which was also reflected in nitrogen and phosphorus removal efficiencies. From microalgal growth and nitrogen removal points of view, the effluent dilutions used in assays 3 and 4 (intermediate dilutions) seem to be the most adequate, as microalgal growth was not inhibited in these conditions and nitrogen mass removal was quite satisfactory, achieving final concentrations of (7.1 ± 0.7) mg_N L^{-1} and (2.81 ± 0.05) mg_N L^{-1}, respectively. Regarding phosphorus removal, concentrations achieved in the last day of culturing in assays 3 and 4 were higher ((4.63 ± 0.04) mg_P L^{-1} and (2.940 ± 0.005) mg_P L^{-1}, respectively) than the one obtained in assay 5 ((1.85 ± 0.02) mg_P L^{-1}). However, the results obtained in assay 5 suggest a growth limitation, mainly related to nitrogen concentration. Accordingly, the obtained results indicate that these values can be further improved by studying different N:P molar ratios, different microalgal cultivation conditions, dilution with other effluents, among others. Improving the remediation performance can significantly contribute to the development of an effective microalgae-based remediation process of pulp and paper industry effluents.

Author Contributions: Conceptualization, B.P., A.L.G., V.J.P.V. and J.C.M.P.; methodology, B.P., A.L.G. and J.C.M.P.; investigation, B.P. and A.F.E.; resources, V.J.P.V. and J.C.M.P.; data curation, B.P., A.L.G., A.F.E., V.J.P.V. and J.C.M.P.; writing (original draft preparation), B.P.; writing (review and editing), B.P., A.L.G., A.F.E., S.M.A.G.U.d.S., A.A.U.d.S., V.J.P.V. and J.C.M.P.; supervision, A.L.G., S.M.A.G.U.d.S., A.A.U.d.S., V.J.P.V. and J.C.M.P.; project administration, S.M.A.G.U.d.S., A.A.U.d.S., V.J.P.V. and J.C.M.P.; funding acquisition, S.M.A.G.U.d.S., A.A.U.d.S., V.J.P.V. and J.C.M.P. All authors have read and agreed to the published version of the manuscript.

Funding: This research was funded by: (i) Base Funding—UIDB/00511/2020 of the Laboratory for Process Engineering, Environment, Biotechnology, and Energy—LEPABE—funded by national funds through the FCT/MCTES (PIDDAC); (ii) Base Funding—UIDB/50020/2020 of the Associate Laboratory LSRE-LCM—funded by national funds through FCT/MCTES (PIDDAC); (iii) Project PTDC/BTA-BTA/31736/2017—POCI-01-0145-FEDER-031736—funded by FEDER funds through COMPETE2020—Programa Operacional Competitividade e Internacionalização (POCI) and with financial support of FCT/MCTES through national funds (PIDDAC); and (iv) the Coordenação de Aperfeiçoamento de Pessoal de Nível Superior—Brasil (CAPES)—Finance Code 001. V.J.P. Vilar acknowledges the FCT Individual Call to Scientific Employment Stimulus 2017 (CEECIND/01317/2017). J.C.M. Pires acknowledges the FCT Investigator 2015 Programme (IF/01341/2015).

Conflicts of Interest: The authors declare no conflict of interest.

References

1. Usha, M.T.; Sarat Chandra, T.; Sarada, R.; Chauhan, V.S. Removal of nutrients and organic pollution load from pulp and paper mill effluent by microalgae in outdoor open pond. *Bioresour. Technol.* **2016**, *214*, 856–860. [CrossRef] [PubMed]
2. Reid, N.; Bowers, T.; Lloyd-Jones, G. Bacterial community composition of a wastewater treatment system reliant on N_2 fixation. *Appl. Microbiol. Biotechnol.* **2008**, *79*, 285–292. [CrossRef]
3. Pokhrel, D.; Viraraghavan, T. Treatment of pulp and paper mill wastewater—A review. *Sci. Total Environ.* **2004**, *333*, 37–58. [CrossRef] [PubMed]
4. Piao, W.; Kim, Y.; Kim, H.; Kim, M.; Kim, C. Life cycle assessment and economic efficiency analysis of integrated management of wastewater treatment plants. *J. Clean. Prod.* **2016**, *113*, 325–337. [CrossRef]
5. Singh, G.; Thomas, P.B. Nutrient removal from membrane bioreactor permeate using microalgae and in a microalgae membrane photoreactor. *Bioresour. Technol.* **2012**, *117*, 80–85. [CrossRef]
6. Zang, Y.; Li, Y.; Wang, C.; Zhang, W.; Xiong, W. Towards more accurate life cycle assessment of biological wastewater treatment plants: A review. *J. Clean. Prod.* **2015**, *107*, 676–692. [CrossRef]
7. Tarlan, E.; Dilek, F.B.; Yetis, U. Effectiveness of algae in the treatment of a wood-based pulp and paper industry wastewater. *Bioresour. Technol.* **2002**, *84*, 1–5. [CrossRef]
8. Li, K.; Liu, Q.; Fang, F.; Luo, R.; Lu, Q.; Zhou, W.; Huo, S.; Cheng, P.; Liu, J.; Addy, M.; et al. Microalgae-based wastewater treatment for nutrients recovery: A review. *Bioresour. Technol.* **2019**, *291*, 121934. [CrossRef]
9. Pires, J.C.M.; Alvim-Ferraz, M.C.M.; Martins, F.G.; Simões, M. Carbon dioxide capture from flue gases using microalgae: Engineering aspects and biorefinery concept. *Renew. Sustain. Energy Rev.* **2012**, *16*, 3043–3053. [CrossRef]
10. Rawat, I.; Kumar, R.R.; Mutanda, T.; Bux, F. Dual role of microalgae: Phycoremediation of domestic wastewater and biomass production for sustainable biofuels production. *Appl. Energy* **2011**, *88*, 3411–3424. [CrossRef]
11. Odjadjare, E.C.; Mutanda, T.; Olaniran, A.O. Potential biotechnological application of microalgae: A critical review. *Crit. Rev. Biotechnol.* **2017**, *37*, 37–52. [CrossRef] [PubMed]
12. Rizwan, M.; Mujtaba, G.; Memon, S.A.; Lee, K.; Rashid, N. Exploring the potential of microalgae for new biotechnology applications and beyond: A review. *Renew. Sustain. Energy Rev.* **2018**, *92*, 394–404. [CrossRef]
13. Gentili, F.G. Microalgal biomass and lipid production in mixed municipal, dairy, pulp and paper wastewater together with added flue gases. *Bioresour. Technol.* **2014**, *169*, 27–32. [CrossRef] [PubMed]
14. OECD. Test. No. 201: Freshwater Alga and Cyanobacteria. In *Growth Inhibition Test*; Organization for Economic Co-operation and Development: Paris, France, 2011. [CrossRef]

15. Rice, E.W.; Bridgewater, L.; Association, A.P.H.; Association, A.W.W.; Federation, W.E. *Standard Methods for the Examination of Water and Wastewater*, 22nd ed.; American Public Health Association: Washington, DC, USA, 2012; p. 724.
16. Silva, N.F.P.; Gonçalves, A.L.; Moreira, F.C.; Silva, T.F.C.V.; Martins, F.G.; Alvim-Ferraz, M.C.M.; Boaventura, R.A.R.; Vilar, V.J.P.; Pires, J.C.M. Towards sustainable microalgal biomass production by phycoremediation of a synthetic wastewater: A kinetic study. *Algal Res.* **2015**, *11*, 350–358. [CrossRef]
17. Larsdotter, K. Wastewater treatment with microalgae—A literature review. *Vatten* **2006**, *62*, 31.
18. Hodaifa, G.; Martínez, M.E.; Sánchez, S. Use of industrial wastewater from olive-oil extraction for biomass production of *Scenedesmus obliquus*. *Bioresour. Technol.* **2008**, *99*, 1111–1117. [CrossRef]
19. Tao, R.; Kinnunen, V.; Praveenkumar, R.; Lakaniemi, A.-M.; Rintala, J.A. Comparison of *Scenedesmus acuminatus* and *Chlorella vulgaris* cultivation in liquid digestates from anaerobic digestion of pulp and paper industry and municipal wastewater treatment sludge. *J. Appl. Phycol.* **2017**, *29*, 2845–2856. [CrossRef]
20. Polishchuk, A.; Valev, D.; Tarvainen, M.; Mishra, S.; Kinnunen, V.; Antal, T.; Yang, B.; Rintala, J.; Tyystjärvi, E. Cultivation of *Nannochloropsis* for eicosapentaenoic acid production in wastewaters of pulp and paper industry. *Bioresour. Technol.* **2015**, *193*, 469–476. [CrossRef]
21. Molinuevo-Salces, B.; Riaño, B.; Hernández, D.; Cruz García-González, M. Microalgae and Wastewater Treatment: Advantages and Disadvantages. In *Microalgae Biotechnology for Development of Biofuel and Wastewater Treatment*; Alam, M.A., Wang, Z., Eds.; Springer Singapore: Singapore, 2019; pp. 505–533.
22. Barros, A.I.; Gonçalves, A.L.; Simões, M.; Pires, J.C.M. Harvesting techniques applied to microalgae: A review. *Renew. Sustain. Energy Rev.* **2015**, *41*, 1489–1500. [CrossRef]
23. Gerde, J.A.; Yao, L.; Lio, J.; Wen, Z.; Wang, T. Microalgae flocculation: Impact of flocculant type, algae species and cell concentration. *Algal Res.* **2014**, *3*, 30–35. [CrossRef]
24. Chisti, Y. Biodiesel from microalgae. *Biotechnol. Adv.* **2007**, *25*, 294–306. [CrossRef] [PubMed]
25. Gonçalves, A.L.; Pires, J.C.M.; Simões, M. A review on the use of microalgal consortia for wastewater treatment. *Algal Res.* **2017**, *24*, 403–415. [CrossRef]

Review

Arsenic Contamination of Groundwater and Its Implications for Drinking Water Quality and Human Health in Under-Developed Countries and Remote Communities—A Review

Samuel B. Adeloju [1,2,*], Shahnoor Khan [1,3] and Antonio F. Patti [1]

1 School of Chemistry, Monash University, Clayton, VIC 3800, Australia; shahnoor.khan.monash@gmail.com (S.K.); tony.patti@monash.edu (A.F.P.)
2 Faculty of Science, Charles Sturt University, Albury, NSW 2640, Australia
3 Ministry of Public Administration, Bangladesh Secretariat, Dhaka 1000, Bangladesh
* Correspondence: sadeloju@csu.edu.au; Tel.: +61-2-6051-9681

Featured Application: Groundwater contamination is a major global issue. A good understanding of the associated chemistry and fate of a contaminant such as arsenic in groundwater is important for minimizing or avoiding potential health, social and economic implications when used as a water source for human consumption.

Abstract: Arsenic is present naturally in many geological formations around the world and has been found to be a major source of contamination of groundwater in some countries. This form of contamination represents a serious threat to health, economic and social well-being, particularly in under-developed countries and remote communities. The chemistry of arsenic and the factors that influence the form(s) in which it may be present and its fate when introduced into the environment is discussed briefly in this review. A global overview of arsenic contamination of groundwater around the world is then discussed. As a case study, the identified and established causes of groundwater contamination by arsenic in Bangladesh is highlighted and a perspective is provided on the consequential health, agricultural, social and economic impacts. In addition, the relevant removal strategies that have been developed and can generally be used to remediate arsenic contamination are discussed. Also, the possible influence of groundwater inorganic compositions, particularly iron and phosphate, on the effectiveness of arsenic removal is discussed. Furthermore, some specific examples of the filter systems developed successfully for domestic arsenic removal from groundwater to provide required potable water for human consumption are discussed. Lastly, important considerations for further improving the performance and effectiveness of these filter systems for domestic use are outlined.

Keywords: arsenic; groundwater; contamination; water quality; domestic filter systems; health effects; treatment methods; Bangladesh

Citation: Adeloju, S.B.; Khan, S.; Patti, A.F. Arsenic Contamination of Groundwater and Its Implications for Drinking Water Quality and Human Health in Under-Developed Countries and Remote Communities—A Review. *Appl. Sci.* **2021**, *11*, 1926. https://doi.org/10.3390/app11041926

Academic Editors: Bart Van der Bruggen and Dino Musmarra

Received: 7 January 2021
Accepted: 16 February 2021
Published: 22 February 2021

1. Introduction

Arsenic (As) is a chemical element which occurs naturally and is commonly present in the earth's crust [1]. It has been found in air, biota, water, soil and rocks [1,2]. In particular, arsenic present in an aqueous medium is of most concern because of its likely detrimental impact on plants, animals and humans. When present in high concentrations in drinking water, arsenic has been found to adversely affect human health [3] and this topic has been the subject of several recent reviews [4–6]. In this comprehensive review, we have additionally provided a detailed evaluation of treatment options for arsenic removal from water, with due consideration for affordable small scale technologies that can be easily implemented in developing countries such as Bangladesh, India, Nepal and Pakistan.

Arsenic has been introduced to the environment through various natural and anthropogenic sources [7]. The primary natural source of arsenic is from As-enriched minerals [2]. Typical examples of natural sources include volcanoes and eroded arsenic bearing rocks, such as arsenopyrite (FeAsS), lollingite ($FeAs_2$), orpiment (As_2S_3) and realgar (AsS) [8]. Under oxidizing and reducing conditions, arsenic can be mobilized at pH 6.5–8.5 [3,9] which is a common pH range in groundwater [9]. On the other hand, the common anthropogenic sources include agriculture, livestock and industrial manufacturing [10]. It is also widely used for the manufacture of glassware, industrial chemicals, copper, lead alloys and pharmaceuticals [10]. Industrial processes, such as smelting of iron ores, mining, pulp and paper production, cement manufacture, burning of fuels and wastes are known sources for the release of arsenic into the environment [8]. However, the focus of this review is on the contamination of groundwater by arsenic from predominantly natural sources.

Arsenic is a group Va element on the periodic table and is classified as a metalloid. In the environment, it exists in several oxidation states as As(3-) (arsine), As(0) (arsenic), As(3+) (arsenite) and As(5+) (arsenate) [11]. In natural waters, the common soluble arsenic species are the inorganic oxyanions of As(III) or As(V) [12]. As(III) species include $As(OH)_3$, $H_2AsO_3^-$, $HAsO_3^{2-}$ and AsO_3^{3-} [13,14]. As(V) species are AsO_4^{3-}, $HAsO_4^{2-}$, $H_2AsO_4^-$ and H_3AsO_4 [13,14]. Arsenic (III) forms complexes preferentially with oxides and nitrogen [14]. On the other hand, As(V) forms complexes with sulfides [14].

There are also organic arsenic species such as arseno-sugars dimethyl arsenic acid (DMAA) and monomethyl arsenic acid (MMAA) [15]. However, in drinking water treatment, these species are not of significant concern due to the limited effect they have on human health, and their ease of elimination from the body [15].

Under reducing anaerobic conditions, the dominant species are trivalent arsenic [9], while in oxygen rich aerobic conditions pentavalent species are dominant [9,11,12]. However, the key considerations for controlling the arsenic species present under various conditions are redox potential (Eh) and pH [9,16–18]. Figure 1 shows that, under oxidizing conditions, $H_2AsO_4^-$ is the dominant arsenate species at low pH (pH < 6.9), whereas $HAsO_4^{2-}$ is the dominant arsenate species at a higher pH, as illustrated in Figure 1c [9,16,17]. On the other hand, $H_3AsO_3^0$ is the dominant arsenite species at pH < 9.2 when operating in reducing conditions, as illustrated in Figure 1b.

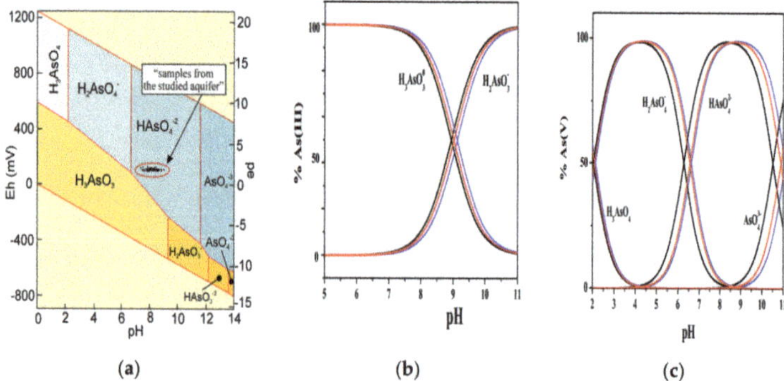

Figure 1. pH dependence of the aqueous speciation of As(III) and As(V) species. (**a**) redox potential (Eh)–pH diagram [9,17], (**b**) arsenite species and (**c**) arsenate species [16]. Operating conditions: 25 °C and 1 bar total pressure. Reproduced with permission from Elsevier and RSC.

The presence of nitrate in groundwater has been proposed as a likely significant contributing mechanism for the oxidation of arsenic to As(V) [18]. Nitrate can act as a terminal electron acceptor and this mechanism is likely to be more prevalent under anoxic conditions.

Reported cases of groundwater contamination by arsenic are more diverse than commonly realized and extend from under-developed to developed countries, including Argentina, Bangladesh, Canada, Chile, China, Hungary, India, Japan, Mexico, Nepal, Poland, Taiwan and the USA [15]. Figure 2 shows where groundwater or surface water contamination by arsenic has been reported. Croatia, Hungary and Serbia are three of the countries in Europe where very high concentrations of arsenic in groundwater have been reported. In Hungary, the arsenic concentrations in groundwater were found to exceed the WHO (10 μg/L) and EC guidelines (7.5 μg/L, EC Directive 2006/118/EC) by several times [19]. The concentrations of arsenic in various groundwater samples collected in Hungary and Romania ranged from <0.5 to 240 μg/L [20]. More distinctly, the highest arsenic concentrations (23 to 208 μg/L, mean 123 μg/L) were obtained in waters dominated by methanogenesis, whereas waters dominated by sulfate reduction gave lower arsenic concentrations (<0.5 to 58 μg/L, mean 11.5 μg/L) [20]. In Serbia, the extent of arsenic contamination is not fully resolved and yet to be determined [15]. As illustrated in Figure 2, the most affected countries in the Americas are Argentina, Chile, Mexico and the United States. In Latin America, the estimated number of people affected by arsenic contamination >50 μg/L is at least four million [21]. Some wells in Argentina, Bolivia and Peru were found to contain extremely high arsenic concentrations at high mg/L levels [21].

Figure 2. Countries affected by arsenic contamination of groundwater or surface waters [15]. Reproduced with permission from IJAET.

The global variation of arsenic concentrations in groundwater is summarized in Table 1. Among the affected countries, Bangladesh and West Bengal in India have the largest population at risk of exposure to arsenic contamination [7]. This is why we chose Bangladesh as a case study to highlight the extent of groundwater contamination by arsenic. The people most affected in Bangladesh and West Bengal reside where there is no access to integrated urban water supply systems, mostly in rural areas. Consequently, millions of people living in rural areas in these countries are regularly drinking water contaminated with arsenic. A relatively large number of people in this region have already been identified as having arsenic related disease symptoms [15].

UNICEF reported that 1.4 million tube wells out of the 4.75 million (around 30%) in Bangladesh contained arsenic concentration above 50 μg/L [22]. A household drinking water quality survey conducted by UNICEF in Bangladesh found that 12.6% of drinking water samples exceeded the country's arsenic standard for drinking water [22]. This represents approximately 22 million Bangladesh people at risk of adverse arsenic exposure [22].

Furthermore, in West Bengal (India), an estimated 6 million people were found to have been exposed to relatively high arsenic concentrations, ranging from 50 to 3,200 μg/L [23,24].

Table 1. Global variation of arsenic concentrations in groundwater.

Country	Region	[As], μg/L	Permissible Limit, μg/L
Afghanistan	Ghazni	10–500	10 (WHO)
* Argentina		10–1000	50 (WHO)
# Australia	Perth Northern New South Wales Victoria (Gold mining regions)	7000 52–337 <1–300,000	10 (WHO)
Bangladesh	Noakhali	<1–4730	50 (WHO)
Brazil	Minas Gerais (Southeastern Brazil)	0.4–350 (Surface water)	10 (WHO)
Cambodia	Prey Veng and Kandal-Mekong delta	Up to 9001–1610	10 (WHO)
Canada	Nova Scotia (Halifax county)	1.5–738.8	10 (WHO)
* Chile		900–1040	50 (WHO)
China	-	50–4440	50 (WHO)
* China Inner Mongolia		1–2400	50 (WHO)
# Croatia (Eastern)	Osijek area	27 (Shallow groundwater) 205–240 (Deep groundwater)	10 (WHO)
Finland	Southwest Finland	17–980	10 (WHO)
# Germany	-	10–150	10 (WHO)
# Ghana	-	<1–175	50 (WHO)
Greece	Fairbanks (mine tailings)	Up to 10,000	10 (WHO)
* Hungary	Pannonian Basin	10–176 0.5–240	10 (WHO)
India	West Bengal Uttar Pradesh	10–3200	50 (WHO)
Italy	Volcano island and Phlegrean Fields	0.1–6940	10 (WHO)
Japan	Fukuoka Prefecture (southern region)	1–293	10 (WHO)
# Lao PDR	Laos	277.8	50 (WHO)
Mexico	Lagunera	8–620	25
Nepal	Rupandehi	Up to 2620	50
# New Zealand	Taupo Volcanic zone	21	10 (WHO)
Pakistan	Muzaffargarh (southwestern Punjab)	Up to 906	50
* Peru		500	50 (WHO)
* Romania		10–176	10 (WHO)
# Slovakia	Banska Bystrica and Nitra	37–39	10 (WHO)
# Spain	-	<1–100	10 (WHO)
Taiwan	-	10–1820	10 (WHO)
Thailand	Ron Phibun	1–>5000	10 (WHO)
# UK	Southwest England	<1–80	10 (WHO)
USA	Tulare Lake	Up to 2600	10 (USEPA)
Vietnam	Red River Delta (North Vietnam) Mekong Delta (South Vietnam)	<1–3050	10 (WHO)

Adapted and modified from [3], * [15] and # [25].

Usually after a few years of extraction of groundwater, a considerable increase in the concentration of arsenic in wells has been observed [15]. Future increases in arsenic contamination issues in drinking water are expected despite the establishment of a stringent standard for arsenic in drinking water and this may spread to other countries [15]. Due to the greater risk that arsenic contamination of groundwater poses to the Bangladeshi people, it is useful to take a closer look at the widespread nature of the problem in this country.

2. Case Study—Arsenic Contamination of Groundwater in Bangladesh

2.1. Background

Bangladesh is a tropical country with a total area of about 147,570 km^2 and an estimated population of 160 million as of 2016 [26]. About 58% of the surface area is arable land and about 11% comprises forests and woodlands. The contribution of the agricultural sector to national GDP (Gross Domestic Product) is about 18%, while about 72% of its people are based in rural settings [26]. The per capita income in 2015–2016 is US$ 1466 [26].

Although there is abundant groundwater in Bangladesh and the productivity of the aquifers are very high, variable water tables are observed in different parts of the country, more commonly shallow and below the ground surface by 1–10 m [27]. Groundwater is generally free from pathogenic microorganisms and, if not contaminated, can be reasonably safe to use. For these reasons groundwater has remained an appealing and readily available reserve for drinking water in many countries [27]. This has resulted in the widespread use of groundwater as the main source of drinking water, particularly as tube wells for the past forty years or more. The tube wells have contributed significantly to the reduction of mortality rate of diarrheal diseases [28]. Bangladesh achieved a remarkable success by providing 97% of the rural population with tube well water [28].

However, the coverage for safe drinking water has been significantly reduced from 97% to 74% as a consequence of the extensive contamination of groundwater with arsenic [28]. Consequently, the mortality rate in Bangladesh was increased as a result of arsenic contamination of the drinking water [29]. About 12.6% of its population are still accessing this contaminated water [22]. This has led to the recognition of arsenic contamination as a catastrophic issue in Bangladesh [22]. To understand the cause of this problem, a knowledge of the geological characteristics of Bangladesh is useful.

2.2. Geology of Bangladesh

Young (Holocene) alluvial and deltaic sediments dominate the geology of Bangladesh. The major river systems of the Bengal Basin are responsible for the deposition of these sediments [27]. Most of these sediments, of several hundreds of meters in thickness, have been deposited within the last 6000–10,000 years [27]. The Basin is now recognized in the world among the most rapidly developed delta systems [27]. In the north, the surface sediments consist mainly of coarse-grained mountain-front alluvial fan deposits, while alluvial sands and silts represent the main composition of sediments from central Bangladesh. The predominant composition of the sediments in the south are deltaic silts and clays. On approaching the southern part of Bangladesh, the deposits tended to be increasingly fine-grained [27]. Figure 3 illustrates the surface geological units of Bangladesh. Evidently, the surface geological characteristics are very diverse in nature.

2.3. Mobilization of Arsenic in Groundwater in Bangladesh

The two most widely discussed hypotheses for the sources of arsenic contamination of groundwater in Bangladesh are the pyrite oxidation hypothesis and the iron oxyhydroxide reduction hypothesis [6,30].

The "Barind and Modhupur Tracts" made up of up-faulted terraces of older (Pleistocene) sediments are the key characteristics of the surface geology in north-central Bangladesh. Unlike the surrounding alluvium deposits, these sediments have experienced significantly more weathering [27]. The younger alluvial sediments in these tracts are present at depths ranging or greater than 50–200 m. However, for the older sediments, the extent of its distribution is still relatively unknown [27]. In south-east Bangladesh, older (Tertiary) sediments dominate the geology of the Chittagong Hill Tracts and their main composition are sandstone, silt and limestone [27]. Within 20–80 m depth, the sediment was found to be very rich in arsenopyrite [31]. The granitic and metamorphic processes of the Himalayas was the original source of the sediment [31]. After the introduction of the sediment into the Ganges delta, it became part of the aquifers [31]. Additional arsenic is retained as an adsorbed coating with ferric oxy-hydroxide on the sediments [32]. It is understood that the transport and deposition of arsenic and ferric oxy-hydroxide in the Ganges delta along with abundant organic matter resulted from the oxidation of arsenopyrite [33,34]. Figure 4 illustrates the arsenic contaminated areas in Bangladesh [24]. Obviously, the arsenic concentrations present in many areas are considerably above 50 μg/L.

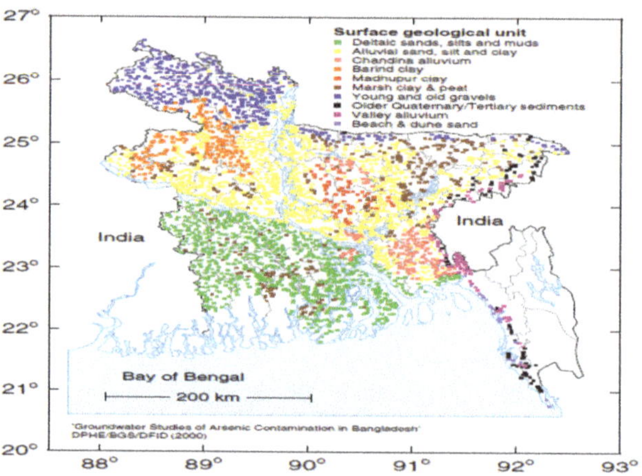

Figure 3. Surface geological units of Bangladesh [27]. Reproduced with permission from BGS.

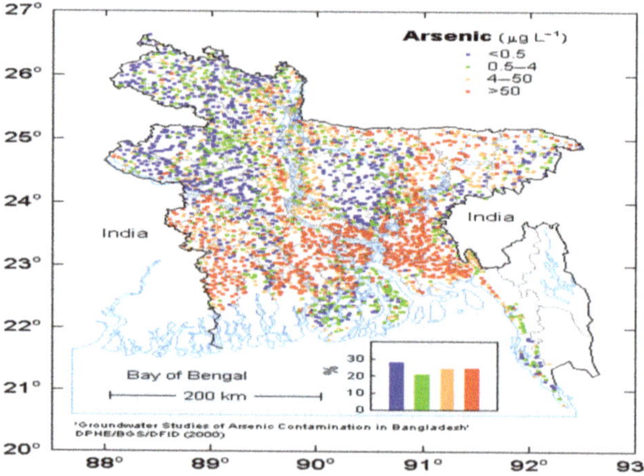

Figure 4. Arsenic contaminated areas in Bangladesh [24]. Reproduced with permission from BGS.

2.3.1. Pyrite Oxidation Hypothesis

Studies have shown that there is a high level of arsenopyrite (FeAsS) in the alluvial regions of Bangladesh [31,35]. The basis of the pyrite oxidation hypothesis is that the oxidation of arsenopyrite resulted in the release of arsenic into groundwater [36–39]. This process is understood to be aided by the invasion of the aquifers by atmospheric oxygen when the water table is lower than these deposits, resulting in its diffusion into the pore space and groundwater. Consequently, the reaction between the arsenopyrite and available oxygen resulted in a water-soluble form of arsenic which was consequently released into the groundwater [30].

The associated reactions for this process are:

$$4FeAsS + 11O_2 + 6H_2O \rightarrow 4\ FeSO_4 + 4\ H_2AsO_3^- + 4H^+$$

$$4FeAsS + 13O_2 + 6H_2O \rightarrow 4\ FeSO_4 + 4\ H_2AsO_4^- + 4H^+$$

The above hypothesis is supported by the absence of arsenic-affected people before irrigation became more intensive at the beginning of the 1980s. Consequently, the total irrigated area which used groundwater increased significantly from 41% to 71% within 1982/1983 and 1996/1997, respectively, while in the same period the use of surface water decreased considerably from 59% to 29% [30].

Based on the pyrite oxidation hypothesis, the lowering of the water table resulted from the considerable use of groundwater, and, in turn, led to groundwater contamination with arsenic. The arsenic mobilization proposed by this hypothesis is widely accepted as the main mechanism for intrusion of arsenic into groundwater in Bangladesh [36,40,41]. However, the hypothesis is not yet fully validated as there is a need for further hydrological and geochemical data [42].

2.3.2. Iron Oxyhydroxide Reduction Hypothesis

Another suggestion proposed as the source of groundwater contamination with arsenic in Bangladesh is based on the so-called iron oxyhydroxide reduction hypothesis. This hypothesis is based on the premise that the source of arsenic was from the Ganges region, upstream of Bangladesh, where arsenic sulfide minerals followed by abrasion were carried by water about 1.6–1.8 million to 10,000 years ago during the late Pleistocene age [43]. As these arsenic-containing minerals travelled down the Ganges, arsenic was adsorbed to iron oxyhydroxide (FeOOH). The arsenic-rich iron oxyhydroxides were deposited at the Gangetic delta, and formed an alluvial aquifer. Various processes, including burial of vegetation, agriculture and floods, led to the introduction of organic carbon into the aquifer [8]. This organic carbon provided food for bacteria and the presence of methane in the water indicates that organic matter is being utilized in the aquifer by anaerobic bacteria [30]. Consequently, the redox potential of the groundwater is lowered by this biological process. Due to this reducing environment, the iron oxyhydroxide is broken down, resulting in the introduction of adsorbed arsenic into the groundwater [36,39,43–45]. The adsorption of arsenic by hydrous iron oxyhydroxide is supported by the following reactions [8]:

$$Fe\ (OH)_3 + H_3AsO_4 \rightarrow FeAsO_4{\cdot}2H_2O + H_2O$$

$$\equiv FeOH^o + AsO_4^{3-} + 3H^+ \rightarrow \equiv FeH_2AsO_4 + H_2O$$

$$\equiv FeOH^o + AsO_4^{3-} + 2H^+ \rightarrow \equiv FeHAsO_4^- + H_2O$$

The reduction of hydrous iron oxides can occur by the following reaction:

$$Fe(OH)_3 + e^- \rightarrow Fe^{2+} + 3OH^-$$

Therefore, the presence of anoxic conditions is necessary for arsenic-rich iron oxyhydroxide to be reduced and, in turn, release arsenic. This process mobilizes iron and its adsorbed load into the groundwater. Some researchers have accepted the arsenic mobilization based on the iron oxyhydroxide reduction hypothesis as the main mechanism for groundwater contamination by arsenic in Bangladesh [27,36,42,46]. However, a more comprehensive sampling and systematic analysis of ferric oxyhydroxide in the areas affected is needed to validate the reduction hypothesis [27,42].

Furthermore, it may also be possible for both of the mechanisms discussed above to contribute to arsenic release. This is in view of the fact that the pyrite oxidation occurs under an oxic condition, while iron oxyhydroxide reduction occurs under an anoxic condition.

3. Effects of Arsenic Contamination

Groundwater contamination by arsenic in Bangladesh has had adverse effects on human health, agriculture, social well-being and the economy of the country. To highlight the seriousness of these impacts, each of these is discussed in more details below.

3.1. Effects on Health

Groundwater contamination by arsenic has resulted in numerous serious health consequences in Bangladesh [47–49]. Globally, arsenic contamination in drinking water has reportedly led to the exposure of about 150 million people leading to serious health effects [50]. Of these, about 110 million people live in Bangladesh, Cambodia, China, India, Laos, Myanmar, Nepal, Pakistan, Taiwan and Vietnam [51]. In addition, many people are affected through consumption of arsenic-contaminated foods produced from the use of arsenic contaminated groundwater [50,51]. It has been found that rice grains contribute about half of the daily intake of arsenic in Bangladesh [52]. A survey conducted within arsenic-affected villages in Bangladesh collected hair, nail, urine and skin-scale samples totaling more than 10,000 which were subsequently analyzed for arsenic [53]. The results of the survey indicated that the arsenic concentrations in 93.8% and 95.1% of nails and urine samples, respectively, exceeded the normal level. For nail, the acceptable arsenic concentration range is 0.43–1.08 mg/kg [54], while for urine the range is 0.005–0.04 mg/day [55]. Furthermore, 83.2% and 97.4% of hair and skin-scale samples, respectively, gave arsenic contents that far exceeded the toxic level of 1 mg/kg [56]. These results therefore revealed a serious impact on health that has arisen from human exposure to arsenic through groundwater contamination in Bangladesh.

A recent comprehensive overview on the adverse effects from inorganic arsenic exposure details the major health issues [57]. Carcinogenic effects are particularly prevalent and the most common consequences resulting from poisoning from drinking arsenic contaminated water are skin diseases, including hyperkeratosis, keratosis, leuco-melanosis and melanosis (hyper pigmentation) [57–64]. Arsenicosis is initially manifested by melanosis which usually occurs all over the body [57,62]. The hardening of the melanosis spots leads to the commencement of keratosis [62]. A long-term exposure to arsenic results in the hyperkeratosis of the palms and soles [59]. Other associated health effects resulting from ingestion of arsenic include cardiac failure, chromosomal abnormality, cirrhosis, diabetes mellitus, gangrene, goiter, hypertension (high blood pressure), liver enlargement, myocardial degeneration, peripheral neuropathy and skin cancers [59,62,63].

Also, the development of cognitive and psychological functions in children has been shown to be affected by extended consumption of arsenic contaminated drinking water [65]. Furthermore, areas where highly contaminated groundwater is used have been shown to be more prone to higher fetal loss and infant deaths [66].

3.2. Effects on Agriculture

Groundwater is the main source of water used for agriculture in Bangladesh. Consequently, when contaminated with arsenic, it can result in the introduction of arsenic into agricultural soils and crops, such as rice and vegetables [51,67,68]. For the same reason, it has also been found in many areas in Bangladesh that high arsenic concentrations are present in the agricultural soils and the usage of groundwater contaminated with arsenic for irrigation has been identified as the main cause [50,52].

A main consequence is the dual arsenic accumulation in vegetables and rice grains from both soils and irrigation water [30,50,51,69–71]. More seriously, the phytotoxic effects of arsenic can result in a significant reduction of crop yields [52]. The variation of arsenic contents of selected foods from different countries is demonstrated by the data in Table 2. In general, arsenic concentrations in these foods are lowest when obtained in countries with little or no exposure to arsenic contamination of water source.

Rice, which is a staple part of the diet in Bangladesh, has been extensively studied in the context of mitigating and reducing the uptake of arsenic during its cultivation. An excellent review on this topic has been recently published [72]. The authors have indicated that by employing water management, physico-chemical and biological strategies, either alone or combined, the uptake of inorganic and methylated arsenic species in rice cultivation can be successfully decreased. These approaches may also be adapted for other crops. The adoption of these strategies is a major challenge for countries like Bangladesh, where socio-economic factors can be a hindrance.

Table 2. Concentrations of arsenic in selected foods from different countries.

Country	Vegetables (μg/kg)	Rice (μg/kg)	Fish/Shrimp (μg/kg)	Other Foods (μg/kg)
Australia		30 (20–40)		
Bangladesh	54.5 (<5–540)	500 (30–1840)	(97–1318)	45.9 (44.9–46.9) Betel leaf
Bangladesh [a]	(70–3990)	496 (58–1830)		
Bangladesh [b]			(214–266)	
China		140 (20–460)		
China [a]		930		
Europe	(<5–87)			
United Kingdom	2 (green veg) 4.9 (other veg)			
USA		250 (30–660)		
West Bengal (India)		140 (20–400)		
West Bengal (India) [a]		250 (140–480)330 (180–430)		

[a] Samples collected from arsenic-affected area and [b] marine species. Adapted and modified from [73].

3.3. Social Effects

The prevalence of arsenic poisoning has triggered many social implications [48,49,74,75]. Relatives, friends and neighbors often ostracize the people affected by arsenic poisoning. There is a general tendency to avoid or discourage the arsenic-affected people from being seen in public. More concerning, school attendance by their children is usually prohibited, while attendance at work places and public meetings is not encouraged for the adults [76].

For women, arsenic poisoning has more dire social consequences [76]. Young women who have had arsenicosis are often constrained from getting married. Even married women who have arsenic related diseases are also socially ostracized. Evidently, the males, females and children who are affected by arsenic poisoning are severely disadvantaged socially [76].

3.4. Economic Impact

There is a direct relationship between the economic impact of suffering from arsenicosis and the social impact experienced from the disease [48,49,74,75,77]. Usually once a family member becomes sick from arsenic poisoning, various coping mechanisms are experienced. This may involve selling of assets, reduction of basic needs, reduction of access to education and the associated burden of financial loans [76]. Consequently, a large-scale poisoning of the population can impact the nation's economy, requiring the Government, in extreme cases, to reduce its social and economic development programs as a basis for dealing with the disaster.

Attempts to minimize or eliminate the impacts of groundwater contamination with arsenic have resulted in various strategies for arsenic removal from groundwater with a varying level of successes. Nevertheless, there is still considerable research being undertaken in this area to achieve efficient and reliable removal of arsenic from groundwater to ensure safe human consumption. Some of these removal processes and strategies are discussed below.

4. Arsenic Removal Processes

A major approach for reducing arsenic poisoning is by treatment of arsenic-contaminated water and the new and emerging treatment technologies have received considerable evaluation and have also been recently reviewed [47,78–80]. The major categories of treatment methods that have been used for this purpose are described below [81]:

(i) Oxidation and filtration process: This usually involves the oxidation of arsenic with inorganic iron and manganese oxides, followed by removal of the residues by filtration [82,83]. Due to the uncharged nature of As(III) complexes, an oxidation step is often employed to oxidize As(III) complexes to As(V) complexes [84].

(ii) Biological oxidation: Microorganisms are used in this case to oxidize As(III) to As(V), followed by As(V) removal with iron and manganese dioxide [85,86].

(iii) Co-precipitation: This involves the addition of a suitable oxidizing agent to oxidize As(III) to As(V), followed by coagulation, sedimentation and filtration [87,88].

1. [(iv)] Adsorption: This approach involves the adsorption of arsenic with activated alumina, activated carbon, hydrated iron oxide, iron based adsorbents and zero valent iron [89–91].

(v) Ion exchange: This involves the use of relevant exchange resins (cation or anion) for separation of arsenic [92,93].

(vi) Membrane technology: This includes the use of processes such as electrodialysis, nanofiltration and reverse osmosis for removal of arsenic [94–98].

The following sub-sections provide more detailed discussion of some of these treatment methods.

4.1. Biological Process

Numerous biological treatment options for removal of metals from drinking water are available [99] and the biological treatment of groundwater for arsenic removal employs naturally occurring microorganisms, such as *Gallionellaferruginea* and *Leptothrixochracea* [100]. The addition of these organisms to the groundwater results in the formation of new iron oxide precipitates on the filter, which subsequently removes arsenic from the water by adsorption. Under optimized conditions, As(III) is oxidized by these microorganisms, enabling removal of up to 95% of arsenic from waters which contain 200 μg/L of arsenic [100]. This process also enabled the removal of As(V). For a comprehensive overview of biological transformations of arsenic, which includes possible detoxification mechanisms, refer to a recent review on this topic [101].

4.2. Precipitative Processes

This category of arsenic removal processes includes coagulation and filtration [87], coagulation-assisted microfiltration [102], enhanced coagulation [103] and lime softening [104]. These are very popular processes that require the addition of either aluminum sulfate, ferric chloride or ferric sulfate as a coagulant to change either the chemical or physical properties of dissolved colloidal or suspended matter [105]. To promote rapid settling out of the particles by gravity, the enhancement of agglomeration is used in some cases, otherwise filtration is used to remove the particles [105]. Agglomeration is usually accomplished by changing the surface charge properties of solids with coagulants to enable formation of a flocculated precipitate. The resulting products of this process are larger particles or flocs that filter or settle out more easily under gravity [106].

Greater than 90% As(V) removal has been successfully achieved with coagulation processes [107]. The use of coagulation with alum, ferric chloride and ferric sulfate for removal of As(III) is far less efficient than the removal of As(V) [107]. To ensure efficient arsenic removal, As(III) is usually oxidized to As(V) prior to coagulation. Effective As(V) removal is usually accomplished at pH 7.6 or less with iron and aluminum coagulants. However, in terms of stability, iron coagulants offer more advantage than aluminum coagulants when operating within a pH range of 5.5 to 8.5 [9]. The key factors that influence the choice of the optimum coagulant dose are the water quality and the arsenic concentration in the treated water. The overall cost of the water treatment process can be significantly increased if further adjustment of the pH is necessary to achieve a more effective removal of arsenic [106].

Some field studies have demonstrated the effectiveness of coagulation/filtration in reducing the arsenic level to below 5 μg/L [106]. However, under optimum operating conditions, the achievement of a residual arsenic level of less than 3 μg/L is possible [107].

4.3. Membrane Processes

Siddique et al. [108] have recently reviewed the application of nanofiltration membrane technologies and their advantages and disadvantages relative to other methods. The membrane processes used for arsenic removal mainly include reverse osmosis [94,95] and electrodialysis [96]. While these processes are effective, they are generally more costly. Nonetheless, they are capable of removing arsenic through filtration, electric repulsion, and adsorption of arsenic-bearing compounds. The size exclusion property of a chosen membrane enables rejection of arsenic compounds larger than its pore size [107]. Besides the size of the compounds, the rejection by the membrane can be influenced by other factors. In some instances, arsenic compounds that are 1–2 orders of magnitude smaller than the membrane pore size have been rejected, thus suggesting that besides physical straining, other removal mechanisms may be involved [107]. Two factors that play significant roles in arsenic rejection by a membrane are shape and chemical characteristics of the arsenic compounds. Removal of arsenic compounds on membranes can also be accomplished through repulsion or surface adsorption depending on the charge and hydrophobicity of the membrane and the feed water [109]. The achievement of removal efficiency of 97% for As(V) and 92% for As(III) in a single pass by reverse osmosis has been reported [107].

4.3.1. Arsenic Removal by Membrane Distillation

As a non-isothermal membrane separation process, membrane distillation (MD) utilizes a microporous hydrophobic membrane with pore size ranging from 0.01 μm to 1 μm [97]. For effective operation, only vapor and non-condensable gases must be present within the membrane pores and the membrane must not be wet [97,98]. Commercially available hydrophobic micro-porous membranes include polytetrafluoroethylene, polyethylene, polypropylene and polyvinylidenefluoride membranes [98]. The simplest and most economical and efficient of the different kinds of MD is direct contact membrane distillation (DCMD) [97,98]. DCMD directly separates the hot feed and the cold permeate with the aid of the membrane. Up to 100% of arsenic has been removed from contaminated groundwater with DCMD [97,98].

4.3.2. Membrane and Adsorption Process Hybrid

New technologies with different processes have also been adopted for further improvement of arsenic removal [2]. Membrane technologies along with low cost adsorptive media have been shown to be effective for the removal of arsenic from water [110]. This has led to the production of a cheap, easy to operate filter system for delivering safe and arsenic-free drinking water. This system consists of three basic components: an organic membrane, a tank/drum in which the membrane is inserted and an adsorptive cartridge made of industrial waste products [110].

4.4. Adsorptive Processes

Adsorptive processes include ion exchange [94,95], iron oxide/hydroxide coated sand [89,90,111,112], iron-hydroxide coated alumina [91], granular ferric hydroxide (GFH) [113,114] and natural iron ores [115]. A desirable feature of the available adsorbents for developing countries is that they should be cheap, readily available and effective [116].

The basis of adsorption methods is the ability to achieve the attraction of soluble arsenic species to the surface of a suitable sorbent. These methods are cost-effective and have attracted a considerable interest for the development of adsorbents that are capable of cheap and efficient arsenic removal from drinking water. The two most commonly used sorbents are based on: (a) iron compounds, including akaganeite (β-FeOOH), amorphous hydrous ferric oxide (FeOOH), goethite (α-FeOOH) and poorly crystalline hydrous ferric oxide [111], and (b) aluminum compounds, including activated alumina γ-Al_2O_3 and gibbsite $Al(OH)_3$ [117].

Among other sorbents considered are carbon from coconut husks [118], carbon from fly ash [119,120], hybrid polymeric materials [121], red mud [122,123], titanium dioxide [124,125], manganese dioxide [126,127], orange peel [128] and fungal biomass-based bio-sorbent [86,129].

4.4.1. Iron Oxides/Oxyhydroxides Based Adsorbents

The effectiveness of various iron compounds for metal ion removal has been demonstrated in a number of studies [35,130]. Those found to be effective sorbents for arsenic removal from aqueous solutions include β-FeOOH, FeOOH, α-FeOOH and poorly crystalline hydrous ferric oxide [80,90,111,112]. Other iron oxides/oxyhydroxides-based sorbents that have been considered include Cerium (IV)-doped iron oxide adsorbent [115], GFH [113,114], iron-hydroxide coated alumina [91], iron oxide-coated polymeric minerals [100,131], iron oxide-coated sand [132], magnetic iron oxide/activated carbon composite [132], iron oxide-coated cement [96], magnetically modified zeolite [90], natural iron ores [115], silica-containing iron(III) oxide [133] and iron containing waste materials such as fly ash and red mud [120].

The adsorption of arsenic by hydrous ferric oxide (FeOOH), particularly akaganeite, ferrihydrite, goethite and GFH, has been demonstrated to be very strong and effective [114]. The breakthrough behavior of a GFH fixed bed filter was investigated for arsenic removal and maximum uptakes of 28.9 mg/g and 42.7 mg/g at pH 7 were achieved with initial low and high concentrations of 0–1 mg/L and 1–8 mg/L, respectively [114]. The use of goethite for arsenic removal also achieved adsorption capacity of 25 mg/g [112]. The ability to use goethite to achieve a high arsenic removal rate was suggested in another study [134]. The achievement of higher arsenic uptakes of 65 mg/g at pH 3.5 and 22 °C [90] and of 120 mg/g at a pH range of 4.5–7 and 25 °C [111] has been successfully demonstrated with the use of a synthetic β-FeOOH.

The iron oxyhydroxides used in these studies have particle sizes in the nanometer range. However, when used in sorption columns, nanoparticles can be problematic due to the difficulty in achieving solid/liquid separation [111]. It is therefore a major challenge to find a natural adsorbent that contains nanoparticles of iron oxides that can remove arsenic effectively and can also be used in column filtration technology.

The use of iron oxide coated sand (IOCS) as an adsorbent for removal of arsenic and iron has been investigated by UNESCO-IHE (The Netherlands) for 20 or more years. It is obtained as a by-product of the iron removal process within the Dutch groundwater treatment plants. The investigation of the use of IOCS obtained from different groundwater treatment plants in the Netherlands demonstrates that high to very high arsenic removal efficiencies were achieved for As(III) and As(V) depending on the iron and manganese content in their coatings. Arsenic removal efficiency of 100 percent was achieved with IOCS which contained 353.8 mg Fe/g IOCS and 17.2 mg Mn/g [135].

4.4.2. Application of Nanoparticles for Arsenic Removal from Water

Various nanomaterials have been employed for water treatment [136]. Their reactivities and high surface areas have attracted a lot of interest in their consideration as novel adsorbents for removal of heavy metals and arsenic [137]. Maiti et al. [138] have reviewed the use of nanomaterials for arsenic removal from water and some of the key nanomaterial technologies are summarized in this section. The implementation of these technologies has been limited by the costs of production and the difficulty in recovery and recycling of the nanoparticles. Of these, iron- and titanium-based nanoparticles are most commonly used for removal of arsenic [139–143].

The nano-adsorbents that have been considered for arsenic removal include nanoparticles of zero-valent iron (nZVI) [144,145], iron oxide nanoparticles α-Fe_2O_3, Fe_3O_4 and α-FeOOH [139,146,147]. The arsenic removal capacity achieved with the iron oxide nanoparticles is influenced by the oxidation state of iron [143].

The oxidation of As(III) to As(V) is supported by the metal core and a thin layer of amorphous iron (oxy)hydroxide present in nanoparticles of nZVI [148]. Although nZVI is a useful adsorbent for arsenic removal, the production of toxic solid wastes from its synthesis is a main disadvantage [149].

Due to their high surface-to-volume ratios [75], iron oxide nanoparticles are 5–10 times more effective for arsenic removal than with the use of micron-sized particles [139]. Ultrafine hematite α-Fe_2O_3 nanoparticles were successfully synthesized and used for treatment of laboratory-prepared and arsenic contaminated natural water [150]. Rapid removal of As(III) and As(V) was achieved, enabling removal of 74% of As(III) within 30 min. The BET (Brunauer–Emmett–Teller) specific surface area was 162 m^2/g and the average particle diameter was 5.0 nm. The achieved adsorption capacity for As(III) was 95 mg/g, while for As(V) it was much lower at 47 mg/g.

The performance of magnetite nanoparticles which have a BET surface area of 69.4 m^2/g and a mean particle diameter of 20 nm has also been investigated for arsenic removal [151]. At pH > 7, the adsorption of As(V) rapidly decreases, while a more consistent adsorption was achieved for As(III) at pH 2–9. The achieved adsorption capacity for As(III) was 8.0 mg/g and was 8.8 mg/g for As(V). It was found that the presence of phosphate interfered with arsenic removal [151] and its removal prior to the treatment is recommended.

The removal of As(V) from water was investigated with a nanocomposite of silica and goethite [146]. The synthesized silica nanoparticles had particle sizes ranging from 150 to 250 nm. The achieved adsorption capacity for the goethite/silica nanocomposite was 17.64 mg/g at pH 3.0. The adsorption of As(V) on the nanocomposite was rapid, reaching equilibrium within 120 min [146].

Arsenic removal and photocatalytic oxidation of As(III) have been investigated in a number of studies with nanocrystalline titanium dioxide (TiO_2) [152–155]. The adsorption of As(III) and As(V) reached equilibrium on this nano-adsorbent within four hours and 80% or more of both species were adsorbed [152]. The photocatalytic efficiency of nanocrystalline TiO_2 in oxidizing As(III) was also demonstrated, achieving full conversion in the presence of sunlight and dissolved oxygen to As(V) within 25 min [152].

Another titanium-based nano-adsorbent for As(III) which does not require oxidation to As(V) is hydrous titanium dioxide nanoparticles ($TiO_2 \times H_2O$) [156]. High maximum adsorption capacities have been achieved with TiO_2 nanoparticles for As(III). An adsorption capacity of 83 mg/g was achieved at a near neutral pH, while at pH 9 it was higher at 96 mg/g [157]. The titanium dioxide nanoparticles provided an effective, low-cost and single-step process for arsenic removal from contaminated water. Nevertheless, due to their particle sizes in the nanometer range, some care and extreme caution must be taken to prevent the possible dispersion of these nanoparticles into the environment [157]. This may be prevented or avoided by either granulation of these nanoparticles into micron-sized particles or loading onto very porous host materials [157].

The reported removal efficiencies achieved by some arsenic removal processes are summarized in Table 3. Evidently, the removal processes based on the use of Fe_2O_3 filter,

As(III) oxidation by (OCl−) and Fe precipitation, enhanced coagulation/filtration with ferric chloride, iron doped activated carbon, hybrid activated alumina, iron based sorbents, layered double hydroxide (LDH), modified zeolites and laterite and limonite achieved removal efficiencies ≥95%, thus demonstrating that processes involving oxidation and filtration, precipitation and adsorption are effective for removal of arsenic from contaminated water. The chosen process therefore depends on the simplicity and convenience of use if to be adopted for providing drinking water for human consumption.

Table 3. Treatment efficiencies reported for some arsenic removal processes.

Treatment Process As(V)	Removal Efficiency	As Concentration
Oxidation and Filtration		
Aeration and filtration	>90%	300 μg As(III)/L
Fe_2O_3 filter	>95%	100–400 μg As(III)/L
As(III) oxidation by (OCl−) and Fe precipitation	>98%	300 μg As(III)/L
Co-precipitation		
Enhanced lime softening	90%	
Enhanced coagulation/filtration with alum	<90%	
Enhanced coagulation/filtration with ferric chloride	95%	
Adsorption		
Iron doped activated carbon	>95%	311 μg As/L
Hybrid activated alumina	>95%	2–20 mg As/L
Iron based sorbents	Up to 98%	
Layered double hydroxide (LDH)	Up to 96%	300 μg As(V)/L
Modified zeolites	Up to 99%	100–400 μg As/L
Modified clays	Up to 80%	0.15 μM As
Laterite and limonite	Up to 95%	500 μg As/L

Adapted and modified from [75].

5. Household Filter for Arsenic Removal for Drinking Water

A major and impactful outcome from the various reported approaches for treatment of groundwater for removal of arsenic is their adoption in designing potable household filters that have been successfully used for treatment of groundwater as a source of domestic drinking water, particularly in under-developed countries. Several designs of arsenic filters have been developed, proposed and evaluated. Yet, there are still emerging new developments in this area.

One of the most successful domestic treatment units is the SONO filter which is a two-bucket system [158]. The composition of the upper bucket includes a composite iron layer (4–5 cm thick) of a mixture of metal iron and iron hydroxides. This layer is then covered with sand layers. Also, sand and charcoal layers are included in the lower bucket for removal of the iron hydroxides and residual organic matter. Figure 5 shows that the arsenic contaminated groundwater passes through coarse sand, composite iron matrix (CIM), brick chips and wood charcoal [158].

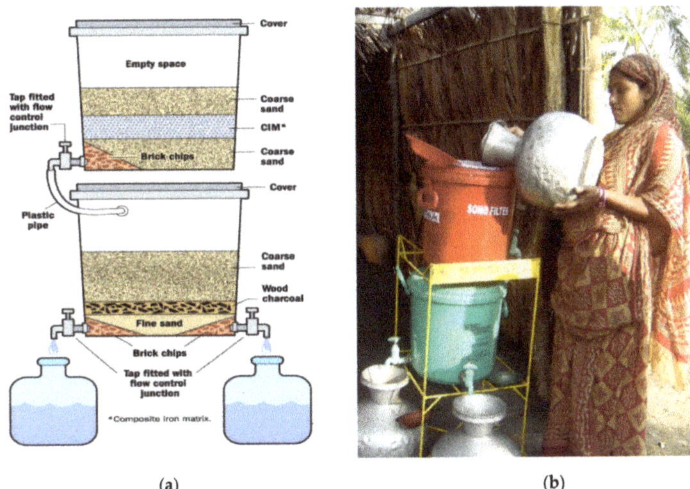

(a) (b)

Figure 5. SONO Filter. (**a**) Schematic and (**b**) in domestic use [158]. Reproduced with permission from Taylor & Francis.

The SONO filter which is manufactured in Bangladesh utilizes a composite iron matrix (CIM) to remove arsenic by complexation reactions on the metal surface as follows [158]:

$$FeOH + H_2AsO_4^- \rightarrow FeHAsO_4^- + H_2O \text{ and}$$

$$FeOH + HAsO_4^{2-} \rightarrow FeAsO_4^{2-} + H_2O$$

These reactions occur without the need for any pre- or post-chemical treatment. Also no regeneration is required and no toxic wastes are produced [158]. The resulting spent material from the arsenic removal process is a solid self-contained iron-arsenate cement which is non-toxic and does not leach out when in contact with rainwater.

The arsenic concentrations in contaminated groundwater used in one study ranged from 5–4000 µg/L, but these were reduced to 3–30 µg/L after treatment [158]. The treated waters from the filter were found to meet the limits of 10 µg/L and 50 µg/L set by WHO and the Bangladeshi government, respectively. Also, the filter is relatively cheap costing about $40 for over five years operation and capable of producing 20–30 L/hour to support one to two families drinking and cooking needs [158]. The approval for use of the filter was granted by government and at the early stages about 30,000 of these filters were provided throughout Bangladesh. Its use subsequently spread to India, Nepal and Pakistan. The National Academy of Engineering has recognized the innovation of the SONO filter for arsenic removal by awarding it the Grainger Challenge Prize for sustainability due to "its affordability, reliability, ease of maintenance, social acceptability, and environmental friendliness [158]".

With regards to ongoing maintenance of SONO filters, an important requirement is the need to flush each bucket with 5 L of hot water to eliminate pathogenic bacteria and minimize coliform counts. The filter is expected to last for five years, but if the flow rate decreases, it can be improved by removing the sand layers for washing and reuse or replacement with new sand.

Another commonly used arsenic filter for treatment of contaminated water for arsenic removal is the Kanchan^{TM} Arsenic Filter (KAF) [159]. This filter was developed in 2003 through a collaborative effort between the Massachusetts Institute of Technology (MIT) and Environment and Public Health Organization on the mitigation of arsenic contamination of groundwater [160].

Figure 6 shows that the components of a KAF include a plastic container, PVC pipe, diffuser basin, brick chips, iron nails, fine sand, coarse sand and gravel [160]. The arsenic contaminated groundwater is poured into the diffuser basin where it comes into contact with the brick particles and iron nails. Upon contact with water and air, the iron nails rust, consequently producing ferric hydroxide particles that quickly adsorb arsenic from the water. The adsorption process is repeated when new iron surfaces are exposed as the outer surfaces scale off. These continuous processes result in the retention of arsenic on the filter components and produce drinkable filtered water. In addition, the KAF is capable of removing pathogens from the water.

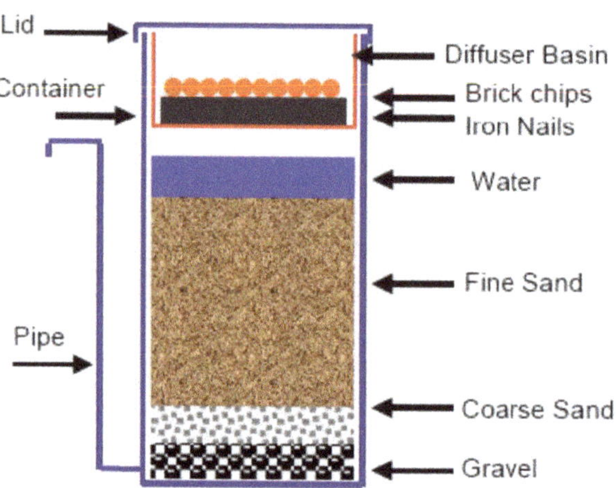

Figure 6. Components of a Kanchan Arsenic Filter [160]. Reproduced with permission from Taylor & Francis.

Factors influencing the performance of a KAF include arsenic concentration, duration of use, filter maintenance, flow rate, monitoring and handling and other water components (such as water hardness, iron, chloride and phosphate concentrations) [161]. The effectiveness of KAFs in reducing the concentration of arsenic in groundwater to less than 50 μg/L has been demonstrated [160]. But the effectiveness of the filter can be affected occasionally by the unexpected and unpredictable variability of groundwater conditions, arsenic concentration and climatic conditions.

One of the more recently developed filters is the Pakistan Arsenic Filter (PAF) [162]. This unit employs three different forms of iron (mesh, nails and slag). The iron is held down in a plastic bucket (25 L) with diffuser plates. A tap is attached to the bucket to enable the adjustment of water level and for taking samples at the completion of the treatment, as shown in Figure 7.

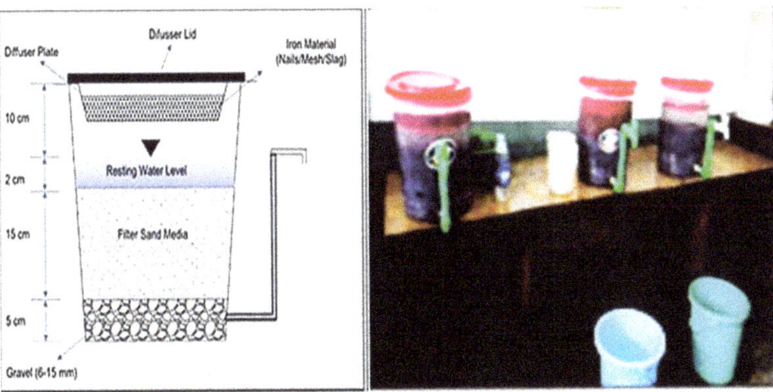

Figure 7. Components of a Pakistan Arsenic Filter (PAF) and its use for arsenic removal from groundwater [162]. Reproduced with permission.

Figure 7 shows the specific components of a PAF. The round gravel layer overlaid by sand layer can be seen at the bottom. The unit used 1 kg of iron (mesh, nails and slag) contained within the diffuser plate in the bucket and the rate of water flow through the filter is 30 L/min which decreased over an 8-week trial period to 20 L/min [162]. The arsenic contaminated groundwater goes through a diffuser plate and then passes through the sand and gravel. Treated water is transferred through a plastic pipe via a tap, as shown in Figure 7. The highest efficiency achieved for arsenic removal was with the use of iron mesh due to its larger surface area which generates more ferric hydroxide. The PAF was produced at a cost of US$ 5 which is affordable and accessible to more people [162]. However, it is only applicable to contaminated water which contains 50 to 100 µg/L arsenic, but it was suggested that the unit can be used in series to achieve improved arsenic removal.

Besides the few examples of arsenic filters highlighted in this paper, there are many other filters used or proposed for removal of arsenic from groundwater for domestic consumption. We have decided to describe only the selected few that have had significant impact for domestic provision of suitable drinking water. Also, it is important to note that there is ongoing research to develop simpler and easier to use filter systems.

6. Conclusions and Future Considerations

This review has demonstrated that arsenic contamination of groundwater impacts the availability of safe and good quality water for domestic use (drinking and cooking) and agriculture in under-developed countries such as Bangladesh, India, Pakistan and Nepal, as well as some parts of more developed countries such as the USA. This can have dire consequences on human health, agriculture, economic and social well-being. The need for a good understanding of the chemistry of arsenic and its fate in the environment has also been demonstrated to be very important for dealing with or minimising the consequences of groundwater contamination by arsenic. The case study about Bangladesh provided a sharp focus on the dire consequences and impact of groundwater contamination by arsenic with serious health, agricultural, social and economic impacts. Furthermore, it highlights that when groundwater contamination by arsenic is identified, rapid efforts need to be directed to reducing/minimizing the associated impacts by adopting and utilising relevant treatment and removal strategies, as well as seeking alternatives for drinking water sources where possible. Socio-economic factors will also play a significant role in the adoption of safer drinking water options. A recent study in Bangladesh illustrated the need for public education combined with the provision of alternative drinking water sources to overcome attitudes and the reluctance of people to switch to better alternatives [163]. The adoption of many of these strategies in the development of various filter systems for arsenic removal has had a great impact on the ability of people to access safe drinking water in some parts of the affected countries and communities. Many of the available potable treatment systems are effective in ensuring the removal and minimisation of arsenic in groundwater dependent communities to acceptable local standards. However, there are still issues with the long-term maintenance of these systems, especially in ensuring ongoing effective removal of arsenic, disinfection, filter replacement and sludge disposal. Ongoing education of local communities on how to maintain these systems, including how often the filters should be replaced and disinfection protocols is necessary to ensure safety, particularly where there is no alternative water source to groundwater. There is therefore still a need for ongoing research on developing more robust and highly effective filter systems for arsenic removal from groundwater for use for domestic consumption, with the aim of further reducing capital and operation costs, improving user friendliness, minimizing maintenance requirements and resolving or eliminating the need for sludge management and disposal. Also, portable devices that can be adopted for household and community monitoring of arsenic concentrations in drinking water are now available [164,165]. Das et al. [164] have developed two low-cost field test kits for detection of arsenic in water. These kits are capable of detecting as low as 10 μg/L of total arsenic in groundwater within 7 min. In addition, there are commercially available quick test systems for arsenic detection in water within 12 min [165]. If adopted, these devices will ensure communities can adequately assess their water safety and, thus, minimize the incidences of arsenic poisoning. Furthermore, the use of various nanomaterials will play a significant role in the development of a next generation efficient filter system. In the long-term, the development of a more affordable version of the ArsenicMaster Whole House Arsenic Water Filtration System [166], shown in Figure 8, which is maintenance-free will go a long way to addressing the various issues with existing filters if it can be designed to suit the arsenic levels found in groundwater in the under-developed countries and remote communities at a reasonable capital cost.

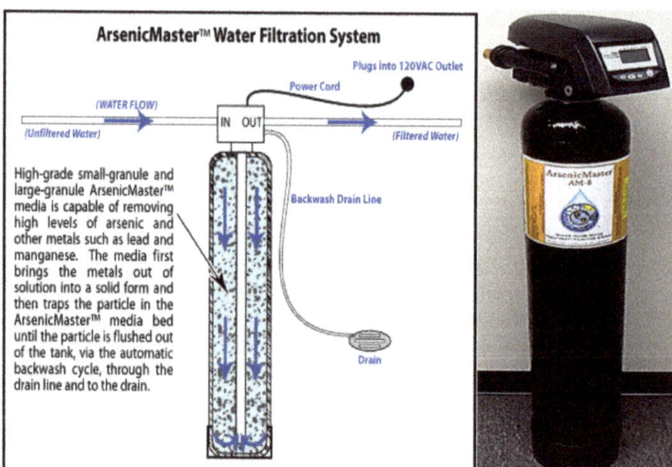

Figure 8. Components of the ArsenicMaster Whole House Arsenic Water Filtration System and a commercially available version [166]. Reproduced with permission.

Author Contributions: Conceptualization, S.B.A. and S.K.; methodology, S.B.A., S.K. and A.F.P.; formal analysis, S.B.A., S.K. and A.F.P.; investigation, S.B.A. and S.K.; resources, S.B.A. and A.F.P.; data curation, S.K and S.B.A.; Writing—Original draft preparation, S.K., S.B.A. and A.F.P.; Writing—Review and editing, S.K., S.B.A. and A.F.P.; visualization, S.B.A.; supervision, S.B.A. and A.F.P.; project administration, S.B.A.; funding acquisition, S.B.A. All authors have read and agreed to the published version of the manuscript.

Funding: This research received no external funding.

Institutional Review Board Statement: Not applicable.

Informed Consent Statement: Not applicable.

Data Availability Statement: Not applicable for a review paper.

Acknowledgments: The provision of postgraduate research scholarships and research facilities by Monash University to S.K. is gratefully acknowledged.

Conflicts of Interest: The authors declare no conflict of interest and that the funders had no role in the design of the study; in the collection, analyses, or interpretation of data; in the writing of the manuscript, or in the decision to publish the results.

References

1. Mandal, B.K.; Suzuki, K.T. Arsenic round the world: A review. *Talanta* **2002**, *58*, 201–235. [CrossRef]
2. Singh, R.; Singh, S.; Parihar, P.; Singh, V.P.; Prasad, S.M. Arsenic contamination, consequences and remediation techniques: A review. *Ecotoxicol. Environ. Saf.* **2015**, *112*, 247–270. [CrossRef] [PubMed]
3. Shankar, S.; Shanker, U.; Shikha. Arsenic contamination of groundwater: A review of sources, prevalence, health risks, and strategies for mitigation. *Sci. World J.* **2014**, *2014*, 1–18. [CrossRef] [PubMed]
4. Shaji, E.; Santosh, M.; Sarath, K.; Prakash, P.; Deepchand, V.; Divya, B. Arsenic contamination of groundwater: A global synopsis with focus on the Indian Peninsula. *Geosci. Front.* **2021**, *12*, 101079. [CrossRef]
5. Ali, W.; Rasool, A.; Junaid, M.; Zhang, H. A comprehensive review on current status, mechanism, and possible sources of arsenic contamination in groundwater: A global perspective with prominence of Pakistan scenario. *Environ. Geochem. Health* **2019**, *41*, 737–760. [CrossRef]
6. Saha, N.; Rahman, M.S. Groundwater hydrogeochemistry and probabilistic health risk assessment through exposure to arsenic-contaminated groundwater of Meghna floodplain, central-east Bangladesh. *Ecotoxicol. Environ. Saf.* **2020**, *206*, 111349. [CrossRef]
7. Mohan, D.; Pittman, C.U., Jr. Arsenic removal from water/wastewater using adsorbents-A critical review. *J. Hazard. Mater.* **2007**, *142*, 1–53. [CrossRef]
8. Ahmed, M.F.; Ali, M.A. *Environmental Chemistry of Arsenic, Arsenic Contamination: Bangladesh Perspective*; International Training Network BUET: Dhaka, Bangladesh, 2003; pp. 21–41.
9. Smedley, P.; Kinniburgh, D. A review of the source, behaviour and distribution of arsenic in natural waters. *Appl. Geochem.* **2002**, *17*, 517–568. [CrossRef]
10. Azcue, J.M.; Nriagu, J.O. Arsenic historical perspective. In *Arsenic in the Environment*; Part, J., Nriagu, J.O., Eds.; John Wiley & Sons: London, UK, 1994; pp. 1–49.
11. Sharma, V.K.; Sohn, M. Aquatic arsenic: Toxicity, speciation, transformations, and remediation. *Environ. Int.* **2009**, *35*, 743–759. [CrossRef]
12. Gupta, D.K.; Srivastava, S.; Huang, H.; Romero-Puertas, M.C.; Sandalio, L.M. *Detoxification of Heavy Metals*; Springer: Cham, Switzerland, 2011; pp. 169–179.
13. Al-Abed, S.R.; Jegadeesan, G.; Purandare, J.; Allen, D. Arsenic release from iron rich mineral processing waste: Influence of pH and redox potential. *Chemosphere* **2007**, *66*, 775–782. [CrossRef] [PubMed]
14. Bodek, I. *Environmental Inorganic Chemistry: Properties, Processes, and Estimation Methods*; Pergamon Press: New York, NY, USA, 1988.
15. Singh, M.K.; Kumar, A. A global problem of arsenic in drinking water and its mitigation—A review. *Int. J. Adv. Eng. Technol.* **2012**, *3*, 196–203.
16. Cassone, G.; Chillé, D.; Foti, C.; Giuffrè, O.; Ponterio, R.; Sponer, J.; Saija, F. Stability of hydrolytic arsenic species in aqueous solutions: As3+ vs. As5+. *Phys. Chem. Chem. Phys.* **2018**, *20*, 23272–23280. [CrossRef]
17. Quinodóz, F.B.; Maldonado, L.; Blarasin, M.; Matteoda, E.; Lutri, V.; Cabrera, A.; Albo, J.G.; Giacobone, D. The development of a conceptual model for arsenic mobilization in a fluvio-eolian aquifer using geochemical and statistical methods. *Environ. Earth Sci.* **2019**, *78*, 206. [CrossRef]
18. Syafiuddin, A.; Boopathy, R.; Hadibarata, T. Challenges and solutions for sustainable groundwater usage: Pollution control and integrated management. *Curr. Pollut. Rep.* **2020**, *6*, 310–327. [CrossRef]
19. Csalagovits, I. Arsenic–bearing artesian waters of Hungary. *Annu. Rep. Geol. Inst. Hung.* **1999**, *1992–1993/II*, 85–92.
20. Rowland, H.A.; Omoregie, E.O.; Millot, R.; Jimenez, C.; Mertens, J.; Baciu, C.; Hug, S.J.; Berg, M. Geochemistry and arsenic behaviour in groundwater resources of the Pannonian Basin (Hungary and Romania). *Appl. Geochem.* **2011**, *26*, 1–17. [CrossRef]
21. Bundschuh, J.; Armienta, M.A.; Birkle, P. *Natural Arsenic in Groundwaters of Latin America*; CRC Press: New York, NY, USA, 2008.
22. UNICEF. Arsenic Mitigation in Bangladesh Fact Sheet. 2010. Available online: http://www.unicef.org/bangladesh/Arsenic.pdf (accessed on 12 December 2020).
23. WHO. *Guidelines for Drinking-Water Quality: Recommendations*, 3rd ed.; World Health Organization: Geneva, Switzerland, 2004; Volume 1.

24. BGS; DPHE. *Arsenic Contamination of Groundwater in Bangladesh*; Final, Report; British Geological Survey: Keyworth, UK; Department of Public Health Engineering: Dhaka, Bangladesh, 2001; Volume 2.
25. Herath, I.; Vithanage, M.; Bundschuh, J.; Maity, J.P.; Bhattacharya, P. Natural arsenic in global groundwaters: Distribution and geochemical triggers for mobilization. *Curr. Pollut. Rep.* **2016**, *2*, 68–89. [CrossRef]
26. BBS. *Per Capaita Income of Bangladesh: Bangladesh Bureau of Staticts*; Ministry of Planning, Government of Bangladesh: Dhaka, Bangladesh, 2016.
27. BGS. *Groundwater Studies for Arsenic Contamination in Bangladesh*; Main report and, supplemental; Government of the Peoples Republic of Bangladesh: Dhaka, Bangladesh; Ministry of Local Government, Rural Development and Cooperatives: Dhaka, Bangladesh; Department of Public Health Engineering: Dhaka, Bangladesh; Mott MacDonald International Ltd.: London, UK, 2000; Volume 1–3.
28. GoB; UNDP. *Millennium Development Goals Needs Assessment & Costing 2009–2015 Bangladesh*; Government of Bangladesh: Dhaka, Bangladesh; UNDP: New York, NY, USA, 2015.
29. Tan, S.N.; Yong, J.W.H.; Ng, Y.F. Arsenic exposure from drinking water and mortality in Bangladesh. *Lancet* **2010**, *376*, 1641–1642. [CrossRef]
30. Ahmed, M.F.; Ahmed, T. Status of remediation of arsenic contamination of groundwater in Bangladesh. *Compr. Water Qual. Purif.* **2014**, *1*, 104–121.
31. Polizzotto, M.L.; Harvey, C.F.; Li, G.; Badruzzman, B.; Ali, A.; Newville, M.; Sutton, S.; Fendorf, S. Solid-phases and desorption processes of arsenic within Bangladesh sediments. *Chem. Geol.* **2006**, *228*, 97–111. [CrossRef]
32. Uddin, A.; Shamsudduha, M.; Saunders, J.; Lee, M.-K.; Ahmed, K.M.; Chowdhury, M. Mineralogical profiling of alluvial sediments from arsenic-affected Ganges–Brahmaputra floodplain in central Bangladesh. *Appl. Geochem.* **2011**, *26*, 470–483. [CrossRef]
33. Harvey, C.F.; Ashfaque, K.N.; Yu, W.; Badruzzaman, A.; Ali, M.A.; Oates, P.M.; Michael, H.A.; Neumann, R.B.; Beckie, R.; Islam, S.; et al. Groundwater dynamics and arsenic contamination in Bangladesh. *Chem. Geol.* **2006**, *228*, 112–136. [CrossRef]
34. McArthur, J.M.; Banerjee, D.M.; Hudson-Edwards, K.A.; Mishra, R.; Purohit, R.; Ravenscroft, P.; Cronin, A.; Howarth, R.J.; Chatterjee, A.; Talukder, T.; et al. Natural organic matter in sedimentary basins and its relation to arsenic in anoxic ground water: The example of West Bengal and its worldwide implications. *Appl. Geochem.* **2004**, *19*, 1255–1293. [CrossRef]
35. Bundschuh, J.; Bhattacharya, P.; Sracek, O.; Mellano, M.F.; Ramírez, A.E.; Storniolo, A.D.R.; Martín, R.A.; Cortés, J.; Litter, M.I.; Jean, J.-S. Arsenic removal from groundwater of the Chaco-Pampean Plain (Argentina) using natural geological materials as adsorbents. *J. Environ. Sci. Heal. Part A* **2011**, *46*, 1297–1310. [CrossRef]
36. Acharyya, S.K.; Shah, B.A. Groundwater arsenic pollution affecting deltaic West Bengal, India. *Curr. Sci.* **2010**, *99*, 1787–1794.
37. Mallick, S.; Rajagopal, N. Groundwater development in the arsenic-affected alluvial belt of West Bengal—Some questions. *Curr. Sci.* **1996**, *70*, 956–958.
38. Mandal, B.K.; Chowdhury, T.R.; Samanta, G.; Mukherjee, D.P.; Chanda, C.R.; Saha, K.C.; Chakraborti, D. Impact of safe water for drinking and cooking on five arsenic-affected families for 2 years in West Bengal, India. *Sci. Total. Environ.* **1998**, *218*, 185–201. [CrossRef]
39. Singh, A.K. Chemistry of arsenic in groundwater of Ganges-Brahmaputra river basin. *Curr. Sci.* **2006**, *91*, 599–606.
40. Fazal, M.A.; Kawachi, T.; Ichion, E. Extent and severity of groundwater arsenic contamination in Bangladesh. *Water Int.* **2001**, *26*, 370–379. [CrossRef]
41. Acharyya, S. *Arsenic in Groundwater—Geological Overview: Consultation on Arsenic in Drinking Water*; World Health Organization (WHO): New Delhi, India, 1997.
42. Safiuddin, M.; Shirazi, S.M.; Yussof, S. Arsenic contamination of groundwater in Bangladesh: A review. *Int. J. Phys. Sci.* **2011**, *6*, 6791–6800. [CrossRef]
43. Nickson, R.; McArthur, J.; Burgess, W.; Ahmed, K.M.; Ravenscroft, P.; Rahmanñ, M. Arsenic poisoning of Bangladesh groundwater. *Nat. Cell Biol.* **1998**, *395*, 338. [CrossRef]
44. Fendorf, S.; Michael, H.A.; Van Geen, A. Spatial and temporal variations of groundwater arsenic in south and southeast Asia. *Science* **2010**, *328*, 1123–1127. [CrossRef]
45. Reza, A.S.; Jean, J.-S.; Yang, H.-J.; Lee, M.-K.; Woodall, B.; Liu, C.-C.; Lee, J.-F.; Luo, S.-D. Occurrence of arsenic in core sediments and groundwater in the Chapai-Nawabganj District, northwestern Bangladesh. *Water Res.* **2010**, *44*, 2021–2037. [CrossRef] [PubMed]
46. McArthur, J.M.; Ravenscroft, P.; Safiulla, S.; Thirlwall, M.F. Arsenic in groundwater: Testing pollution mechanisms for sedimentary aquifers in Bangladesh. *Water Resour. Res.* **2001**, *37*, 109–117. [CrossRef]
47. Maity, J.P.; Nath, B.; Kar, S.; Chen, C.-Y.; Banerjee, S.; Jean, J.-S.; Liu, M.-Y.; Centeno, J.A.; Bhattacharya, P.; Chang, C.L.; et al. Arsenic-induced health crisis in peri-urban Moyna and Ardebok villages, West Bengal, India: An exposure assessment study. *Environ. Geochem. Health* **2012**, *34*, 563–574. [CrossRef] [PubMed]
48. Islam, M.; Islam, F. *Arsenic Contamination in Groundwater in Bangladesh: An Environmental and Social Disaster*; IWA Water Wiki: London, UK, 2010; Available online: www.iwawaterwiki.org (accessed on 16 December 2020).
49. Safiuddin, M.; Karim, M. Groundwater arsenic contamination in Bangladesh: Causes, effects and remediation. In Proceedings of the 1st International Conference and Annual Paper Meet on Civil Engineering, Chittagong, Bangladesh, 2–3 November 2016; Volume 43, pp. 79–84.

50. Huq, S.I.; Joardar, J.; Parvin, S.; Correll, R.; Naidu, R. Arsenic contamination in food-chain: Transfer of arsenic into food materials through groundwater irrigation. *J. Heal. Popul. Nutr.* **2006**, *24*, 305–316.
51. Khan, S.I.; Ahmed, A.M.; Yunus, M.; Rahman, M.; Hore, S.K.; Vahter, M.; Wahed, M. Arsenic and cadmium in food-chain in Bangladesh—An exploratory study. *J. Heal. Popul. Nutr.* **2010**, *28*, 578–584. [CrossRef]
52. *ITNC Position Paper on Bangladesh Response to Arsenic Contamination of Groundwater*; International Training Network Centre, Bangladesh University of Engineering and Technology (BUET): Dhaka, Bangladesh, 2008.
53. SOES; DCH. *Groundwater Arsenic Contamination in Bangladesh*; A summarized survey, report; The School of Environmental Studies, Jadavpur University: Kolkata, India; Dhaka Community Hospital: Dhaka, Bangladesh, 2000.
54. Dhar, R.K.; Biswas, B.K.; Samanta, G.; Mandal, B.K.; Chakraborti, D.; Roy, S.; Jafar, A.; Islam, A.; Ara, G.; Kabir, S.; et al. Groundwater arsenic calamity in Bangladesh. *Curr. Sci.* **1997**, *73*, 48–59.
55. Farmer, J.G.; Johnson, L.R. Assessment of occupational exposure to inorganic arsenic based on urinary concentrations and speciation of arsenic. *Occup. Environ. Med.* **1990**, *47*, 342–348. [CrossRef]
56. Arnold, H.; Odam, R.; James, W. *Diseases of the Skin: Clinical Dermatology*, 8th ed.; WB Sanders: Philadelphia, PA, USA, 1990.
57. Tchounwou, P.B.; Yedjou, C.G.; Udensi, U.K.; Pacurari, M.; Stevens, J.J.; Patlolla, A.K.; Noubissi, F.; Kumar, S. State of the science review of the health effects of inorganic arsenic: Perspectives for future research. *Environ. Toxicol.* **2019**, *34*, 188–202. [CrossRef]
58. WHO. *Guidelines for Drinking Water Quality*; WHO: Geneva, Switzerland, 2011; Volume 4, pp. 315–318.
59. Dastgiri, S.; Mosaferi, M.; Fizi, M.A.; Olfati, N.; Zolali, S.; Pouladi, N.; Azarfam, P. Arsenic exposure, dermatological lesions, hypertension, and chromosomal abnormalities among people in a rural community of northwest Iran. *J. Health Popul. Nutr.* **2010**, *28*, 14–22. [PubMed]
60. Lazaroff, C. Arsenic Standard Delayed by Call for More Studies; Environment News Service: 2001. Available online: www.ens-newswire.com (accessed on 10 October 2011).
61. Smith, A. *Report and Action Plan for Arsenic in Drinking Water Focusing on Health, Bangladesh*; Assignment Report (WHO Project BAN CWS 001); World Health Organization (WHO): New Delhi, India, 1997; p. 9.
62. Yunus, M.; Sohel, N.; Hore, S.K.; Rahman, M. Arsenic exposure and adverse health effects: A review of recent findings from arsenic and health studies in Matlab, Bangladesh. *Kaohsiung J. Med Sci.* **2011**, *27*, 371–376. [CrossRef] [PubMed]
63. Khalequzzaman, M.; Faruque, F.S.; Mitra, A.K. Assessment of Arsenic Contamination of Groundwater and Health Problems in Bangladesh. *Int. J. Environ. Res. Public Health* **2005**, *2*, 204–213. [CrossRef]
64. Karim, M.M. Arsenic in groundwater and health problems in Bangladesh. *Water Res.* **2000**, *34*, 304–310. [CrossRef]
65. Asadullah, M.N.; Chaudhury, N. Poisoning the mind: Arsenic contamination of drinking water wells and children's educational achievement in rural Bangladesh. *Econ. Educ. Rev.* **2011**, *30*, 873–888. [CrossRef]
66. Sohel, N.; Vahter, M.; Ali, M.; Rahman, M.; Rahman, A.; Streatfield, P.K.; Kanaroglou, P.S.; Persson, L. Åke spatial patterns of fetal loss and infant death in an arsenic-affected area in Bangladesh. *Int. J. Heal. Geogr.* **2010**, *9*, 53. [CrossRef]
67. Bhattacharya, P.; Samal, A.C.; Majumdar, J.; Santra, S.C. Arsenic contamination in rice, wheat, pulses, and vegetables: A study in an arsenic affected area of West Bengal, India. *Water Air Soil Pollut.* **2010**, *213*, 3–13. [CrossRef]
68. Martin, M.; Ferdousi, R.; Hossain, K.M.J.; Barberis, E. Arsenic from groundwater to paddy fields in Bangladesh: Solid-liquid partition, sorption and mobility. *Water Air and Soil Pollut.* **2010**, *212*, 27–36. [CrossRef]
69. Rahman, M.A.; Hasegawa, H. High levels of inorganic arsenic in rice in areas where arsenic-contaminated water is used for irrigation and cooking. *Sci. Total Environ.* **2011**, *409*, 4645–4655. [CrossRef]
70. Saha, N.; Zaman, M.R. Concentration of selected toxic metals in groundwater and some cereals grown in Shibganj area of Chapai Nawabganj, Rajshahi, Bangladesh. *Curr. Sci.* **2011**, *101*, 427–431.
71. Williams, P.; Islam, M.; Hussain, S.; Meharg, A. Arsenic absorption by rice. Behavior of Arsenic in Aquifers, Soils and Plants. (Conference Proceedings), Dhaka, Bangladesh, 16–18th January 2005.
72. Kumarathilaka, P.; Seneweera, S.; Ok, Y.S.; Meharg, A.A.; Bundschuh, J. Mitigation of arsenic accumulation in rice: An agronomical, physicochemical, and biological approach—A critical review. *Crit. Rev. Environ. Sci. Technol.* **2020**, *50*, 31–71. [CrossRef]
73. Jiang, J.-Q.; Ashekuzzaman, S.M.; Jiang, A.; Sharifuzzaman, S.M.; Chowdhury, S.R. Arsenic contaminated groundwater and its treatment options in Bangladesh. *Int. J. Environ. Res. Public Health* **2013**, *10*, 18–46. [CrossRef] [PubMed]
74. Ahmed, A.A.M.; Alam, J.B.; Ahmed, A.A.M. Evaluation of socioeconomic impact of arsenic contamination in Bangladesh. *J. Toxicol. Environ. Health Sci.* **2011**, *3*, 298–307.
75. Mahmood, S.; Halder, A. The socioeconomic impact of arsenic poisoning in Bangladesh. *J. Toxicol. Environ. Health Sci.* **2011**, *3*, 65–73.
76. Ahmad, S.A.; Sayed, M.H.; Khan, M.H.; Karim, M.N.; Haque, M.A.; Bhuiyan, M.S. Sociocultural aspects of arsenicArsenicosis in Bangladesh: Community perspective. *J. Environ. Sci. Health A Toxicol. Hazard Subst Environ. Eng.* **2007**, *42*, 1945–1958. [CrossRef]
77. Rahmana, M.A.; Rahmanc, A.; Kaiser Khan, M.Z.; Renzahod, A.M.N. Human health risks and socio-economic perspectives of arsenic exposure in Bangladesh: A scoping review. *Ecotoxicol. Environ. Saf.* **2018**, *150*, 335–343. [CrossRef]
78. Guha Mazumder, D.; Dasgupta, U.B. Chronic arsenic toxicity: Studies in West Bengal, India. *Kaohsiung J. Med. Sci.* **2011**, *27*, 360–370. [CrossRef] [PubMed]
79. Alka, S.; Shahir, S.; Ibrahim, N.; Ndejiko, M.J.; Vo, D.-V.N.; Manan, F.A. Arsenic removal technologies and future trends: A mini review. *J. Clean. Prod.* **2021**, *278*, 123805. [CrossRef]

80. Weerasundara, L.; Ok, Y.-S.; Bundschuh, J. Selective removal of arsenic in water: A critical review. *Environ. Pollut.* **2021**, *268*, 115668. [CrossRef] [PubMed]
81. Jain, C.K.; Singh, R.D. Technological options for the removal of arsenic with special reference to South East Asia. *J. Environ. Manag.* **2012**, *107*, 1–18. [CrossRef]
82. Ghurye, G.; Clifford, D.; Tripp, A. Iron coagulation and direct microfiltration to remove arsenic from groundwater. *J. Am. Water Work. Assoc.* **2004**, *96*, 143–152. [CrossRef]
83. Leupin, O.X.; Hug, S.J. Oxidation and removal of arsenic (III) from aerated groundwater by filtration through sand and zero-valent iron. *Water Res.* **2005**, *39*, 1729–1740. [CrossRef]
84. Bissen, M.; Frimmel, F.H. Arsenic—A review. Part II: Oxidation of arsenic and its removal in water treatment. *Acta Hydrochim. Hydrobiol.* **2003**, *31*, 97–107. [CrossRef]
85. Katsoyiannis, I.A.; Zouboulis, A.I. Application of biological processes for the removal of arsenic from groundwaters. *Water Res.* **2004**, *38*, 17–26. [CrossRef]
86. Pokhrel, D.; Viraraghavan, T. Arsenic removal from an aqueous solution by a modified fungal biomass. *Water Res.* **2006**, *40*, 549–552. [CrossRef]
87. Hering, J.G.; Chen, P.-Y.; Wilkie, J.A.; Elimelech, M. Arsenic removal from drinking water during coagulation. *J. Environ. Eng.* **1997**, *123*, 800–807. [CrossRef]
88. Chen, H.-W.; Frey, M.M.; Clifford, D.; McNeill, L.S.; Edwards, M. Arsenic treatment considerations. *J. Am. Water Work. Assoc.* **1999**, *91*, 74–85. [CrossRef]
89. Lakshmipathiraj, P.; Narasimhan, B.R.V.; Prabhakar, S.; Raju, G.B. Adsorption of arsenate on synthetic goethite from aqueous solutions. *J. Hazard. Mater.* **2006**, *136*, 281–287. [CrossRef] [PubMed]
90. Vaclavikova, M.; Matik, M.; Jakabsky, S.; Hredzak, S. Preparation and sorption properties of Fe-nanomaterials for removal of arsenic from waters. In *Book of Abstract of NATO CCMS on Clean Products and Processes*; NATO: Lillehammer, Norway, 2005; p. 13.
91. Hlavay, J.; Polyak, K. Determination of surface properties of ironhydrocide-coated alumina adsorbent prepared for removal of arsenic from drinking water. *J. Colloid Interface Sci.* **2005**, *284*, 71–77. [CrossRef]
92. Jekel, M. Removal of arsenic in drinking water treatment. In *Arsenic in the Environment. Part I: Cycling and Characterization*; Nriagu, J.O., Ed.; Wiley: New York, NY, USA, 1994; p. 119.
93. Rau, I.; Gonzalo, A.; Valiente, M. Arsenic(V) adsorption by immobilized iron mediation. Modeling of the adsorption process and influence of interfering anions. *React. Funct. Polym.* **2003**, *54*, 85–94. [CrossRef]
94. Ning, R.Y. Arsenic removal by reverse osmosis. *Desalination* **2002**, *143*, 237–241. [CrossRef]
95. Kang, M.; Kawasaki, M.; Tamada, S.; Kamei, T.; Magara, Y. Effect of pH on the removal of arsenic and antimony using reverse osmosis membranes. *Desalination* **2000**, *131*, 293–298. [CrossRef]
96. Kundu, S.; Gupta, A. Analysis and modeling of fixed bed column operations on As(V) removal by adsorption onto iron oxide-coated cement (IOCC). *J. Colloid Interface Sci.* **2005**, *290*, 52–60. [CrossRef]
97. Manna, A.K.; Sen, M.; Martin, A.R.; Pal, P. Removal of arsenic from contaminated groundwater by solar-driven membrane distillation. *Environ. Pollut.* **2010**, *158*, 805–811. [CrossRef]
98. Yarlagadda, S.; Gude, V.G.; Camacho, L.M.; Pinappu, S.; Deng, S. Potable water recovery from As, U, and F contaminated ground waters by direct contact membrane distillation process. *J. Hazard. Mater.* **2011**, *192*, 1388–1394. [CrossRef]
99. Hasan, H.A.; Muhammad, M.H.; Ismail, N.I. A review of biological drinking water treatment technologies for contaminants removal from polluted water resources. *J. Water Process Eng.* **2020**, *33*, 101035. [CrossRef]
100. Wang, S.; Zhao, X. On the potential of biological treatment for arsenic contaminated soils and groundwater. *J. Environ. Manag.* **2009**, *90*, 2367–2376. [CrossRef]
101. Mazumder, P.; Sharma, S.K.; Taki, K.; Kalamdhad, A.S.; Kumar, M. Microbes involved in arsenic mobilization and respiration: A review on isolation, identification, isolates and implications. *Environ. Geochem. Health* **2020**, *42*, 3443–3469. [CrossRef]
102. Han, B.; Runnels, T.; Zimbron, J.; Wickamasinghe, R. Arsenic removal from drinking water by floculation and microfiltration. *Desalination* **2002**, *145*, 293–298. [CrossRef]
103. Kumar, P.R.; Chaudhari, S.; Khilar, K.C.; Mahajan, S. Removal of arsenic from water by electrocoagulation. *Chemosphere* **2004**, *55*, 1245–1252. [CrossRef]
104. Dutta, A.; Chaudhuri, M. Removal of arsenic from ground water by lime softening with powdered coal additive. *J. Water SRT-Aqua.* **1991**, *40*, 25–29.
105. Petrusevski, B.; Sharma, S.K.; Schippers, J.C. *Groundwater Treatment. Part II, UNESCO-IHE Lecture Notes*; UNESCO: Delft, The Netherlands, 2003.
106. Schippers, J.C. *Surface Water Treatment, UNESCO-IHE Lectures Notes*; UNESCO: Delft, The Netherlands, 2003.
107. US-EPA Office of Ground Water and Drinking Water. *Implementation Guidance for the Arsenic Rule*; EPA report-816-D-02-005; EPA: Cincinnati, OH, USA, 2002.
108. Siddique, T.A.; Dutta, N.K.; Choudhury, N.R. Nanofiltration for arsenic removal: Challenges, recent developments, and perspectives. *Nanomaterials* **2020**, *10*, 1323. [CrossRef] [PubMed]
109. Xu, Y.; Nakajima, T.; Ohki, A. Adsorption and removal of arsenic (V) from drinking water by aluminium loaded Shirazuzeolite. *J. Hazard. Mater. B* **2002**, *92*, 275–278. [CrossRef]

110. Vigneswaran, S.; Nguyen, T.V. Water Filtration System Wins $500k Technology Against Poverty Prize. School of Environmental and Environmental Engineering, UTS. 2017. Available online: https://www.ecovoice.com.au/uts-water-filtration-system-wins-technology-against-poverty-500k-prize/ (accessed on 15 December 2017).
111. Deliyanni, E.; Bakoyannakis, D.; Zouboulis, A.; Matis, K. Sorption of As(V) ions by akaganéite-type nanocrystals. *Chemosphere* **2003**, *50*, 155–163. [CrossRef]
112. Matis, K.A.; Lehmann, M.; Zouboulis, A.I. Modelling sorption of metals from aqueous solution onto mineral particles: The case of arsenic ions and goethite ore. In *Natural Microporous Materials in Environmental Technology*; Springer International Publishing: New York, NY, USA, 1999; pp. 463–472.
113. Badruzzaman, M.; Westerhoff, P.; Knappe, D.R. Intraparticle diffusion and adsorption of arsenate onto granular ferric hydroxide (GFH). *Water Res.* **2004**, *38*, 4002–4012. [CrossRef] [PubMed]
114. Sperlich, A.; Werner, A.; Genz, A.; Amy, C.; Worch, E.; Jekel, I. Breakthrongh behavior of granular ferric hydroxide (C PH) fixed-bed adsorption filters: Modeling and experimental approaches. *Water Res.* **2005**, *39*, 1190–1198. [CrossRef]
115. Zhang, W.; Singh, P.; Paling, E.; Delides, S. Arsenic removal from contaminated water by natural iron ores. *Miner. Eng.* **2004**, *17*, 517–524. [CrossRef]
116. Asere, T.G.; Stevens, C.V.; Du Laing, G. Use of (modified) natural adsorbents for arsenic remediation: A review. *Sci. Total Environ.* **2019**, *676*, 706–720. [CrossRef]
117. Singh, T.S.; Pant, K. Equilibrium, kinetics and thermodynamic studies for adsorption of As(III) on activated alumina. *Sep. Purif. Technol.* **2004**, *36*, 139–147. [CrossRef]
118. Manju, G.; Raji, C.; Anirudhan, T. Evaluation of coconut husk carbon for the removal of As from water. *Water Res.* **1998**, *32*, 3062–3070. [CrossRef]
119. Pattanayak, J.; Mondal, K.; Mathew, S.; Lalvani, S. A parametric evaluation of the removal of As(V) and As(III) by carbon-based adsorbents. *Carbon* **2000**, *38*, 589–596. [CrossRef]
120. Bertocchi, A.F.; Ghiani, M.; Peretti, R.; Zucca, A. Red mud and fly ash for remediation of mine sites contaminated with As, Cd, Cu, Pb and Zn. *J. Hazard. Mater.* **2006**, *134*, 112–119. [CrossRef]
121. Demarco, M.J.; Sengupta, A.K.; Greenleaf, J.E. Arsenic removal using a polymeric/inorganic hybrid sorbent. *Water Res.* **2003**, *37*, 164–176. [CrossRef]
122. Genç-Fuhrman, H.; Bregnhøj, H.; McConchie, D. Arsenate removal from water using sand–red mud columns. *Water Res.* **2005**, *39*, 2944–2954. [CrossRef]
123. Altundoğan, H.; Altundoğan, S.; Tümen, F.; Bildik, M. Arsenic adsorption from aqueous solutions by activated red mud. *Waste Manag.* **2002**, *22*, 357–363. [CrossRef]
124. Dutta, P.K.; Ray, A.K.; Sharma, V.K.; Millero, F.J. Adsorption of arsenate and arsenite on titanium dioxide suspensions. *J. Colloid Interface Sci.* **2004**, *278*, 270–275. [CrossRef] [PubMed]
125. Jing, C.; Meng, X.; Liu, S.; Baidas, S.; Patraju, R.; Christodoulatus, C.; Korfiatis, C. Surface complexation of organic arsenic on nanocrystalline titanium dioxide. *J. Colloid Interface Sci.* **2005**, *290*, 14–21. [CrossRef] [PubMed]
126. Deschamps, E.; Ciminelli, V.S.; Höll, W.H. Removal of As(III) and As(V) from water using a natural Fe and Mn enriched sample. *Water Res.* **2005**, *39*, 5212–5220. [CrossRef]
127. Lenoble, V.; Laclautre, C.; Serpaud, B.; Deluchat, V.; Bollinger, J.-C. As(V) retention and As(III) simultaneous oxidation and removal on a MnO_2-loaded polystyrene resin. *Sci. Total Environ.* **2004**, *326*, 197–207. [CrossRef]
128. Ghimire, K.; Inoue, K.; Yamaguchi, H.; Makino, K.; Miyajima, T. Adsorptive separation of arsenate and arsenite anions from aqueous medium by using orange waste. *Water Res.* **2003**, *34*, 4945–4953. [CrossRef]
129. Loukidou, M.; Matis, K.; Zouboulls, A.; Liakopoulou-Kyriakidou, M. Removal of As(V) from wastewaters by chemically modified fungal biomass. *Water Res.* **2003**, *37*, 4544–4552. [CrossRef]
130. Maiti, A.; Thakur, B.K.; Basu, J.K.; De, S. Comparison of treated laterite as arsenic adsorbent from different locations and performance of best filter under field conditions. *J. Hazard. Mater.* **2013**, *262*, 1176–1186. [CrossRef] [PubMed]
131. Thirunavukkarasu, O.S.; Viraraghavan, T.; Subramanian, K.S. Arsenic removal from drinking water using iron oxide-coated sand. *Water Air Soil Pollut.* **2003**, *142*, 95–111. [CrossRef]
132. Sahira Joshi, S.; Sharma, M.; Kumari, A.; Surendra Shrestha, S.; Shrestha, B. Arsenic removal from water by adsorption onto iron oxide/nano-porous carbon magnetic composite. *Appl. Sci.* **2019**, *9*, 3732. [CrossRef]
133. Zeng, L. A method for preparing silica-containing iron(III) oxide adsorbents for arsenic removal. *Water Res.* **2003**, *37*, 4351–4358. [CrossRef]
134. Mamindy-Pajany, Y.; Hurel, C.; Marmier, N.; Roméo, M. Arsenic adsorption onto hematite and goethite. *Comptes Rendus Chim.* **2009**, *12*, 876–881. [CrossRef]
135. Omeroglu, P. Use of Iron Coated Sand for Arsenic Removal. M.Sc. Thesis, SEE 131, UNESCO-IHE, Delft, The Netherlands, 2001.
136. Mondal, P.; Bhowmick, S.; Chatterjee, D.; Figoli, A.; Van der Bruggen, B. Remediation of inorganic arsenic in groundwater for safe water supply: A critical assessment of technological solutions. *Chemosphere* **2013**, *92*, 157–170. [CrossRef]
137. Hristovski, K.; Baumgardner, A.; Westerhoff, P. Selecting metal oxide nanomaterials for arsenic removal in fixed bed columns: From nanopowders to aggregated nanoparticle media. *J. Hazard. Mater.* **2007**, *147*, 265–274. [CrossRef]
138. Maiti, A.; Mishra, S.; Chaudhary, M. Nanoscale materials for arsenic removal from water. In *Nanoscale Materials in Water Purification*; Elsevier BV: Amsterdam, The Netherlands, 2019; pp. 707–733.

139. Habuda-Stanić, M.; Nujić, M. Arsenic removal by nanoparticles: A review. *Environ. Sci. Pollut. Res.* **2015**, *22*, 8094–8123. [CrossRef]
140. Wong, W.; Wong, H.Y.; Badruzzaman, A.B.M.; Goh, H.H.; Zaman, M. Recent advances in exploitation of nanomaterial for arsenic removal from water: A review. *Nanotechnology* **2016**, *28*, 042001. [CrossRef]
141. Hua, M.; Zhang, S.; Pan, B.; Zhang, W.; Lv, L.; Zhang, Q. Heavy metal removal from water/wastewater by nanosized metal oxides: A review. *J. Hazard. Mater.* **2012**, *211–212*, 317–331. [CrossRef]
142. Qu, X.; Alvarez, P.J.; Li, Q. Applications of nanotechnology in water and wastewater treatment. *Water Res.* **2013**, *47*, 3931–3946. [CrossRef]
143. Tang, S.C.N.; Lo, I.M.C. Magnetic nanoparticles: Essential factors for sustainable environmental applications. *Water Res.* **2013**, *47*, 2613–2632. [CrossRef] [PubMed]
144. Kanel, S.R.; Manning, B.; Charlet, L.; Choi, H. Removal of arsenic(III) from groundwater by nanoscale zero-valent iron. *Environ. Sci. Technol.* **2005**, *39*, 1291–1298. [CrossRef]
145. Jegadeesan, G.; Mondal, K.; Lalvani, S.B. Arsenate remediation using nanosized modified zerovalent iron particles. *Environ. Prog.* **2005**, *24*, 289–296. [CrossRef]
146. Attinti, R.; Sarkar, D.; Barrett, K.R.; Datta, R. Adsorption of arsenic(V) from aqueous solutions by goethite/silica nanocomposite. *Int. J. Environ. Sci. Technol.* **2015**, *12*, 3905–3914. [CrossRef]
147. Adegoke, H.I.; Adekola, F.A.; Fatoki, O.S.; Ximba, B.J. A comparative study of sorption of As (V) ions on nanoparticle hematite, goethite and magnetite. *Nanotechnology* **2014**, *1*, 184–187.
148. Ramos, M.A.V.; Yan, W.; Li, X.-Q.; Koel, B.E.; Zhang, W.-X. Simultaneous oxidation and reduction of arsenic by zero-valent Iron nanoparticles: Understanding the significance of the core−shell structure. *J. Phys. Chem. C* **2009**, *113*, 14591–14594. [CrossRef]
149. Litter, M.I.; Morgada, M.E.; Bundschuh, J. Possible treatments for arsenic removal in Latin American waters for human consumption. *Environ. Pollut.* **2010**, *158*, 1105–1118. [CrossRef]
150. Tang, W.; Li, Q.; Gao, S.; Shang, J.K. Arsenic(III,V) removal from aqueous solution by ultrafine α-Fe_2O_3 nanoparticles synthesized from solvent thermal method. *J. Hazard. Mater.* **2011**, *192*, 131–138. [CrossRef] [PubMed]
151. Chowdhury, S.R.; Yanful, E.K. Arsenic removal from aqueous solutions by adsorption on magnetite nanoparticles. *Water Environ. J.* **2010**, *25*, 429–437. [CrossRef]
152. Pena, M.E.; Korfiatis, G.P.; Patel, M.; Lippincott, L.; Meng, X. Adsorption of As(V) and As(III) by nanocrystalline titanium dioxide. *Water Res.* **2005**, *39*, 2327–2337. [CrossRef]
153. Jegadeesan, G.; Al-Abed, S.R.; Sundaram, V.; Choi, H.; Scheckel, K.G.; Dionysiou, D.D. Arsenic sorption on TiO_2 nanoparticles: Size and crystallinity effects. *Water Res.* **2010**, *44*, 965–973. [CrossRef]
154. Nabi, D.; Aslam, I.; Qazi, I.A. Evaluation of the adsorption potential of titanium dioxide nanoparticles for arsenic removal. *J. Environ. Sci.* **2009**, *21*, 402–408. [CrossRef]
155. Ashraf, S.; Siddiqa, A.; Shahida, S.; Qaisar, S. Titanium-based nanocomposite materials for arsenic removal from water: A review. *Heliyon* **2019**, *5*, e01577. [CrossRef]
156. Guan, X.; Du, J.; Meng, X.; Sun, Y.; Sun, B.; Hu, Q. Application of titanium dioxide in arsenic removal from water: A review. *J. Hazard. Mater.* **2012**, *215–216*, 1–16. [CrossRef] [PubMed]
157. Xu, Z.; Li, Q.; Gao, S.; Shang, J.K. As(III) removal by hydrous titanium dioxide prepared from one-step hydrolysis of aqueous $TiCl_4$ solution. *Water Res.* **2010**, *44*, 5713–5721. [CrossRef] [PubMed]
158. Hussam, A.; Munir, A.K.M. A simple and effective arsenic filter based on composite iron matrix: Development and deployment studies for groundwater of Bangladesh. *J. Environ. Sci. Health Part A* **2007**, *42*, 1869–1878. [CrossRef] [PubMed]
159. Gianotti, R. *Filtering Out Arsenic in Nepal*; MIT Komaza Magazine: Cambridge, MA, USA, 2011; pp. 17–19.
160. Ngai, T.K.; Shrestha, R.R.; Dangol, B.; Maharjan, M.; Murcott, S.E. Design for sustainable development—Household drinking water filter for arsenic and pathogen treatment in Nepal. *J. Environ. Sci. Health Part A* **2007**, *42*, 1879–1888. [CrossRef]
161. Singh, A.; Smith, L.S.; Shrestha, S.; Maden, N. Efficacy of arsenic filtration by kanchan arsenic filter in Nepal. *J. Water Health* **2014**, *12*, 596–599. [CrossRef]
162. Hayder, S.; Ahmed, T.; Tariq, M. Arsenic: A low-cost household level treatment for rural settings in developing countries. *J. Pollut. Eff. Control.* **2018**, *6*, 1–5. [CrossRef]
163. Naus, F.L.; Burer, K.; Van Laerhoven, F.; Griffioen, J.; Ahmed, K.M.; Schot, P. Why do people remain attached to unsafe drinking water options? Quantitative evidence from Southwestern Bangladesh. *Water* **2020**, *12*, 342. [CrossRef]
164. Das, J.; Sarkar, P.; Panda, J.; Pal, P. Low-cost field test kits for arsenic detection in water. *J. Environ. Sci. Health Part A* **2013**, *49*, 108–115. [CrossRef] [PubMed]
165. Industrial Test Systems Quick 481396-5 Arsenic for Water Quality Testing. Available online: https://www.amazon.com/Industrial-Test-Systems-481396-5-Arsenic/dp/B07V9C3W5Y (accessed on 18 January 2021).
166. Vitasalus Inc., Troy, MI, USA. Available online: http://www.equinox-products.com/ArsenicMaster.htm (accessed on 20 December 2020).

MDPI

Article

The Effect of Superabsorbent Polymers on the Microstructure and Self-Healing Properties of Cementitious-Based Composite Materials

Irene A. Kanellopoulou, Ioannis A. Kartsonakis and Costas A. Charitidis *

School of Chemical Engineering, R-Nano Lab, Laboratory of Advanced, Composite, Nanomaterials and Nanotechnology, National Technical University of Athens, 9 Heroon Polytechniou str., Zografou Campus, 15773 Athens, Greece; ikan@chemeng.ntua.gr (I.A.K.); ikartso@chemeng.ntua.gr (I.A.K.)
* Correspondence: charitidis@chemeng.ntua.gr; Tel.: +30-210-772-4046

Featured Application: Superabsorbent polymers of novel structure have been used in cementitious-based composite materials improving their self-healing behavior by an index of 60%.

Abstract: Cementitious structures have prevailed worldwide and are expected to exhibit further growth in the future. Nevertheless, cement cracking is an issue that needs to be addressed in order to enhance structure durability and sustainability especially when exposed to aggressive environments. The purpose of this work was to examine the impact of the Superabsorbent Polymers (SAPs) incorporation into cementitious composite materials (mortars) with respect to their structure (hybrid structure consisting of organic core—inorganic shell) and evaluate the microstructure and self-healing properties of the obtained mortars. The applied SAPs were tailored to maintain their functionality in the cementitious environment. Control and mortar/SAPs specimens with two different SAPs concentrations (1 and 2% bwoc) were molded and their mechanical properties were determined according to EN 196-1, while their microstructure and self-healing behavior were evaluated via microCT. Compressive strength, a key property for mortars, which often degrades with SAPs incorporation, in this work, practically remained intact for all specimens. This is coherent with the porosity reduction and the narrower range of pore size distribution for the mortar/SAPs specimens as determined via microCT. Moreover, the self-healing behavior of mortar-SAPs specimens was enhanced up to 60% compared to control specimens. Conclusively, the overall SAPs functionality in cementitious-based materials was optimized.

Keywords: mechanical properties; microstructure; self-healing; SAP; microCT; cementitious materials; mortar

Citation: Kanellopoulou, I.A.; Kartsonakis, I.A.; Charitidis, C.A. The Effect of Superabsorbent Polymers on the Microstructure and Self-Healing Properties of Cementitious-Based Composite Materials. *Appl. Sci.* **2021**, *11*, 700. https://doi.org/10.3390/app11020700

Received: 30 November 2020
Accepted: 10 January 2021
Published: 13 January 2021

1. Introduction

Cementitious materials have been widely used over the years in the construction sector due to the cement abundance, its low cost and excellent durability [1,2]. In 2017 the global cement production came up to 4.1 billion tones. China and India which currently represent two of the most rapidly growing countries worldwide produced 57% and 7% of the global cement production in that year, while 6% of the same production is attributed to Europe [3]. It is estimated that the cement production will double in the decades to come [2,4].

Nevertheless, cement is prone to cracking as a result of both inherent material properties such as low tensile strength and application related factors such as tension inducement during the infrastructure service life thus compromising its integrity and durability [1,5–7]. Aggressive environment conditions (i.e., wide ambient temperature fluctuation, rich presence of ions, pH levels etc.) combined with external loading favor crack propagation often leading to the formation of an interconnected crack network that allows corrosive factors to penetrate the structure and assault the reinforcement thus leading to its deterioration. This raises huge safety issues and makes maintenance and repair high priority concerns [6–8].

Even though autogenous crack self-healing in cementitious materials has been known for centuries, its effectiveness highly depends on numerous factors, namely the crack width and age, water abundance upon the crack formation, environment conditions such as pH and presence of ions etc. [6,9–11]. Conventional retrofitting/repair methods applied up to now mainly consist of textile reinforced mortar/concrete (TRM/TRC), the textiles used mainly being based on carbon, glass and aramid fibers, polymer crack injection and polymer modified concrete [1,9,10]. These methods have proven to be effective but in several cases, the cost is prohibitive and/or these methods are difficult or even impossible to apply because of the location of the damaged spot on the infrastructure [6,9].

As a result, internal curing promoting methods have gained a lot interest in the last two decades [1,9]. These methods focus on (i) maintaining continuous water provision and thus continuous cement hydration and (ii) restraining cement self-desiccation [1]. Over the years, the incorporation of several internal curing agents in cementitious materials such as Light Weight Aggregates (LWA) [1,2,5,9,12,13], Superabsorbent Polymers (SAPs) [9,10,14–26], Rice Husk Ash (RHA) [1,5,27], bottom ash [1,5,28], fly ash [1,2,5,9,13,29], cenospheres [1,5,30], crushed returned concrete fine aggregates (CCA) [1,13] and wood fibers [1,13] has been tested. In this application field, SAPs seem to have very promising results and gain more research interest [10].

The SAPs are 3-D polymer networks that due to their hydrophilic nature absorb huge amounts of water (even thousands of times their own dry weight), while due to the network crosslinking they retain their structure and are not dissolved [1,14,15,19,22,24,31]. It is confirmed that their total water absorption level is inversely dependent on environment related factors such as the presence of ions and pH values [10,18,22,24].

Due to their properties, SAPs are used in a vast variety of applications such as hygienic products, agriculture, drug delivery systems, sealing, pharmaceuticals, biomedical applications, tissue engineering, biosensors and the construction field [32]. In the construction field, attention has been drawn to different strategies to obtain coatings with debonding properties [33]. One of these strategies is to incorporate SAPs into an intermediate primer layer between the substrate and the top coatings. The trigger mechanism relies on the fact that with a pH variation, the SAPs can enhance their shape due to the water absorption resulting in the reduction of the attachment between the primer layer and both the top coating and substrate, enabling the detachment of the top coating from the corresponding substrate. Moreover, in cementitious pastes a large number of ions are present (Ca^{2-}, K^+, Na^+, $SO_4{}^{2-}$, OH^- etc.) and pH of the mix water ranges between 11 and 13.4 [10,15,22,24,34].

The incorporation of SAPs in cementitious materials initially results in water absorbance during the cement paste mixing procedure and act as water reservoirs that will make water available during cement curing and hardening. In this way, SAPs contribute to maintaining higher levels of relative humidity thus mitigating early age shrinkage and at the same time favoring extended hydrating reactions which lead to denser cement microstructure by reducing the capillary pores in the cementitious matrix which leads to improved structure strength [9,10,15–17,20,23,24]. Furthermore, upon crack formation and in the presence of water, SAPs reabsorb water on a separate event, swell and allow an immediate crack self-healing effect while the promotion of additional hydrating reactions of remaining unhydrated cementitious phases provide a crack self-healing effect, thus acting as internal curing agents [9–11,14–17,23]. On the other hand, when water is released from SAPs particles in the cement matrix, SAPs deswell forming voids around them in the scale of macropores. These macropores are likely to act as strain inducers and be accounted for any strength loss detected in the final structure, a competitive effect to the internal curing promotion that was previously described [9,10,15,17,21,31,35].

It must be clarified that the overall SAPs behavior in cementitious materials highly depends on the nature of SAPs used and their characteristics such as structure, absorption/desorption behavior, morphology (particles shape, size and size distribution) and dosage in the cementitious matrix, as well as water to cement (w/c) ratio and the incorporation procedure adopted [9,15,17,31,35]. In most cases SAPs absorption/desorption has been examined in extracted or synthetic solutions. Nevertheless, there have also been some studies on the in-situ evaluation of this behavior when SAPs are incorporated in a porous cementitious matrix. The results from these studies revealed that SAPs desorption when in contact with a porous, cementitious material, is effected by the bonding between SAPs particles and the cementitious matrix and is governed by diffusion between SAPs particles and capillary sorption in the matrix [36–38]. Consequently, it is imperative to investigate the correlation between SAPs used and the response of the cementitious system with respect to its mechanical properties and its microstructure.

The vast majority of commercial SAPs are copolymeric networks based on acrylic acid or acrylamide that may or may not have been partially neutralized [10,11,14,15,17,18,20–23,31,35]. Because of the diversity of the parameters that have to be satisfied in the construction applications to enhance their internal curing action, SAP particles should preferentially have a homogeneous spherical shape, size in the submicron area so that the voids left behind after their deswelling are smaller and don't affect the structure mechanical properties and chemical affinity to the cement matrix so that they are more easily dispersed homogeneously in the cement paste [23].

The aim of this work was to examine the impact of the incorporation of tailored SAPs with respect to their structure (hybrid structure consisting of organic core—inorganic shell) on mortars mechanical strength in terms of flexural and compressive strength, microstructure and self-healing behavior. The synthesis and characterization of the incorporated SAPs have been thoroughly discussed in previous authors' work. More specifically, the SAP particles used were spherical in the submicron range based on poly (methacrylic acid) crosslinked with ethylene glycol dimethacrylate which were synthesized via radical polymerization and later encapsulated with CaO-SiO_2 inorganic shell via the sol-gel method [25,39]. The incorporation of SAPs in mortars was conducted in two different dosages, 1% and 2% by weight of cement (bwoc). The results obtained in this work revealed that the flexural strength improved by 3%, while the compressive strength remained practically intact for the mortar/SAP composite materials compared to the control specimens. Moreover, the total and closed porosity of the mortar/SAPs specimens were reduced by about 0.5% and 2.5% for mortar-SAPs-1 and mortar-SAPs-2, respectively, while self-healing behavior was enhanced for both SAPs concentrations (in the case of mortar-SAPs-1 by 60% and in the case of mortar-SAPs-2 by 10% compared with mortar-reference specimens).

The added value of this work resides in the optimization of the SAPs functionality in cementitious-based materials and the improvement of the cementitious materials self-healing properties due to tailored SAPs structure, which are easy to fabricate via the combination of the sol-gel process, radical polymerization and the coprecipitation method. Finally, a new approach for the quantitative evaluation of mortars self-healing behavior was proposed.

2. Materials and Methods

2.1. Materials

Sand grade in accordance to CEN, EN 196-1 standard, cement CEM I 52.5 N and in-house synthesized SAPs were used to manufacture conventional mortar specimens. As mentioned earlier, the SAPs used were synthesized according to previous authors' work [39]. In Figure 1 the novel SAPs structure which consists of a hybrid organic core of poly(methacrylic acid) crosslinked with ethylene glycol dimethacrylate encapsulated with a composite inorganic shell of silicon-calcium oxide, P(MAA-co-EGDMA)@CaO-SiO_2 (Figure 1a,b) is shown. Their size ranges from 190 to 320 nm (Figure 1c) while their maximum water absorbance ratio in cement slurry filtrate is determined 1100% their initial dry weight. Before their incorporation in mortars, SAPs are ground to fine powder form, in order to dismantle agglomerates and enhance their performance.

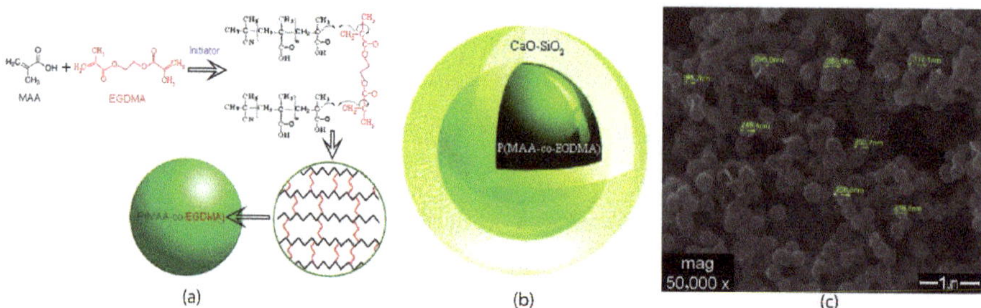

Figure 1. Polymerization reaction to form the organic core P(MAA-co-EGDMA (**a**), schematic representation of organic core—inorganic shell (**b**), SEM image of the synthesized SAPs/P(MAA-co-EGDMA)@CaO-SiO_2 (**c**), revealing the homogenous spherical morphology of SAPs particles in the nanoscale.

2.2. Methods

2.2.1. Manufacture of Mortar/SAPs Specimens

Two specimen series of mortar/SAPs composites were manufactured, prismatic and cylindrical. Prismatic test specimens were fabricated according to the EN 196-1 [40] using a Teflon mold, while cylindrical specimens were using a medical disposable syringe [41–43]. The specimens were cast from a batch of mortar paste with a water/cement ratio (w/c) of 0.50. In accordance to literature review and previous authors' work on cement/SAPs composites [7,10,11,15,16,20,21,23,31,35,39], SAPs were incorporated in mortar in two different dosages, 1 and 2% by weight of cement (bwoc). The SAPs incorporation in the mortar paste was accomplished in succession to the sand fraction and the obtained mortar paste was mixed until the visually monitored dispersion was acceptable. In all cases mixing time needed was more than 30 s. A mortar mixer (MATEST-E094) was used for the dry mechanical mixing of the mixture components and a jolting apparatus (MATEST-E130) to compact the specimens. The molded samples were stored in the moist air room for 24 h and after demolding they were submerged horizontally in a suitable container at 20 °C for 28 days. The composition of the specimens is shown in Table 1.

Table 1. Composition of the mortar specimens.

	Cement (g)	Sand (g)	Water (g)	SAPs (g)
Mortar-reference	450	1350	225	0.0
Mortar-SAPs-1	450	1350	225	4.5
Mortar-SAPs-2	450	1350	225	9.0

2.2.2. Mechanical Properties of Mortar/SAPs Specimens

The flexural strength of mortar/SAPs composites was evaluated using prism specimens with dimensions 40 mm × 40 mm × 160 mm according to EN 196-1 using a universal testing machine of capacity 300 KN (Instron 300DX-B1-C4-G6C) [40]. For each SAPs dosage as well as for control samples, triplicates were manufactured, demolded and tested after 28 days curing in water. The prism halves from the flexural strength tests were used for the compressive strength tests, which were also performed according to EN 196-1.

2.2.3. Microstructure of Mortar/SAPs Specimens

X-ray micro computed tomography (microCT) was utilized to evaluate the mortar/SAPs composites microstructure. This technique is based on the correlation between X-rays absorption, material density and atomic number. High-density materials absorb

X-rays more profoundly and produce light grey projection images. On the other hand, low-density materials are visualized as darker projection images. During the scanning process, angular projections of the specimen were acquired and saved. After the angular projections acquisition was completed, their reconstruction followed. The reconstruction process was executed by the "NRecon" visualization program via the implementation of Feldkamp algorithm. As a result, a 3D reconstructed model of the scanned specimen was produced. Moreover, quantitative parameters were determined using densiometry and morphometry evaluation, the latter based on image segmentation (black and white) which was done via a global threshold method, the Otsu method [44].

In this work, the specimens were scanned using SkyScan 1272 X-ray micro-tomograph at the age of 28 days. The specimens' geometry was chosen to be cylindrical; their diameter was 10 mm and their height ranged between 20 and 30 mm. In order to enhance contrast in microCT images and improve grey scale histogram segmentation iodine was utilized as a contrast agent. Therefore, all specimens were treated with a 3% iodine solution in ethanol prior to their scans. More specifically, at the age of 28 days, the cylindrical specimens were submerged in the iodine solution for 48 h and then, they were dried in an oven at 80 °C for 24 h. If the specimens were not immediately scanned, they were stored in a desiccator [45]. The acquisition settings of the scans are presented in Table 2. In addition, for each scan the flat field correction was applied.

Table 2. Acquisition settings of the X-ray micro-tomography scans.

Acquisition Settings		Value
Source Voltage	kV	60
Source Current	uA	120
Image Pixel Size	um	9
Filter		Al 0.5 + Cu 0.038
Rotation Step	deg	0.2
Rotation	deg	180

The reconstruction and the porosity analysis were performed using NRecon (version 1.6.6.0) and CTAnalyzer (version 1.13) softwares, respectively. The selected volumes of interest were 441, 105 and 350 mm^3 for mortar-reference, mortar-SAPs-1% and mortar-SAPs-2%, respectively. Figure 2 shows representative 2D and 3D reconstructed images of the scanned specimens ((a), (b) mortar-reference; (c), (d) mortar-SAPs-1%; (e), (f) mortar-SAPs-2%).

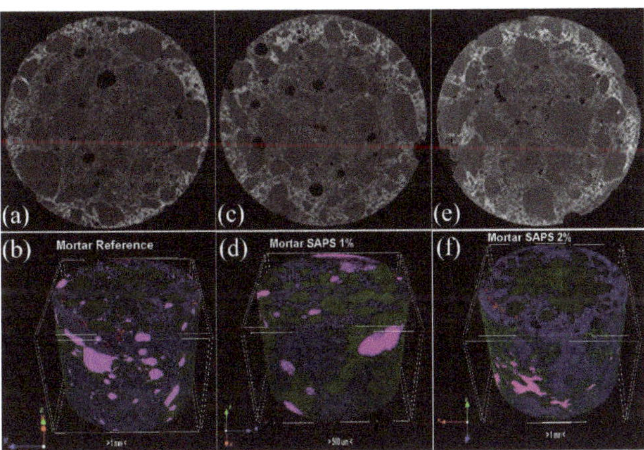

Figure 2. 2D and 3D reconstructed images of the scanned specimens of: mortar-reference (**a**,**b**); mortar-SAPs-1% (**c**,**d**); mortar-SAPs-2% (**e**,**f**).

Moreover, the challenging issues of material phase segmentation and their quantitative analysis were addressed via 2D and 3D reconstructed images data, and more specifically via the grey scale intensity thresholds. In particular, even though microCT does not support chemical analysis, differences in brightness are recorded in the form of Grey Scale Histograms (GSH) due to density and atomic mass variations for different materials. As a result, phase segmentation depends on the thresholding method followed in each case [46]. In this work, GSH were obtained during the scan data processing via CTAnalyzer Software. The Grey Scale Values ranged from "0" to "255". The lower grey scale values correspond to black color in reconstructed images and were attributed to the lack of material or pores. On the other hand, the higher values correspond to white and were attributed to unhydrated cement phases and/or SAPs aggregates, while the intermediary grey scale values are depicted as grey and were attributed to different hydrated cement phases [44,46–50]. In this work, taking into consideration that all hydrated cement phases, mainly Calcium Silicate Hydrates(C-S-H) are expected to show peaks at similar grey scale values, a deconvolution procedure was engaged to identify and quantify the different material phases (pores, hydrated cement phases, unhydrated cement phases and SAPs) assuming that the grey scale values distribution for each phase is Gaussian. Then, a fitting model distribution, comprising a set of four (n = 4) Gauss distributions was numerically fitted to the GSH using an algorithm implemented in Software "Magic Plot" (student version 2.5.1).

2.2.4. Self-Healing Evaluation of Mortar/SAPs Specimens

Cementitious structures can be damaged in a variety of ways, the most common being cracking. MicroCT has been utilized as a laboratory-scale method to evaluate their damage extent and interpret healing mechanisms. In the past more conventional methods, such as SEM (Scanning Electron Microscopy), have been used to evaluate crack morphology and mitigation but they can provide insight only on the specimen surface, whereas microCT can provide useful information on the bulk of the specimen and consequently evaluate internal cracks and internal self-healing [46].

In this work, after 28 days of curing the cylindrical specimens that were previously examined via microCT were precracked under compressive load using a hydraulic press. During their compression, the load was applied in a controlled and smooth manner but the specimens were completely split in two halves in all cases. Therefore, prior to this procedure they were wrapped tightly in a polypropylene based film in order to avoid the complete separation of the two halves but instead to form a crack as shown in Figure 3. Afterwards, the specimen circumference and base were coated with an epoxy resin (a mixture of Sinmast J 158 (component A) and Sinmast S2 liquid primer (component B) by Sintecno in a ratio A:B 77:23) in order to secure the two halves together. The top surfaces of the cylinders were untreated, so that the formed crack could interact with healing agents (water). The epoxy resin was cured at ambient temperature for 1 day and then at 60 °C for 2 h.

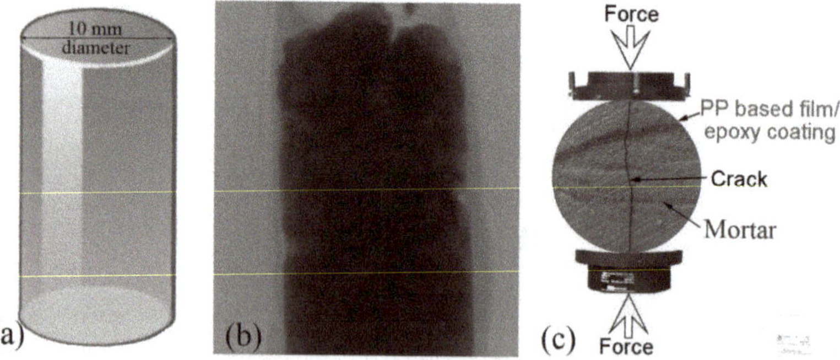

Figure 3. Schematic representation of mortar cylinder (**a**), specimen image from microCT CCD camera (**b**), crack inducement in cylindrical specimens via compression load enforcement (**c**).

In order to estimate their self-healing behavior, the cracked specimens were submerged in water at ambient temperature and they were evaluated at different time slots (0 and 8 days). The self-healing efficiency of mortar/SAPs composites was quantified and visualized via microCT analysis. Prior to microCT evaluation the specimens were dried at 50 °C for three days and were stored in a desiccator. The cracked mortar specimens were scanned using the SkyScan 1272 desktop microCT at 25.0 μm pixel resolution with 0.5 mm aluminum filter. The scanned images were reconstructed via NRecon software, while three-dimensional evaluation was conducted by CTAnalyzer software following a methodology proposed by Nicole Y.C Yu et al., appropriately adjusted for cracked mortar specimens [51]. The main outcomes discussed in this work were the evolution of connectivity density and the percent object volume versus healing time. These parameters were adopted because they allow the quantitative evaluation of crack healing in 3D (not only in 2D), with respect to crack closure in terms of changes in morphology and density and are embedded in the CTAnalyzer software.

3. Results

3.1. Mechanical Properties of Mortar/SAPs Specimens

During the flexural strength test an abrupt, brittle failure was observed in mortars mainly due to deformation localization by the coalescence of narrow microcracks that ultimately lead to macrocracks that expand to the entire specimen [52]. The flexural strength evaluation of the mortar-based composites with SAPs in dosages 1 and 2% bwoc as well as control mortar specimens at the age of 28 days is demonstrated in Figure 4a,b. According to Figure 4b, a slight increment of about 3% can be detected on the flexural strength of the mortar-based composites that contain SAPs in dosages 1% and 2% bwoc, in comparison to the control specimens which practically reveals that the flexural strength remains intact after the SAPs incorporation in both concentrations.

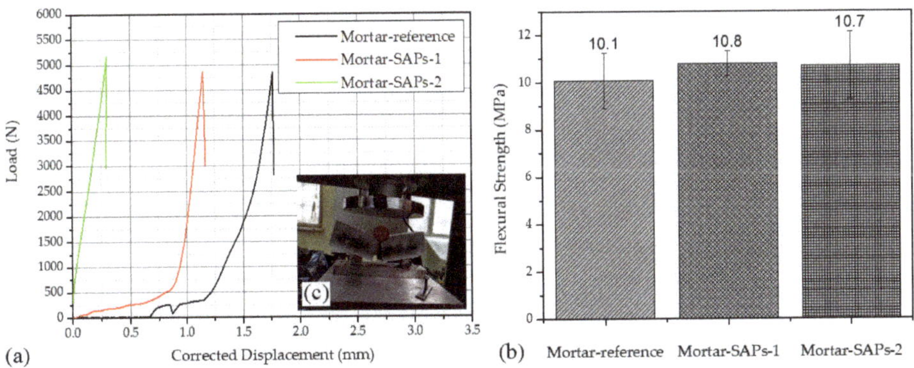

Figure 4. Maximum values of loads for mortar specimens at 28 days of age (**a**), flexural strength of mortar specimens at 28 days of age (**b**), optical image of specimen after flexural test (**c**).

On the other hand, the compressive strength of the mortar based composite materials with the SAPs incorporation for the same mortar age (28 days) and w/c ratio (0.5) is depicted in Figure 5a,b. More specifically, Figure 5a depicts the load applied during the test as a function of the head displacement and Figure 5b shows the calculated compressive strength of the mortar based composite materials with the SAPs incorporation for the same mortar age (28 days) and w/c ratio (0.5). As shown in Figure 5b the compressive strength of the mortar-based composite materials practically remains intact for both SAPs dosages (1% and 2% bwoc) in comparison with the control materials.

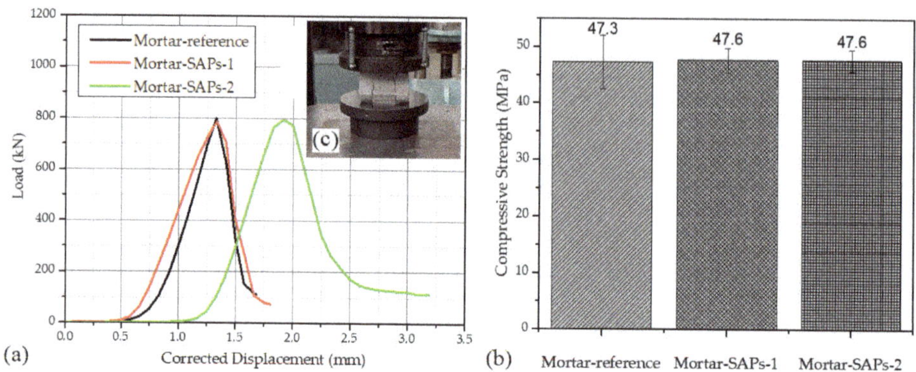

Figure 5. Maximum values of loads for mortar specimens at 28 days of age (**a**), compressive strength of mortar specimens at 28 days of age (**b**), optical image of specimen after compressive test (**c**).

Figure 6 reveals the relationship between the flexural and compressive strength of the mortar-based control and composite materials using the corresponding average values at the age of 28 days. It can be observed that a direct relationship (R^2 = 0.97818) between them exists. Similar behaviors have been reported by other research works for reference mortars [53,54].

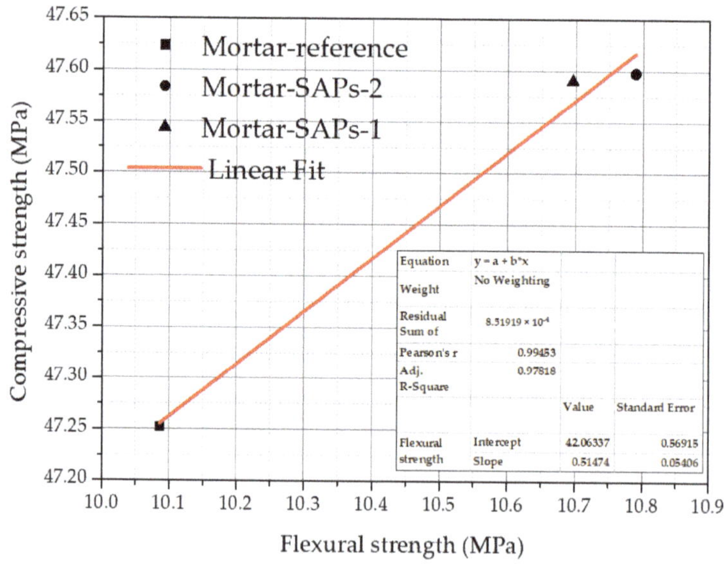

Figure 6. Relationship between compressive and flexural strength of mortar-based materials.

3.2. Microstructure of Mortar/SAPs Specimens

3.2.1. Microstructure Analysis

In this work, the effect of SAPs incorporation on the mortar microstructure in means of porosity and phase segmentation was examined via microCT. In cementitious materials, porosity affects greatly key properties such as mechanical and transport properties [46]. Moreover, the dispersion quality of SAPs in the cementitious matrix can be evaluated through porosity determination. The porosity of cementitious materials is usually divided into gel pores (ranging from a few nanometers to 0.2 μm), capillary pores (ranging from

0.2 to 10 μm), and air voids (above 10 μm) [50,55,56]. In this work, the pore analysis from micro-CT applies only to the pores with diameters larger than 9 μm. The specimen size dictates the minimum distance that can be reached from the X-ray source during scanning and consequently the accuracy of the method. Therefore, the pore analysis from micro-CT scans includes partially capillary pores and the air voids. Air voids are mainly of interest in this work, as in literature, it is reported that the incorporation of SAPs in cementitious matrixes creates air voids around them as a result of the water absorption/desorption by them combined with their poor dispersion in cement and therefore the formation of large SAPs agglomerates [9,10,15,21,31,35,38]. The statistical pore analysis of mortar specimens and the corresponding pore size distribution are exhibited in Table 3 and Figure 7, respectively. The statistical pore analysis parameters calculated are delineated as follows.

Table 3. Statistical pore analysis of mortar specimens.

Property	Unit	Mortar-Reference	Mortar-SAPs-1	Mortar-SAPs-2
Volume of closed porosity	mm^3	10.60	1.87	1.18
Closed porosity	%	2.440	1.800	0.341
Volume of open porosity	mm^3	5.43	1.49	2.54
Open porosity	%	1.230	1.420	0.725
Volume of total porosity	mm^3	16.10	3.36	3.72
Total porosity	%	3.64	3.19	1.06

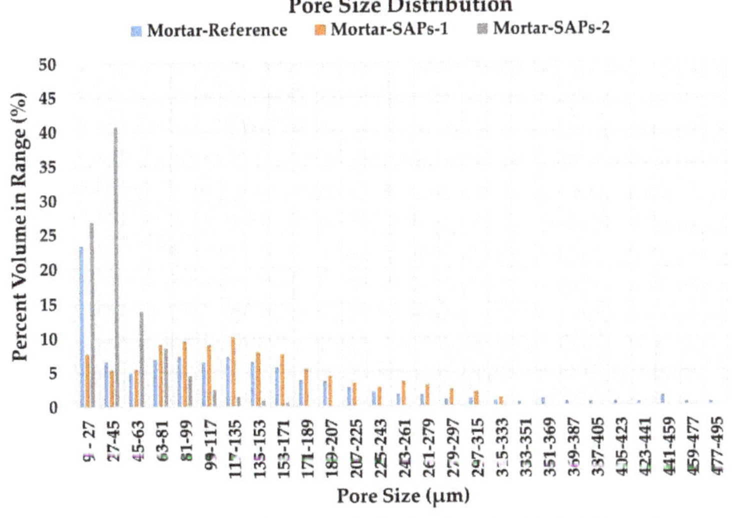

Figure 7. Pore size distribution in mortar-reference, mortar-SAPs-1 and mortar-SAPs-2.

- "Volume of closed porosity" (mm^3) is the total volume of all closed pores within the Volume of Interest (VOI), when a closed pore in 3D is a connected assemblage of space (black) voxels that is fully surrounded on all sides in 3D by solid (white) voxels.
- "Closed porosity" (%) is the volume of closed pores (as defined above) as a percent of the total of solid plus closed pore volume, within the VOI.
- "Volume of open porosity" (mm^3) is the total volume of all open pores within the VOI. An open pore is defined as any space located within a solid object or between solid objects, which has any connection in 3D to the space outside the object or objects.

- "Open porosity" (%) is the volume of open pores (as defined above) as a percent of the total VOI volume
- "Volume of total porosity" (mm^3) is the total volume of all open and closed pores within the VOI
- "Total porosity" (%) is the volume of all open plus closed pores (as defined above) as a percent of the total VOI volume.

3.2.2. Image Segmentation and Phase Identification

There are a number of different approaches on the matter of image segmentation and phase identification via microCT Analysis. One that stands out as it is widely used is the global thresholding method in GSH. The GSH with the Gaussian deconvolution procedure performed for the mortars examined in this work, are given in Figure 8.

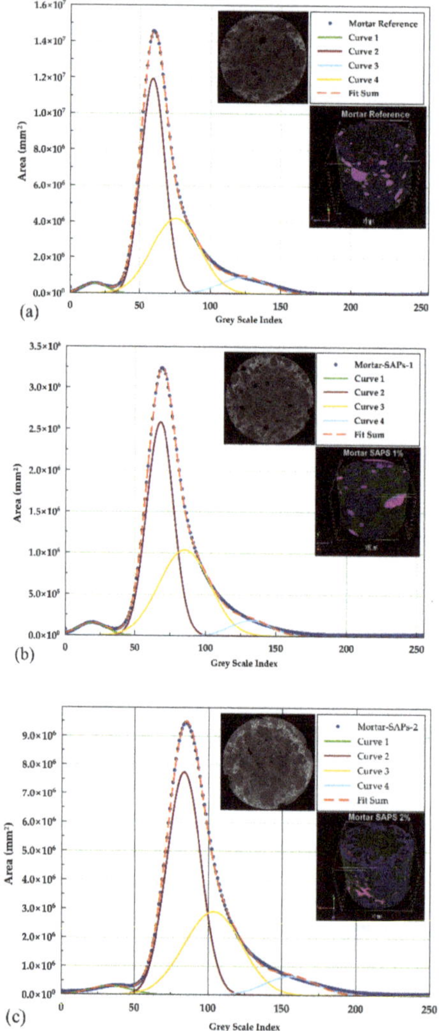

Figure 8. Grey Scale Histogram with the Gaussian deconvolution fit for mortar-reference (**a**), mortar-SAPs-1 (**b**), mortar-SAPs-2 (**c**).

The deconvolution of the initial GSH as described earlier, produced four distinct curves in each case, curves 1 to 4 and the corresponding peaks (Figure 8). Curve 1—Peak 1 were attributed to pores, curve 2 and curve 3 were assigned to hydrated cement phases and curve 4 was ascribed to unhydrated cement phases and SAPs. We observed that for all specimens the main peak in the GSH was deconvoluted in two separate peaks which were attributed to different hydrated products probably owed to the main hydrated cement products which are Calcium Silicate Hydrates, C-S-H and Calcium Hydroxide, $Ca(OH)_2$.

Then, using the GSH data, the quantitative determination of the different phases identified in the mortar specimens was performed and the corresponding results were tabulated in Table 4. Taking into consideration the data in Table 4, it is observed that in the case of mortar-SAPs-1 the unhydrated cement products together with the incorporated SAPs (1% bwoc) represented only the 5% of the total material when the corresponding values for both mortar-reference and mortar-SAPs-2 were 36%. This indicates that the progress of the hydration reactions for this material was remarkably enhanced compared to mortar-reference and mortar-SAPs-2, as a result of SAPs incorporation in the cementitious matrix. This is attributed to more effective SAPs dispersion in the cementitious matrix and therefore enhancement of their functionality to extend hydration reactions during cement curing, thus promoting this as the optimal SAPs concentration in mortars in respect to microstructure evaluation.

Table 4. Quantitative determination of the different cement phases identified in the mortars.

% Phase	Mortar-Reference	Mortar-SAPs-1	Mortar-SAPs-2
Peak 1	2	3	3
Peak 2	54	54	55
Peak 3	8	38	7
Peak 4	36	5	36

3.2.3. Self-Healing Evaluation

In Figure 9 the crack surface and the larger voids in respect to cracks and large holes in the bulk of the specimens immediately after they were cracked (0 days) and after 8 days of healing treatment, are shown for the mortar specimens examined in this work.

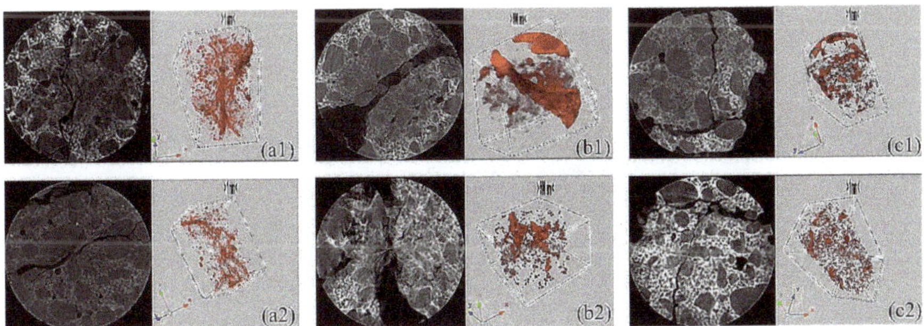

Figure 9. Initial microCT images depicting voids (in respect to both large pores and cracks) in red color for mortar-reference (**a$_1$**), mortar-SAPs-1 (**b$_1$**) and mortar-SAPs-2 (**c$_1$**) and corresponding images after 8 days of healing treatment for mortar-reference (**a$_2$**), mortar-SAPs-1 (**b$_2$**) and mortar-SAPs-2 (**c$_2$**).

MicroCT imaging allows both to visualize the self-healing process, but also to quantitatively analyze it in terms of changes in morphology and density using methods and functions embedded in CTAnalyzer software properly adjusted for mortar specimens. The methodology followed in this work is described as follows.

The region surrounding a crack is rich in products with a vast variety of thicknesses ranging from thick intact mortar to fresh self-healing products that can be thinner or thicker structures. Binarizing or segmenting the structures in this region can therefore be compromised. An effective solution, which has the effect of artificially diminishing the attenuation of thin structures, is the method of adaptive thresholding in CTAnalyzer custom processing procedure. Then, the (Region of Interest) ROI shrink-wrap and stretch over holes functions were performed. As a result, the (Volume of Interest) VOI was wrapped around the boundary of complex and porous objects such as thin self-healing products and the wrapped VOI was prevented from penetrating into the porous spaces of the object but instead only the complex outer margins were marked. Figures 10 and 11 depict the resulting banalization-segmentation images after applying the adaptive thresholding method and shrink-wrap function for mortar-reference, mortar-SAPs-1 and mortar-SAPs-2 specimens at 0 and 8 days of healing treatment, accordingly.

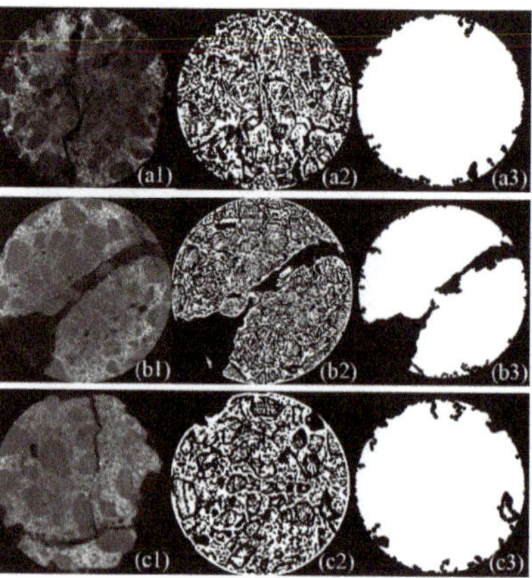

Figure 10. Initial microCT images for mortar-reference (**a$_1$**), mortar-SAPs-1 (**b$_1$**) and mortar-SAPs-2 (**c$_1$**), resulting images for mortar-reference (**a$_2$**), mortar-SAPs-1 (**b$_2$**) and mortar-SAPs-2 (**c$_2$**) and ROI images for mortar-reference (**a$_3$**), mortar-SAPs-1 (**b$_3$**) and mortar-SAPs-2 (**c$_3$**) after adaptive thresholding method and shrink-wrap function are applied before healing treatment.

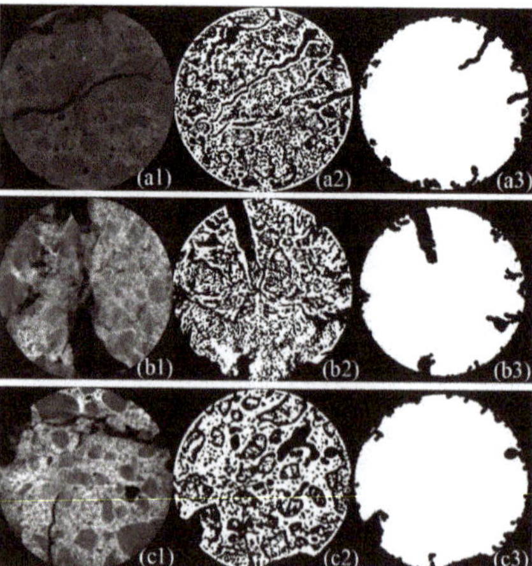

Figure 11. Initial microCT images for mortar-reference (**a$_1$**), mortar-SAPs-1 (**b$_1$**) and mortar-SAPs-2 (**c$_1$**), resulting images for mortar-reference (**a$_2$**), mortar-SAPs-1 (**b$_2$**) and mortar-SAPs-2 (**c$_2$**) and ROI images for mortar-reference (**a$_3$**), mortar-SAPs-1 (**b$_3$**) and mortar-SAPs-2 (**c$_3$**) after adaptive thresholding method and shrink-wrap function are applied after 8 days of healing treatment.

Then, a full 3D analysis was run. The morphometric parameters calculated to quantitatively evaluate self-healing of the cracked mortar specimens were (a) percent object volume (%) and (b) connectivity density (mm^{-3}) [51,57,58]. The first parameter shows the specimen volume variations, while the second one is sensitive to structure complexity changes versus healing time. The corresponding results are shown in Figure 12.

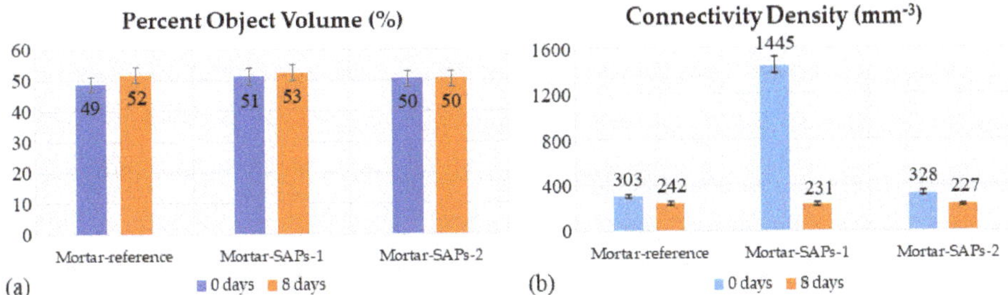

Figure 12. Calculated morphometric parameters to evaluate the self-healing process of mortars. Percent object volume (**a**), connectivity density (**b**). The error bars refer to percent error.

4. Discussion

The overall effect of SAPs incorporation in cementitious materials highly depends on the SAPs properties. In this study, in house synthesized SAPs were used with tailored properties [39]. Firstly, different SAP particle sizes introduce different effects. For the same SAP concentration and w/c ratio, the larger the SAP particle size, the lower the determined strength was in means of flexural and compressive strength. This is directly correlated with the formation of larger SAP voids left behind in the cement structure when SAPs release the amount of water they absorbed during mixing [21]. The size of the SAPs used in this work laid in the submicron scale as shown in Figure 1c, whereas the particle sizes of SAPs tested by other researchers were several micrometers [8–10,15,18,21,23].

Additionally, SAPs chemical constitution triggers different behaviors in the cementitious matrix. As mentioned earlier, most commercially available SAPs are copolymeric networks based on acrylic acid or acrylamide that may or may not have been partially neutralized [10,11,14,15,17,18,20–23,31,35] which present a low affinity with cement and as a result form large aggregates when incorporated in it. In this work, SAPs that have been tested had a hybrid organic core—inorganic shell structure. The introduction of an inorganic coating on the organic SAP core has been proposed by Kanellopoulou I. et al. (2019), via the encapsulation of the polymeric core with an inorganic CaO-SiO_2 shell via solution-gelation technique [31]. Other research groups have investigated the behavior of coated SAPs via the Wurster process but the compensation of the strength reduction achieved was only partial [9].

On the contrary, as indicated by the total porosity reduction and the fact that no mechanical strength degradation was determined with respect to flexural and compressive strength for the mortar/SAPs composite materials, the tailored SAPs in this work presented enhanced chemical affinity with the cementitious matrix. More specifically, due to the enhancement of the chemical affinity between SAPs and cement, large SAP agglomerates were not formed and as a result the size of the SAP voids (porosity) when water was released from their particles were smaller. Additionally, the limited water absorption capacity of the coated SAPs compared to the corresponding capacity of the uncoated particles also favored the size limitation of the SAP voids. The total porosity reduction in both cases of SAPs incorporation in mortars as shown in Table 3 indicates that:

- SAPs were homogeneously dispersed in mortars due to enhanced chemical affinity with the cementitious matrix and

- SAPs promoted hydration reactions in the cement matrix, thus forming denser and consequently more durable structures, since stress inducing points (voids) in the cement matrix were smaller. This is coherent with the mechanical properties behavior discussed later.

The slight flexural strength increment (~3%) of the mortar-SAPs specimens in comparison to the control specimens was evaluated taking into account as corroborating evidence the aforementioned porosity parameters and was attributed to the water absorbed by SAPs during the mixing process, which became available later during cement curing, thus promoting and accelerating hydration reactions. The space surrounding SAP particles was enriched with hydrated cement phases. As a result, mitigation of autogenous shrinkage and early age cracking were observed, while the densification of the cement microstructure led to the enhancement of the mortar/SAPs composites flexural strength [8,52,54,59].

Moreover, compressive strength is considered a key concrete property since it is directly correlated with concrete quality. Nevertheless, it must be clarified that the determined value of each measurement set highly depends on a variety of parameters namely the w/c, additives incorporation and curing conditions, i.e., humidity levels and temperature profiles. Curing conditions can affect drastically the compressive strength of a cementitious material. More specifically, compressive strength degradation has been correlated with low moisture levels during the first day of curing or high temperatures in the initial curing state which is responsible for lower quality hydration products [52]. Furthermore, w/c ratios used in mortars formulation have also been directly correlated with the voids formed in the concrete matrix and consequently with their mechanical strength. More specifically, increased w/c ratios have been known to lead to increased voids in the cementitious matrix and thus to the degradation of mechanical properties [1]. The w/c ratio used in this work (0.5) is high compared to the corresponding ratios examined by other research groups. For example the w/c ratios in cementitious composites with SAPs were 0.30–0.35 in the research of Sun et al. [8], 0.40 and 0.50 in the research of De Belie et al. [9], 0.30, 0.40, 0.50 in the research of Lee et al. [10], 0.35, 0.40 and 0.50 in the research of Farzanian et al. [15] and 0.40 in the research of Kim [54]. Nonetheless, since the effective w/c ratio is not the same as the total w/c ratio, when SAPs are incorporated in mortars, excess of water must be added to counterbalance the amount of water absorbed by SAPs. Considering the dosage of SAPs used in this work (1 and 2% bwoc) the high w/c ratio of 0.50 was chosen. The w/c ratio chosen combined with the curing conditions of the mortar specimens and notably, low moisture during the first day after casting as well as low temperature during the 28 days specimen curing the value of the compressive strength of the control specimens was calculated below 52.5 MPa.

In literature, it is often reported that cementitious materials containing SAPs show poorer compressive performance compared to the corresponding materials without SAPs during all curing periods and for SAPs dosages even lower than those examined in this work, while this behavior became more pronounced for increased SAPs dosages. [1,7,9,10,15,17]. On the contrary, in this work compressive strength was not influenced by the SAPs incorporation in mortars.

The relationship between the determined flexural and compressive strength for the mortar specimens as depicted in Figure 6 was also indicative of the fact that SAPs incorporation in the mortars affected flexural and compressive strength in a similar manner and that their incorporation did not negatively influence the mortar microstructure and hence the mortar strength was not degraded [53,54].

According to the results in Table 3, the percentage of total porosity was reduced by 0.5% and 2.5% in the cases of mortar-SAPs-1 and mortar-SAPs-2, respectively. Moreover, the percentage of open porosity was also reduced in the case of mortar-SAPs-2 compared to mortar-reference by 0.5%. Open porosity represents the pores which are in direct contact with the specimen environment thus allowing humidity and/or other harmful factors (e.g., corrosive factors) to penetrate the bulk of the specimen and cause the material deterioration. As a result, the open porosity reduction in cementitious materials promotes their mechanical integrity and their sustainability.

Additionally, as shown in Figure 7, the size of the pores in the mortar-reference specimens was greater than that of the mortar-SAPs-1 and mortar-SAPs-2 specimens, while the smallest pores were recorded in the case of mortar-SAPs-2. The pore sizes in the mortar-reference specimens showed a wide distribution ranging from 9 to almost 500 μm. On the contrary, the pore sizes in mortar specimens in which SAPs had been incorporated showed a narrower distribution ranging from 9 to 333 μm and from 9 to 171 μm for mortar-SAPs-1 and mortar-SAPs-2, respectively. The more uniform the pore profile the more homogeneous the materials, which is consistent with improved mechanical properties.

Furthermore, the Gaussian deconvolution procedure of the GSH of all the mortar specimens examined in this work, showed that the minimum unhydrated cement phases (5%) were found in the case of the mortar-SAPs-1 specimens, whereas the corresponding phases in mortar-reference (36%) and mortar-SAPs-2 (35%) were almost identical (see Table 4). Firstly, this indicates that SAPs were more homogeneously dispersed in the cementitious matrix for the concentration of 1% bwoc thus allowing them to function as water reservoirs and cement hydration promoting agents and therefore led to denser and more durable cementitious structures. Secondly, it is safe to assume that even greater SAPs concentrations, if needed to render different functionalities to the mortars, will not compromise the mortar behavior.

In mortars, apart from one single crack several other microcracks are formed and distributed in the bulk of the specimens, when they are subjected to compressive loads, which sometimes may not be easy to detect using conventional methods. On the contrary, by running a full 3D analysis via microCT, apart from crack healing, the 3D volume of the specimen was also be thoroughly evaluated. As mentioned earlier, the quantitative analysis of the self-healing progress in pre-cracked mortar specimens took place via microCT Analysis. More specifically, a morphological and an architectural parameter were calculated for all mortar specimens before and after eight days of healing treatment. These parameters were:

- Morphometric parameter, "Percent Object Volume" (%) which shows the percent of the specimen volume variation during the healing treatment. The value of this parameter increases with healing time, since healing products that are formed cover progressively the crack volume in part, thus representing the healing progress with time [51,57,58].
- Architecture parameter, "Connectivity Density" (mm^{-3}) comprising a sensitive indicator of the change in texture and complexity associated with the healing progression. More specifically Connectivity Density is an indicator of complex, very highly porous structures. The value of this parameter decreases with healing time, since crack healing provides denser structures within the crack volume [58,60].

In order to estimate the relative self-healing enhancement of mortar/SAPs specimens versus the control specimens, the relative self-healing enhancement index h_i (%) was calculated [61], based on the results of Connectivity Density since this architecture parameter is very sensitive to texture and complexity changes.

$$h_i(\%) = \left(1 - \frac{CD_{i,n}}{CD_{i,0}}\right) \times 100, \tag{1}$$

where:

h_i: relative self-healing enhancement index for specimen, i
$CD_{i,n}$: value of Connectivity Density for specimen, i after n days of self-healing treatment
$CD_{i,0}$: value of Connectivity Density for specimen, i before self-healing treatment (at 0 days)

More specifically, when no self-healing is observed h_i value is "0", as the final and the initial value of connectivity density are the same. On the other hand, when self-healing is promoted, h_i values become higher and approach the value "100", since the ratio $CD_{i,n}/CD_{i,0}$ approaches "0" as $CD_{i,n}$ decreases. The corresponding values of h_i for the mortar/SAPs and control specimens are shown in Table 5 and it is revealed that the self-healing index was 84% and 31% for mortar-SAPs-1 and mortar-SAPs-2, respectively, while it was 20% for mortar-reference specimen. Comparatively, self-healing was enhanced by about 60% and 10% for mortar-SAPs-1 and mortar-SAPs-2, respectively compared to the mortar-reference specimen. These results reveal that healing efficiency was optimized for the specimen mortar-SAPs-1. This is attributed to the more effective dispersion of SAPs in the cementitious matrix for the SAPs concentration 1% bwoc and therefore the enhancement of their functionality as self-healing agents. This conclusion is in agreement with that drawn by the phase identification in mortars via image segmentation obtained by microCT scans.

Table 5. Indexes h_i for the mortar/SAPs specimens and control specimens.

Specimen	h_i (%)
Mortar-reference	20
Mortar-SAPs-1	84
Mortar-SAPs-2	31

At this point, it must be taken into consideration that the specimen mortar-SAPs-1 was more severely damaged during the artificial inducement of the crack by the application of compressive load, leading to a very high initial value for connectivity density.

Our future goal is to expand this preliminary study on the evaluation of the self-healing behavior of mortars containing the proposed SAPs for longer treatment durations and attain a more profound insight on the healing mechanisms.

5. Conclusions

The presented work, took into consideration the overall behavior of the mortar-SAPs composites manufactured and examined. It also comprises the determination of their mechanical properties, their microstructure evaluation, as well as the evaluation of their self-healing behavior yielding a series of conclusions. Even though it is often reported in literature that the incorporation of SAPs in mortars causes degradation in mechanical properties and specifically in compressive strength [1,7,9,10,15,17], within the manuscript it is shown that, the incorporation of tailored SAPs with respect to their structure (hybrid organic core—inorganic shell structure, spherical shape in the submicron scale) did not negatively influence neither the flexural nor the compressive strength of the mortars. This is directly correlated with the microstructure and porosity evaluation of the mortars, which took place via microCT analysis. In particular, the total porosity was reduced by about 0.5% and 2.5% for mortar-SAPs-1 and mortar-SAPs-2, respectively, while the range of the pore size distribution became narrower for both SAPs concentrations compared to the control specimens. As a result, these SAPs enhanced cement hydration reactions when incorporated in the mortars without introducing more stress inducing sites (macropores left behind in the matrix because of the deswelling of SAPs particles during cement curing) and consequently not compromising the mortars strength. Moreover, the Gaussian deconvolution procedure of the GSH of all the mortar specimens examined in this work, showed that the minimum unhydrated cement phases (5%) after 28 days of aging, were found in the case of mortar-SAPs-1, which also revealed the more pronounced self-healing behavior.

Conclusively, the overall SAPs functionality in cementitious-based materials was optimized while, the SAP concentration of 1% bwoc was promoted as the premium one in reference to mortar composite strength, microstructure and self-healing enhancement.

Author Contributions: Conceptualization, I.A.K. (Irene A. Kanellopoulou) and I.A.K. (Ioannis A. Kartsonakis); methodology, I.A.K. (Irene A. Kanellopoulou); software, I.A.K. (Irene A. Kanellopoulou) and I.A.K. (Ioannis A. Kartsonakis); validation, I.A.K. (Irene A. Kanellopoulou), I.A.K.

(Ioannis A. Kartsonakis) and C.A.C.; formal analysis, I.A.K. (Irene A. Kanellopoulou) and I.A.K. (Ioannis A. Kartsonakis); investigation, I.A.K. (Irene A. Kanellopoulou); resources, I.A.K. (Irene A. Kanellopoulou) and I.A.K. (Ioannis A. Kartsonakis); data curation, I.A.K. (Irene A. Kanellopoulou); writing—original draft preparation, I.A.K. (Irene A. Kanellopoulou) and I.A.K. (Ioannis A. Kartsonakis); writing—review and editing, I.A.K. (Ioannis A. Kartsonakis) and C.A.C.; visualization, I.A.K. (Irene A. Kanellopoulou) and I.A.K. (Ioannis A. Kartsonakis); supervision, C.A.C.; project administration, C.A.C.; funding acquisition, C.A.C. All authors have read and agreed to the published version of the manuscript.

Funding: These results are part of the projects that have received funding from the European Unions' HORIZON 2020 research and innovation program under grant agreement no. 685445 (LORCENIS) and under grant agreement no. 814505 (DECOAT).

Institutional Review Board Statement: Not applicable.

Informed Consent Statement: Not applicable.

Data Availability Statement: Data sharing not applicable.

Conflicts of Interest: The authors declare no conflict of interest.

References

1. He, Z.; Shen, A.; Guo, Y.; Lyu, Z.; Li, D.; Qin, X.; Zhao, M.; Wang, Z. Cement-based materials modified with superabsorbent polymers: A review. *Constr. Build. Mater.* **2019**, *225*, 569–590. [CrossRef]
2. Zhang, W.; Zheng, Q.; Ashour, A.; Han, B. Self-healing concrete composites for sustainable infrastructures: A review. *Compos. Part B Eng.* **2020**, *189*, 107892. [CrossRef]
3. The European Cement Association (CEMBUREAU). *Activity Report*; The European Cement Association (CEMBUREAU): Belgium, Brussels, 2018; pp. 1–53.
4. Sanjuán, M.Á.; Andrade, C.; Mora, P.; Zaragoza, A. Carbon Dioxide Uptake by Mortars and Concretes Made with Portuguese Cements. *Appl. Sci.* **2020**, *10*, 646. [CrossRef]
5. Liu, J.; Shi, C.; Ma, X.; Khayat, K.H.; Zhang, J.; Wang, D. An overview on the effect of internal curing on shrinkage of high performance cement-based materials. *Constr. Build. Mater.* **2017**, *146*, 702–712. [CrossRef]
6. Ferrara, L.; Van Mullem, T.; Alonso, M.C.; Antonaci, P.; Borg, R.P.; Cuenca, E.; Jefferson, A.; Ng, P.-L.; Peled, A.; Roig-Flores, M.; et al. Experimental characterization of the self-healing capacity of cement based materials and its effects on the material performance: A state of the art report by COST Action SARCOS WG2. *Constr. Build. Mater.* **2018**, *167*, 115–142. [CrossRef]
7. Ma, S.; Huang, C.; Baah, P.; Nantung, T.; Lu, N. The influence of water-to-cement ratio and superabsorbent polymers (SAPs) on solid-like behaviors of fresh cement pastes. *Constr. Build. Mater.* **2021**, *275*, 122160. [CrossRef]
8. Sun, B.; Wu, H.; Song, W.; Li, Z.; Yu, J. Design methodology and mechanical properties of Superabsorbent Polymer (SAP) cement-based materials. *Constr. Build. Mater.* **2019**, *204*, 440–449. [CrossRef]
9. De Belie, N.; Gruyaert, E.; Al-Tabbaa, A.; Antonaci, P.; Baera, C.; Bajare, D.; Darquennes, A.; Davies, R.; Ferrara, L.; Jefferson, T.; et al. A Review of Self-Healing Concrete for Damage Management of Structures. *Adv. Mater. Interfaces* **2018**, *5*, 1800074. [CrossRef]
10. Lee, H.X.D.; Wong, H.S.; Buenfeld, N.R. Self-sealing of cracks in concrete using superabsorbent polymers. *Cem. Concr. Res.* **2016**, *79*, 194–208. [CrossRef]
11. Snoeck, D.; De Belie, N. Autogenous Healing in Strain-Hardening Cementitious Materials With and Without Superabsorbent Polymers: An 8-Year Study. *Front. Mater.* **2019**, *6*, 48. [CrossRef]
12. Sisomphon, K.; Copuroglu, O.; Fraaij, A. Application of encapsulated lightweight aggregate impregnated with sodium monofluorophosphate as a self-healing agent in blast furnace slag mortar. *HERON* **2011**, *56*, 13–32.
13. Bentz, D.P.; Weiss, W.J. *Internal Curing A 2010 State-of-the-Art Review*; National Institute of Standards and Technology: Gaithersburg, MD, USA, 2011. [CrossRef]
14. Lee, H.X.D.; Wong, H.S.; Buenfeld, N.R. Potential of superabsorbent polymer for self-sealing cracks in concrete. *Adv. Appl. Ceram.* **2013**, *109*, 296–302. [CrossRef]
15. Farzanian, K.; Pimenta Teixeira, K.; Perdigão Rocha, I.; De Sa Carneiro, L.; Ghahremaninezhad, A. The mechanical strength, degree of hydration, and electrical resistivity of cement pastes modified with superabsorbent polymers. *Constr. Build. Mater.* **2016**, *109*, 156–165. [CrossRef]
16. Snoeck, D. Superabsorbent polymers to seal and heal cracks in cementitious materials. *RILEM Tech. Lett.* **2018**, *3*, 32–38. [CrossRef]
17. Assmann, A. Physical Properties of Concrete Modified with Superabsorbent Polymers. Ph.D. Thesis, Universität Stuttgart, Stuttgart, Germany, 2013.
18. Schröfl, C.; Snoeck, D.; Mechtcherine, V. A review of characterisation methods for superabsorbent polymer (SAP) samples to be used in cement-based construction materials: Report of the RILEM TC 260-RSC. *Mater. Struct.* **2017**, *50*, 197. [CrossRef]

19. Snoeck, D.; Van Tittelboom, K.; Steuperaert, S.; Dubruel, P.; De Belie, N. Self-healing cementitious materials by the combination of microfibres and superabsorbent polymers. *J. Intell. Mater. Syst. Struct.* **2012**, *25*, 13–24. [CrossRef]
20. Snoeck, D.; Dewanckele, J.; Cnudde, V.; De Belie, N. X-ray computed microtomography to study autogenous healing of cementitious materials promoted by superabsorbent polymers. *Cem. Concr. Compos.* **2016**, *65*, 83–93. [CrossRef]
21. Ma, X.; Liu, J.; Wu, Z.; Shi, C. Effects of SAP on the properties and pore structure of high performance cement-based materials. *Constr. Build. Mater.* **2017**, *131*, 476–484. [CrossRef]
22. Mechtcherine, V. Use of superabsorbent polymers (SAP) as concrete additive. *RILEM Tech. Lett.* **2016**, *1*, 81. [CrossRef]
23. Pelto, J.; Leivo, M.; Gruyaert, E.; Debbaut, B.; Snoeck, D.; De Belie, N. Application of encapsulated superabsorbent polymers in cementitious materials for stimulated autogenous healing. *Smart Mater. Struct.* **2017**, *26*, 105043. [CrossRef]
24. Jensen, O.M.; Hansen, P.F. Water-entrained cement-based materials II. Experimental observations. *Cem. Concr. Res.* **2002**, *32*, 973–978. [CrossRef]
25. Pantelakis, S.; Kanellopoulou, I.; Karaxi, E.K.; Karatza, A.; Kartsonakis, I.A.; Charitidis, C.; Koubias, S. Hybrid superabsorbent polymer networks (SAPs) encapsulated with SiO_2 for structural applications. *MATEC Web Conf.* **2018**, *188*, 01025. [CrossRef]
26. Karatzas, A.; Bilalis, P.; Kartsonakis, I.A.; Kordas, G.C. Reversible spherical organic water microtraps. *J. Non Cryst. Solids* **2012**, *358*, 443–445. [CrossRef]
27. Van Tuan, N.; Ye, G.; van Breugel, K.; Copuroglu, O. Hydration and microstructure of ultra high performance concrete incorporating rice husk ash. *Cem. Concr. Res.* **2011**, *41*, 1104–1111. [CrossRef]
28. Wyrzykowski, M.; Ghourchian, S.; Sinthupinyo, S.; Chitvoranund, N.; Chintana, T.; Lura, P. Internal curing of high performance mortars with bottom ash. *Cem. Concr. Compos.* **2016**, *71*, 1–9. [CrossRef]
29. Jozwiak-Niedzwiedzka, D. Microscopic observations of self-healing products in calcareous fly ash mortars. *Microsc. Res. Tech.* **2015**, *78*, 22–29. [CrossRef] [PubMed]
30. Liu, F.; Wang, J.; Qian, X.; Hollingsworth, J. Internal curing of high performance concrete using cenospheres. *Cem. Concr. Res.* **2017**, *95*, 39–46. [CrossRef]
31. De Meyst, L.; Mannekens, E.; Araujo, M.; Snoeck, D.; Van Tittelboom, K.; Van Vlierberghe, S.; De Belie, N. Parameter Study of Superabsorbent Polymers (SAPs) for Use in Durable Concrete Structures. *Materials* **2019**, *12*, 1541. [CrossRef]
32. Ahmed, E.M. Hydrogel: Preparation, characterization, and applications: A review. *J. Adv. Res.* **2015**, *6*, 105–121. [CrossRef]
33. Goulis, P.; Kartsonakis, I.A.; Charitidis, C.A. Synthesis and Characterization of a Core-Shell Copolymer with Different Glass Transition Temperatures. *Fibers* **2020**, *8*, 71. [CrossRef]
34. Krafcik, M.J.; Erk, K.A. Characterization of superabsorbent poly (sodium-acrylate acrylamide) hydrogels and influence of chemical structure on internally cured mortar. *Mater. Struct.* **2016**, *49*, 4765–4778. [CrossRef]
35. Tan, Y.; Chen, H.; Wang, Z.; Xue, C.; He, R. Performances of Cement Mortar Incorporating Superabsorbent Polymer (SAP) Using Different Dosing Methods. *Materials* **2019**, *12*, 1619. [CrossRef] [PubMed]
36. Farzanian, K.; Ghahremaninezhad, A. The effect of the capillary forces on the desorption of hydrogels in contact with a porous cementitious material. *Mater. Struct.* **2017**, *50*, 216. [CrossRef]
37. Farzanian, K.; Ghahremaninezhad, A. Desorption of superabsorbent hydrogels with varied chemical compositions in cementitious materials. *Mater. Struct.* **2018**, *51*, 3. [CrossRef]
38. Farzanian, K.; Ghahremaninezhad, A. On the Interaction between Superabsorbent Hydrogels and Blended Mixtures with Supplementary Cementitious Materials. *Adv. Civ. Eng. Mater.* **2018**, *7*, 20180073. [CrossRef]
39. Kanellopoulou, I.; Karaxi, E.K.; Karatza, A.; Kartsonakis, I.A.; Charitidis, C.A. Effect of submicron admixtures on mechanical and self-healing properties of cement-based composites. *Fatigue Fract. Eng. Mater. Struct.* **2019**, *42*, 1494–1509. [CrossRef]
40. European Committee for Standardization. *Methods of Testing Cement—Part 1: Determination of Strength*; European Committee for Standardization: Belgium, Brussels, 2005; Volume BS EN 196-1, pp. 2–33.
41. Gallucci, E.; Scrivener, K.; Groso, A.; Stampanoni, M.; Margaritondo, G. 3D experimental investigation of the microstructure of cement pastes using synchrotron X-ray microtomography (μCT). *Cem. Concr. Res.* **2007**, *37*, 360–368. [CrossRef]
42. Nestle, N.; Dakkouri, M.; Geiger, O.; Freude, D.; Kaerger, J. Blastrurnace slag cements: A construction material with very unusual nuclear spin relaxation behavior during hardening. *J. Appl. Phys.* **2000**, *88*, 4269–4273. [CrossRef]
43. Wang, Y.C.; Li, Z.Y.; Wang, S.H.; Yang, W.G.; Liu, W.; Li, L.Y.; Tang, L.P.; Xing, F. Analysis methodology of XCT results for testing ingress of substances in hardened cement paste: Explained with chloride immersion test. *Constr. Build. Mater.* **2019**, *229*, 116839. [CrossRef]
44. Zhang, M.; He, Y.; Ye, G.; Lange, D.A.; Breugel, K.V. Computational investigation on mass diffusivity in Portland cement paste based on X-ray computed microtomography (μCT) image. *Constr. Build. Mater.* **2012**, *27*, 472–481. [CrossRef]
45. Deboot, J.T. Investigation of X-Ray Computed Tomography for Portland Cement Phase Quantification. Ph.D. Thesis, Oregon State University, Corvallis, OR, USA, 2018.
46. du Plessis, A.; Boshoff, W.P. A review of X-ray computed tomography of concrete and asphalt construction materials. *Constr. Build. Mater.* **2019**, *199*, 637–651. [CrossRef]
47. MacLeod, A.J.N.; Collins, F.G.; Duan, W.; Gates, W.P. Quantitative microstructural characterisation of Portland cement-carbon nanotube composites using electron and x-ray microscopy. *Cem. Concr. Res.* **2019**, *123*, 105767. [CrossRef]
48. Wyrzykowski, M.; Assmann, A.; Hesse, C.; Lura, P. Microstructure development and autogenous shrinkage of mortars with C-S-H seeding and internal curing. *Cem. Concr. Res.* **2020**, *129*, 105967. [CrossRef]

49. Guntoro, P.I.; Ghorbani, Y.; Koch, P.-H.; Rosenkranz, J. X-ray Microcomputed Tomography (μCT) for Mineral Characterization: A Review of Data Analysis Methods. *Minerals* **2019**, *9*, 183. [CrossRef]
50. Bossa, N.; Chaurand, P.; Vicente, J.; Borschneck, D.; Levard, C.; Aguerre-Chariol, O.; Rose, J. Micro- and nano-X-ray computed-tomography: A step forward in the characterization of the pore network of a leached cement paste. *Cem. Concr. Res.* **2015**, *67*, 138–147. [CrossRef]
51. Yu, N.Y.; Schindeler, A.; Peacock, L.; Mikulec, K.; Fitzpatrick, J.; Ruys, A.J.; Cooper-White, J.J.; Little, D.G. Modulation of anabolic and catabolic responses via a porous polymer scaffold manufactured using thermally induced phase separation. *Eur. Cell. Mater.* **2013**, *25*, 190–203. [CrossRef]
52. Soliman, A.M. Early-Age Shrinkage of Ultra High-Performance Concrete: Mitigation and Compensating Mechanisms. Ph.D. Thesis, University of Western Ontario, London, KY, Canada, 2011.
53. Medeiros, M.H.F.; Helene, P.; Selmo, S. Influence of EVA and acrylate polymers on some mechanical properties of cementitious repair mortars. *Constr. Build. Mater.* **2009**, *23*, 2527–2533. [CrossRef]
54. Kim, M.O. Influence of Polymer Types on the Mechanical Properties of Polymer-Modified Cement Mortars. *Appl. Sci.* **2020**, *10*, 1061. [CrossRef]
55. Pipilikaki, P.; Beazi-Katsioti, M. The assessment of porosity and pore size distribution of limestone Portland cement pastes. *Constr. Build. Mater.* **2009**, *23*, 1966–1970. [CrossRef]
56. Diamond, S. The microstructure of cement paste and concrete—A visual primer. *Cem. Concr. Compos.* **2004**, *26*, 919–933. [CrossRef]
57. Blazejczyk, A. Morphometric Analysis of One-Component Polyurethane Foams Applicable in the Building Sector via X-ray Computed Microtomography. *Materials* **2018**, *11*, 1717. [CrossRef] [PubMed]
58. Panmekiate, S.; Ngonphloy, N.; Charoenkarn, T.; Faruangsaeng, T.; Pauwels, R. Comparison of mandibular bone microarchitecture between micro-CT and CBCT images. *Dentomaxillofac. Radiol.* **2015**, *44*, 20140322. [CrossRef] [PubMed]
59. Sosa, I.; Thomas, C.; Polanco, J.A.; Setién, J.; Tamayo, P. High Performance Self-Compacting Concrete with Electric Arc Furnace Slag Aggregate and Cupola Slag Powder. *Appl. Sci.* **2020**, *10*, 773. [CrossRef]
60. Yu, Y.; Zhang, Y.X. Effect of capillary connectivity and crack density on the diffusivity of cementitious materials. *Int. J. Mech. Sci.* **2018**, *144*, 849–857. [CrossRef]
61. Ferrara, L.; Krelani, V.; Moretti, F. On the use of crystalline admixtures in cement based construction materials: From porosity reducers to promoters of self healing. *Smart Mater. Struct.* **2016**, *25*, 084002. [CrossRef]

Article

Preparation and Characterization of Licorice-Chitosan Coatings for Postharvest Treatment of Fresh Strawberries

Somaris E. Quintana [1,2,*], Olimpia Llalla [1], Luis A. García-Zapateiro [2], Mónica R. García-Risco [1] and Tiziana Fornari [1]

1 Research Institute of Food Science (CIAL), CEI UAM+CSIC, Autonomous University of Madrid, 28049 Madrid, Spain; ollallac@unam.edu.pe (O.L.); monica.rodriguez@uam.es (M.R.G.-R.); tiziana.fornari@uam.es (T.F.)

2 Research Group of Complex Fluid Engineering and Food Rheology, University of Cartagena, Cartagena 130015, Colombia; lgarciaz@unicartagena.edu.co

* Correspondence: somaris.quintana@predoc.uam.es; Tel.: +34-910-017-900

Received: 31 October 2020; Accepted: 24 November 2020; Published: 26 November 2020

Abstract: Several plant extracts are being investigated to produce edible coatings, mainly due to their antioxidant and antimicrobial activities. In this study, licorice root extracts were produced by ultrasound-assisted extraction and were combined with chitosan to elaborate edible coatings. Different solvents and temperatures were used in the extraction process, and the antioxidant and antimicrobial activity of the extracts were assessed. The most bioactive extracts were selected for the development of the edible coatings. The rheological properties of the coatings were studied, and they were applied on strawberry to evaluate their physicochemical and microbiological properties. The addition of licorice extract to chitosan resulted in positive effects on the rheological properties of the coatings: the incorporation of phytochemicals to chitosan decreased the shear stress and improved the restructuring ability of the coating solutions. The films presented a reduction of the Burger model parameter, indicating a reduction of rigidity. Furthermore, the strawberry coated with chitosan and licorice extract maintained good quality parameters during storage and showed the best microbiological preservation in comparison with controls. Hence, the use of chitosan with licorice extract is a potential strategy to produce edible coating for improving the postharvest quality of fruits.

Keywords: licorice; chitosan; edible coating; strawberry shelf life; rheological properties; ultrasound-assisted extraction

---

1. Introduction

An increase in consumer requirements for safe food has led to the development of new improved packaging systems, including active, intelligent and edible materials. The use of edible biopolymers in food-packaging applications has become an alternative due to their film-forming properties and environmentally friendly behavior [1]. Edible coatings and films are different: edible films are used as wrapping packaging materials, while edible coatings can be directly applied on the surface of food products [2]. Edible coatings are natural, safe and ecologically friendly substitutes, suitable to be applied to reduce water transfer, gaseous exchange and oxidation of fresh products [3]. The edible coatings should preserve the quality, nutritional value and texture of food products by reducing moisture loss and oxygen effects, while maintaining adherence, without altering the original taste and odor [4]. Furthermore, the combination of natural food grade substances in the coating should improve the physical properties of the formed films [5]. In fact, the rheological properties of the film-forming solution, such as thickness, spreadability and uniformity of the liquid coating layer, and the film performance, can be significantly affected by the type and composition of the coating constituents, such as polysaccharides, protein, hydrocolloid or composite materials. In general, a reduction in the solution's viscosity provides a processing advantage during high-shear processing operations, whereas high apparent viscosity at low-shear rates provides a better application by dipping [4].

Chitosan is a cationic biopolymer that can be mainly obtained from the deacetylation process of chitin, a renewable natural resource [2] which is widely used in the coating of fresh products. Biological properties have been attributed to chitosan, such as antibacterial and antifungal properties [6]. Then, the application of this substance has been applied in several areas such as agriculture, cosmetic, pharmacy and also food industry [7].

Chitosan has been studied and used for coating fish, meat, fruits and vegetables, to improve the quality and extend the food shelf life [8]. In addition, chitosan can be combined with other functional compounds with antioxidant, antimicrobial or other activities, to enhance the quality and efficacy of the coating [9]. Antioxidants, antimicrobial agents, coloring agents, flavors, nutrients, prebiotics and probiotics are examples of functional compounds that can be incorporated into edible chitosan-based coatings [10]. Particularly, for the case of strawberry (*Fragaria × ananassa*), some works studied the application of chitosan-based coatings [11,12] but a relatively high moisture permeability was reported [2]. Then, the incorporation of natural bioactive compounds such as nisin, natamycin, pomegranate, grape seed extracts [13], aloe vera gel [14], *Thymus capitatus* [15] and *Mentha spicata* essential oils [16] and green tea extract [17], were studied, with the aim of enhancing the performance of the edible coating and the preservation of strawberry. This fruit has been intensively studied, because it is highly consumed [18] and is a relevant source of bioactive compounds such as vitamin C, vitamin E, β-carotene and anthocyanins [19]. Nevertheless, several factors can reduce the fruit quality and limit its commercialization, such as intrinsic physiology, physiological and mechanical lesion, fungal infections and postharvest decay [20–22].

On the other hand, licorice is a traditional therapeutic herb, which grows in various parts of the world, and is well recognized due to its bioactive properties, including antioxidant, anti-fungal, anti-ulcer, anti-inflammatory, as well as anticancer and antiviral. Licorice was commonly used in traditional medicine recipes for the treatment of respiratory complaints, inflammatory disorder and liver problems [23,24]. Accordingly, licorice extracts, particular those obtained from the root, were studied as natural food preservatives. Jiang et al. [25] investigated the efficacy of the addition of licorice extract in the preparation of meat patty to inhibit lipid oxidation during refrigerated and frozen storage. Qui et al. [26] combined chitosan citric acid and aqueous licorice extract to preserve the quality of sea bass (*Lateolabrax japonicas*) fillets by preventing lipid oxidation and microbial growth and thus extending the fish shelf life. Mandanipour et al. [27] evaluated the influence of an ethanolic licorice extract combined with chitosan for controlling blue mold and extending the shelf life in the postharvest storage of apples.

Different methods have been employed for the production of extracts. Among these, the ultrasound-assisted extraction (UAE) is a versatile, flexible and efficient technique employed for the extraction of bioactive compounds [28] due to its high reproducibility, and it is very appreciated for the reduction of solvent consumption [29]. Application of ultrasounds causes the implosion of the solvent cavitation bubbles, leading to the disruption of the vegetal cell membranes. This action facilitates the penetration of solvent into the cells, thereby improving mass transfer and increasing the releasing bioactive compounds.

In this study, ultrasound-assisted extraction of licorice root was studied, using solvents of different polarities, to examine the bioactive composition and the antioxidant and antimicrobial activity of the extracts. Selected extracts were tested for the elaboration of chitosan-based edible coatings. The rheological properties of the coating solutions were studied to assess their ability to form films. Furthermore, the rheological properties of the films were analyzed in order to determine the influence of the incorporation of licorice extract. Finally, a study to evaluate the effect of the application of the edible coatings on strawberry was accomplished, considering the physicochemical and microbiological characteristics of the berries during storage.

2. Materials and Methods

2.1. Chemicals and Fruit

Sodium carbonate anhydrous (99.5% purity) and ethanol (99.5% purity) were purchased from Panreac (Barcelona, Spain). Folin–Ciocalteu reagent, Gallic acid standard (>98% purity), (±)-6-hydroxy-2.5.7.8-tetramethyll-chromane-2-carboxylic acid (Trolox, 97% purity), 2.2-diphenyl-1-pycrilhydrazyl (DPPH, 95% purity), chloramphenicol (≥98% purity), low molecular weight chitosan deacetylated chitin, tween-20, tween-80, acetic acid (≥99.5% FCC, Food Grade), glycerol (>99%, FCC, FG), isoliquiritigenin, liquiritin, glycyrrhizic acid ammonium salt, liquiritigenin and glabridin, were purchased from Sigma–Aldrich (St. Louis, MO, USA). BBL Mueller Hinton II Broth and Difco Wilkins-Chalgren Agar were purchased from Becton, Dickinson and Company (France). Sodium hydroxide (ACS, Reag. Ph Eur, ISO) was purchased from EMSURE®, Merck (Germany).

Strawberries were purchased from a local market. They were brought to the lab for the experimental studies and used immediately. Strawberries were selected for uniformity of color, shape and size and with the absence of physical defects or decay.

2.2. Preparation of Licorice Extracts

Roots of licorice were obtained from Murciana herbalist's (Murcia, Spain) and ground using a Premill 250 hammer mill (Lleal S.A., Granollers, Spain). Then, UAE using an ultrasonic probe (Branson Digital Sonifier 550 model, Danbury, CT, USA) with an electric power of 550 W and frequency of 60 kHz was accomplished. The extractions were carried out for 15 min with a plant/solvent ratio of 1:10, at temperatures of 25 and 50 °C. Four different solvents were used: ethanol, methanol, ethanol:water (50:50) and methanol:water (50:50) (Table 1). The extracts obtained were stored at −20 °C until their use.

Table 1. Extraction yields (% mass extracted related to the mass of licorice root) of licorice root ultrasound-assisted extraction (UAE).

Sample	Solvent	Temperature °C	Yield %
ELe25	Ethanol	25	3.85 ± 0.46^{d}
MLe25	Methanol	25	9.08 ± 2.77^{c}
EWLe25	Ethanol:water (50:50 *v/v*)	25	25.14 ± 1.07^{a}
MWLe25	Methanol:water (50:50 *v/v*)	25	15.69 ± 2.93^{b}
ELe50	Ethanol	50	3.69 ± 0.14^{d}
MLe50	Methanol	50	9.94 ± 2.08^{c}
EWLe50	Ethanol:water (50:50 *v/v*)	50	24.75 ± 0.76^{a}
MWLe50	Methanol:water (50:50 *v/v*)	50	23.36 ± 0.18^{a}

Values with different letters ($^{a-d}$) differ significantly ($p < 0.05$).

2.3. Content of Total Phenolic Compounds (TPC) and Antioxidant Activity

The content of TPC of licorice extract was determined following the method of Folin–Ciocalteu [30]. Initially, 10 μL of extract was mixed with 600 μL of milliQ water and 50 μL of Folin–Ciocalteu reagent. The content was thoroughly mixed and after 3 min, 150 μL of sodium carbonate solution (20% mass) and 190 μL of milliQ water were added to the mixture. After 2 h at room temperature in darkness, the absorbance was measured at 760 nm using a Genesys 10S UV-Vis spectrophotometer (Thermo Fischer Scientific Inc., MA, Waltham, USA). The results were expressed as GAE (mg of gallic acid equivalents/g of extract).

The ability of licorice extracts to scavenge DPPH free radicals was determined according to the method described by Brand-Williams et al. [31]. For the reaction, 25 μL of samples were added to 975 μL of DPPH radical in ethanol (6.1×10^{-5}), which was prepared daily. The reaction took place at room temperature, in the dark, until it reached a plateau. Then, the absorbance was measured at 515 nm in a Genesys 10S UV-Vis spectrophotometer (Thermo Fischer scientific, Waltham, MA, USA). The DPPH concentration in the reaction medium was calculated from a calibration curve (linear regression). A control sample, containing the same volume of solvent instead of the extract, was used to measure the maximum DPPH absorbance. Trolox was used as a reference standard, so results were expressed as TEAC values (μmol Trolox/g extract). All analyses were done in triplicate.

2.4. Determination of Antimicrobial Activity of Licorice Extracts

The licorice extracts were tested against a Gram-positive bacterium, *Staphylococcus aureus* ATCC 25923, and a Gram-negative bacteria, *Escherichia coli* ATCC 25922, following the methods described by Quintana et al. [32]. These tests were carried out to assess a general comparison of the antimicrobial capacity of the extracts and decide which of them could be the most effective for the edible coating. A broth microdilution method was used for the determination of IC_{50} values (i.e., the concentration required to obtain 50% inhibition of growth after 24 h of incubation at 37 °C). All tests were performed in Mueller–Hinton broth supplemented with 0.5% tween-20. The inoculum of bacterial strains was prepared from overnight Mueller–Hinton broth cultures at 37 °C. Test strains were suspended in Muller–Hinton (bacteria) broth to give a final density 10^7 CFU(Colony forming units)/mL. The extract and fractions were diluted in ethanol ranging from 1 to 50 mg/mL. The 96-microwell plates were prepared by dispensing into each well 185 μL of culture broth, 10 μl of the different extract's dilutions, antibiotic solution (chloramphenicol as positive control) or solvent (ethanol as negative control), and 5 μL of the inoculums. In addition, blanks were prepared by adding 190 μL of broth medium to the solvent or licorice extracts wells. The final volume of each well was 200 μl. After dispensing the inoculum, the plates were read in an Infinite 200 PRO plate reader (TECAN, Trading AG, Männedorf, Switzerland) spectrophotometer at 620 nm for T0 (Zero Time). Then, the plates were incubated at 37 °C for 24 h and the absorbance was read for TF (Final Time). Each test was performed in triplicate and repeated twice.

2.5. HPLC Analysis

High Performance Liquid Chromatography (HPLC) analysis was performed as previously described by the authors [33] to identify and quantify main bioactive compounds of licorice, i.e., isoliquiritigenin, liquiritin, glycyrrhizic acid, liquiritigenin and glabridin. A Prominence-i LC-2030C 3D Plus (Shimadzu) unit equipped with a quaternary solvent delivery system, an autosampler, diode-array detection (DAD) detector and a RP-C18 (250 × 4.6 mm; 3 μm) column was used. The column temperature was set at 25 °C. The mobile phase was 0.026% aqueous H_3PO_4 (*v*/*v*) (A) and acetonitrile, applying the following gradient elution: 80–75% (0–20 min), 75–66% (20–30 min), 66–50% (30–50 min), 50–40 (50–60min) and 40% (60–80min) of A. After 5 min, the initial conditions were achieved. The flow rate was 0.7 mL/min and was kept constant during analysis. Injection volume was 20 μL and detection was accomplished at 254, 280 and 370 nm. Calibration curves with the standards were used to determine the content of these bioactive compounds in the different extracts.

2.6. Preparation of Coating Forming Solutions

Edible coatings were prepared with chitosan and the addition of different amounts (1% and 5%) of the licorice extract (LE). Chitosan solution was prepared following the procedures described by Ali et al. [34] with some modifications. Briefly, 4.0 g of chitosan was dissolved in 200 mL of distilled water containing 1.0 mL of glacial acetic acid. The solution was heated with constant stirring for 12 h. The pH of the solution was adjusted to 5.5 with 1 N NaOH, and 0.2 mL of tween-80 was added as emulsifier. Then, 1% or 5% of licorice extract was added and homogenized using an Ultra Turrax homogenizer (T18 basic IKA, Staufen, Germany) at 7500 rpm for 3 min.

2.6.1. Rheological Properties of Coating Solutions and Films

Rheological assay was done in order to evaluate the steady and viscoelastic properties of edible coating solutions employing a modular advanced rheometer system Mars 60, Haake (Thermo-Scientific, Karlsruhe Germany), equipped with a coaxial cylinder (inner radius 12.54 mm, outer radius 11.60 mm, cylinder length 37.6 mm).

Steady-State Shear Test

The continuous shear test was performed at a temperature of 25 °C, over a shear rate in the range of 10^{-3} to 10^{3} s^{-1}, to study the influence of the natural extract on the rheological behavior. Experimental data were fitted to the rheological Power Law model, according to Equation (1):

$$\sigma = K\dot{\gamma}^{n} \tag{1}$$

where σ is the shear stress (Pa), $\dot{\gamma}$ is the shear rate (s^{-1}), K is the consistency coefficient ($Pa{\cdot}s^{n}$) and n is the flux index (dimensionless).

Oscillatory Test

The stress amplitude sweep test was carried out within the range of 0.01 to 1000 Pa and with angular frequency of 0.1 Hz on all samples, in order to determine the linear viscoelastic regime (LVR). Based on the results of these experiments, the frequency sweep was done at 0.1 Pa, to keep the stress in the LVR, within the range of 0.01 to 100 rad·s^{-1}. Thermo-viscoelasticity properties were investigated in a ramp temperature from 20 to 80 °C, under constant frequency (0.1 Hz) in the LVR and at a heating rate of 5 °C/min. Dynamic data were obtained in oscillatory shear experiments. The data recorded include the storage modulus (G′) which provides the elastic component, the loss modulus (G″) which is related to the viscous component of the material and the loss tangent (Tan δ) which is the ratio G″/G′ and provides the ratio of elastic to viscous response of the material under consideration.

Preparation of the Film

In order to evaluate the influence of the addition of licorice extracts to chitosan on the ability of the coating solutions to form films, the casting method [35] was used to prepare the films. The coating solutions were casted in sample holder followed by drying at 47 °C for 12 h. The dried films were peeled and stored in a desiccator containing silica gel to prevent moisture absorption. Films of 0.5 mm of thickness were obtained.

Creep and Recovery

The rheological measurement of films was conducted using a parallel plate geometry (diameter 35 min, gap 0.5 mm). The storage modulus (G′) and loss modulus (G″) were measured from 0.1 to 20 rad·s^{-1} at 25 °C in the LVR. Creep and recovery tests were performed at 25 °C for each sample (with the same size). The tests were recorded at constant stress amplitude at 25 °C for 180 s, followed by release of the stress and recovery for 180 s. Creep curves were analyzed according to the Burgers model with one Kelvin-Voigt element [36] (Equation (2)):

$$J = J_o + J_1\left(1 - exp\left(\frac{-t}{\lambda_{ret}}\right)\right) + \frac{t}{\mu_o} \tag{2}$$

where J is compliance, t is the time, J_o is the instantaneous compliance (Pa^{-1}), J_1 is the retarded compliance (Pa^{-1}), λ_{ret} is the retardation time of Kelvin-Voight element (s) and μ_o is Newtonian viscosity (Pa·s).

2.7. Application of Coatings on Strawberry and Quality Parameters

The strawberries were dipped into three different edible coating samples (chitosan, chitosan + 1% LE, and chitosan + 5% LE) for 1 min. Then, the fruits were air-dried, packed in commercial corrugated boxes and stored at 4.0 ± 1 °C. Uncoated strawberries were used as control. Twenty-five berries for each coating treatment were used and the experiments were performed in duplicate. Quality characteristics of control and coated fruits were determined during storing at 4.0 °C.

2.7.1. Fungal Decay Percentage

Strawberries were visually evaluated for the presence of mold growth during the storage time (10 days). Any strawberry with visible spoilage was considered to be decayed. Fungal decay percentage was calculated by using the following equation:

$$Fungal\ decay\ (\%) = \frac{number\ of\ decayed\ fruits}{total\ number\ of\ fruits} \times 100 \tag{3}$$

2.7.2. Determination of Weight Loss

Strawberries were weighed just after coating and air drying. Then, berries' weight was monitored during 10 days after coating. Weight loss was calculated as the percentage of loss related to initial weight (Equation (4)):

$$Weight\ loss\ \% = \frac{Initial\ weight - Final\ weight}{Initial\ weight} \times 100 \tag{4}$$

2.7.3. Determination of pH and Titratable Acidity (TA)

Following the procedures describe by the Association of Official Agricultural Chemist -AOAC method [37], the pH was measured using a pH-meter. TA was determined by titrating the diluted juice (5 g fruit diluted in 95 mL distilled water) up to pH 8.2 using 0.1N NaOH. The results were expressed as the percent of citric acid (Equation (5)):

$$\mathrm{TA} = \frac{\mathrm{V(NaOH)} \cdot 0.1 \cdot 0.064}{\mathrm{m}} \times 100\ (\%) \tag{5}$$

where V(NaOH) is the volume (mL) of NaOH consumed for titration, 0.1 is the molarity of the NaOH solution, 0.064 is a conversion factor for citric acid and m is the mass of the aliquot sample taken for analysis. Measurements were done in triplicate.

2.7.4. Content of Total Phenolic Compounds (TPC)

Strawberries' pulp was finely chopped and 5 g of it was extracted and homogenized with 10 mL of methanol. After a cleaning-up step via centrifugation (5 min at 4500 rpm and 25 °C) and filtration, the supernatants were recovered and allowed to stand at room temperature for evaporation of solvent. The experiments were done in triplicate. Total phenolic content of the extracted pulp was determined using the Folin–Ciocalteu method [30].

2.7.5. Microbiological Analysis of Strawberries

The microbiological analysis of strawberries during storage was measured according to Haiji et al. [38] with slight modifications. The mesophilic and psychrophilic bacteria and yeast count were considered the most comprehensive method to evaluate the microbiological quality of the strawberries. On day 1 and until day 10 of storage, 10 g of fruit was aseptically transferred into stomacher bags, and 100 mL of sterile saline solution was added to each sample. Serial decimal dilutions of the homogenized sample were prepared. Determination of total aerobic bacteria was carried out on Plate Count Agar (37 °C for 48 h). The enumeration results were signified as log CFU (colony forming units)/g.

2.8. Statistical Analysis

Two replicates were prepared for each coating treatment and for each day. Each sample was measured in triplicate. All data were expressed as mean ± standard error, and the statistical analysis of data was performed using R software version 3.6.2. The significant difference between the treatments (three coating treatments) and days were determined using one-way analysis of variance (ANOVA) with Tukey's HSD (honestly-significant difference) test, grouping at the 95% confidence level.

3. Results and Discussion

3.1. Extraction Yield of Licorice Extracts and Quantification of Bioactive Compounds

Extraction yields are reported in Table 1, together with the deviations obtained in duplicate experiments. Taking into account the solvents used, extraction yield increased in the following order: ethanol, methanol, methanol:water (50:50) and ethanol:water (50:50). Based on the dielectric constants of the solvents (ethanol 24.3, methanol 32.7 and water 78.4, at 25 °C), it can be established that the addition of 50% water to the organic solvents and thus, the increase of solvent polarity, resulted in a significant increase of extraction yield ($p < 0.05$). Nevertheless, extraction temperature did not affect yield significantly ($p > 0.05$), at least in the range of temperatures explored (25 and 50 °C).

While the addition of 50% water to methanol produced nearly a 2-fold increase of extraction yield, the addition of 50% water to ethanol resulted in a 6.5-fold yield increase. Then, the combination of a medium polar solvent (ethanol) with a high polar solvent (water) was the most suitable alternative to obtain high yields in the extraction of licorice root. These results are in accordance to other reported extraction studies, such as the work of Nieto et al. [39] concerning the extraction of grape stems, Arranz et al. [40] in the extraction of marjoram with different ethanol:water mixtures, or Kaderides et al. [41] in the UAE extraction of pomegranate peels, reporting an increase in the extraction yield with an increase in solvent polarity. Moreover, it was found that the combination of solvents is more efficient for extraction of phenolic compounds than a single solvent [42].

The main well-known bioactive compounds of licorice (glycyrrhizin, glabridin, liquiritigenin, isoliquiritigenin and liquiritin) were identified and quantified by HPLC analysis and the results obtained are reported in Table 2. Values in the range of 79.0 to 157.9 mg of these bioactive compounds were determined per gram of the different extracts obtained. Except in the case of ethanolic extracts, the higher concentrations were determined for glycyrrhizin, followed by glabridin, and in considerably lower concentrations, liquiritigenin, isoliquiritigenin and liquiritin.

In the case of ethanolic extracts, the concentration of glabridin was higher than glycyrrhizin, while significantly lower concentrations were observed for the rest of the components. Glabridin, an isoflavan, and glycyrrhizin, a triterpene glycoside, are the most abundant and distinctive compounds of licorice roots with several recognized pharmacological properties [43,44]. While glabridin is nearly insoluble in water, the glycosylated sugars present in glycyrrhizin provides some polarity to this compound and thus is better extracted using water. Figure 1 shows the concentration of these compounds in the extracts (mg/g) obtained at 25 °C as a function of the dielectric constant of the solvent used. In the case of water:organic solvent mixtures, the dielectric constant was estimated as the weighted average of the pure solvent dielectric constants. As it can be observed in Figure 1, the extraction of glycyrrhizin increased considerably as the solvent dielectric constant increased, while the opposite effect is observed for the non-polar compound glabridin. Similar results were observed in the case of licorice root UAE extracts obtained at 50 °C.

Table 2. Concentration (mg/g extract) of bioactive compounds identified in licorice UAE extracts (High Performance Liquid Chromatography—HPLC analysis).

Sample	Liquiritin	Glabridin	Glycyrrhizin	Isoliquiritigenin	Liquiritigenin
ELe25	0.60 ± 0.01 ^{a}	55.80 ± 5.03 ^{a}	17.30 ± 1.56 ^{c}	0.87 ±0.03 ^{a}	0.95 ± 0.10 ^{c}
MLe25	0.60 ± 0.01 ^{a}	21.35 ± 0.72 ^{b}	53.45 ± 0.85 ^{b,c}	0.60 ± 0.01 ^{a}	2.00 ± 0.01 ^{b}
EWLe25	0.70 ± 0.07 ^{a}	2.73 ± 0.12 ^{c}	159.50 ± 12.51 ^{a}	0.40 ± 0.01 ^{a}	3.95 ± 0.30 ^{a}
MWLe25	0.80 ± 0.01 ^{a}	1.00 ± 0.02 ^{c}	123.85 ± 10.11 ^{a,b}	0.40 ± 0.01 ^{a}	3.30 ± 0.35 ^{a,b}
ELe50	1.10 ± 0.12 ^{a}	55.13 ± 2.77 ^{a}	13.60 ± 0.23 ^{c}	0.60 ± 0.01 ^{a}	1.00 ± 0.01 ^{c}
MLe50	1.00 ± 0.06 ^{a}	24.40 ± 0.90 ^{b}	96.80 ± 4.27 ^{b,c}	1.60 ± 0.09 ^{a}	2.60 ± 0.22 ^{b}
EWLe50	0.60 ± 0.01 ^{a}	2.45 ± 0.03 ^{c}	145.10 ± 10.23 ^{a}	0.40 ± 0.01 ^{a}	3.45 ± 0.30 ^{a}
MWLe50	0.80 ± 0.01 ^{a}	0.80 ± 0.07 ^{c}	133.40 ± 15.24 ^{a,b}	0.40 ± 0.00 ^{a}	3.60 ± 0.46 ^{a,b}

Values with different letters ($^{a–c}$) in the column and row differ significantly ($p < 0.05$).

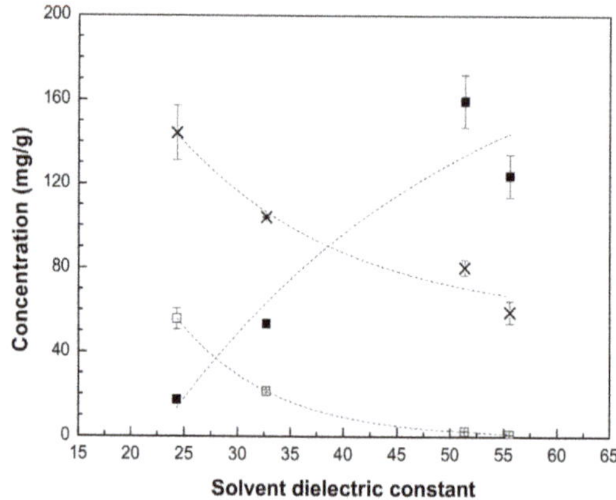

Figure 1. Concentration of glabridin (□), glycyrrhizin (■) and (X) total phenolic compounds (TPC) in UAE licorice extracts as a function of the estimated dielectric constant of the solvent (extraction temperature 25 °C); (—) trend line.

3.2. TPC, Antioxidant and Antimicrobial Activity of Licorice Extracts

3.2.1. TPC and Antioxidant Activity

The TPC of licorice extracts (mg GAE/g), together with their antioxidant (DPPH test) and antimicrobial activities (IC_{50} values), are reported in Table 3. The extracts with the higher content of TPC were obtained with ethanol (ELe25 and ELe50), followed by methanol (MLe25 and MLe50), ethanol:water (EWLe25 and EWLe50) and methanol:water (MWLe25 and MWLe50), with values in the range 60–160 mg GAE/g extract. Figure 1 shows that the TPC values of the extracts obtained at 25 °C decreased when the dielectric constant of the solvent increased. Similar results were obtained at the extraction temperature of 50 °C. Then, it can be established that the addition of water to the organic solvents (methanol or ethanol) enhanced the extraction yields, while the TPC concentration decreased, maybe due to the extraction of other polar compounds, as suggested by Spingo et al. [45]. For example, the glycyrrhizic acid (GA) molecule has several hydroxyl groups and thus, it is easily soluble when extracted with polar solvents [46]. Then, higher concentrations of GA were obtained in the extracts produced using ethanol:water (EWLe25 and EWLe50) and methanol:water (MWLe25 and MWLe50).

Table 3. Total phenolic compounds (TPC), antioxidant activity (DPPH free radicals scavenging assay expressed as TEAC value) and antimicrobial activity (IC_{50} mg/mL) of licorice UAE extracts.

Sample	TPC (mg GAE/g)	DPPH (μmol Trolox/g)	*E. coli* IC_{50} (mg/mL)	*S. aureus* IC_{50} (mg/mL)
ELe25	144.25 ± 3.03 ^{a}	447.37 ± 1.83 ^{a}	>2.5	>2.5
MLe25	104.05 ± 0.22 ^{a,b}	210.77 ± 0.17 ^{a,b}	>2.5	>2.5
EWLe25	80.36 ± 3.67 ^{b}	86.92 ± 0.99 ^{b}	1.21 ± 0.03^{b}	1.38 ± 0.11 ^{a}
MWLe25	59.09 ± 5.40 ^{b}	68.01 ± 1.26 ^{b}	1.09 ± 0.10^{ab}	1.56 ± 0.06 ^{c}
ELe50	162.20 ± 1.30 ^{a}	546. 52 ± 2.16 ^{a}	>2.5	>2.5
MLe50	126.20 ± 2.88 ^{a,b}	277.50 ± 0.24 ^{a,b}	> 2.5	>2.5
EWLe50	85.71 ± 3.07 ^{b}	102.52 ± 0.49 ^{b}	0.84 ± 0.04^{b}	1.43 ± 0.06 ^{d}
MWLe50	79.14 ± 2.16 ^{a,b}	62.18 ± 1.28 ^{b}	0.97 ± 0.03^{b}	1.93 ± 0.01 ^{b}

Values with different letters ($^{a–d}$) in the column and row differ significantly ($p < 0.05$).

Concerning the effect of temperature, an increase of the extraction temperature (25 to 50 °C) increases the TPC content, showing significant differences ($p < 0.05$). A similar positive effect of temperature on total polyphenols recovery has been previously reported, for example in the extraction of olive leaves [47]. However, it should be noted that extremely high extraction temperature may promote possible degradation of phenolic compounds, or may enhance solvent loss through vaporization [48].

As reported elsewhere, phenolic compounds significantly contribute to the antioxidant activity of plant extracts [49,50]. It is generally stated that the higher the TPC value, the higher the antioxidant activity. In this work, a positive relationship between the antioxidant activity (evaluated by the DPPH assay) and the TPC content was obtained, considering all licorice UAE extracts produced. Figure 2 shows the linear correlation ($R^2 = 0.974$) obtained between TPC (mg GAE/g) and TEAC (μmol Trolox/g) values. Regarding the effect of temperature and following the observed tendency in the case of TPC content, for all solvents used, the increase of extraction temperature increased the antioxidant activity of the extract. As mentioned before, the extracts obtained using 50% water were those with the lower TPC values and thus, these extracts are those with the lower antioxidant activity (Figure 2). This result strengthens the hypothesis that the addition of water to the organic solvent enhanced the extraction of polar but non-antioxidant compounds.

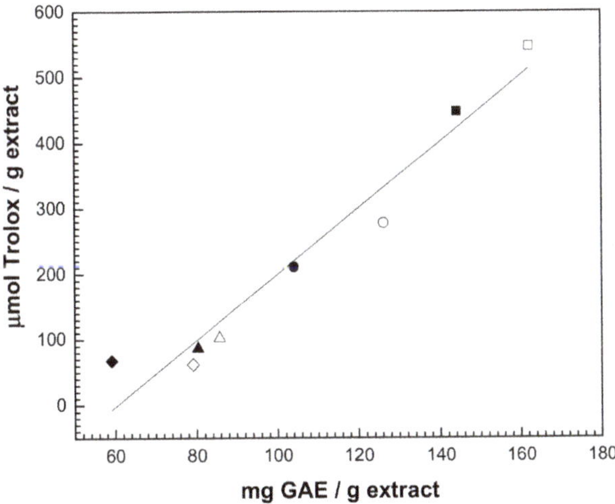

Figure 2. Correlation between antioxidant activity (μmol Trolox/g) and total phenolic compounds (mg GAE/g) of licorice UAE extracts (Table 3). T = 25 °C: (■) ethanol, (•) methanol, (▲) ethanol:water and (◆) methanol:water. T = 50 °C: (□) ethanol, (○) methanol, (Δ) ethanol:water and (◊) methanol:water.

Among the identified and quantified licorice bioactive compounds, only in the case of glabridin, the less polar identified polyphenol, did an increase of its concentration show a positive effect on the antioxidant activity of the extract.

3.2.2. Antimicrobial Activity

Table 3 reports the IC_{50} values (concentration of the extract necessary to attain 50% inhibition) of the licorice UAE extracts tested against a Gram-positive bacteria, *Staphylococcus aureus* ATCC 25923, and a Gram-negative bacteria, *Escherichia coli* ATCC 25922. The IC_{50} values corresponding to Chloramphenicol, an antibiotic useful for the treatment of a number of bacterial infections, are 0.09 mg/mL for *S. aureus* bacteria and 0.08 mg/mL for *E. coli*. Only extracts obtained using mixtures of water and organic solvent (methanol or ethanol) exhibit antimicrobial activity. In comparison with Chloramphenicol, the IC_{50} values of these extracts are around one order of magnitude larger for both types of bacteria. The observed antimicrobial activity may be related to the high glycyrrhizin concentration observed in these extracts (>130 mg/g) due to the recognized antimicrobial activity of this compound [51] and licorice extracts [32].

3.2.3. Selection of Licorice Extract for Preparing Edible Coatings

Considering the Food and Drug Administration recommendation for the use of non-toxic and environmentally safe solvents, such as ethanol [52], and the antimicrobial and antioxidant activity of samples, the extract produced with ethanol:water (50:50) at 50 °C (EWLe50) was selected and produced in sufficient amounts in order to prepare the edible coatings and carry out the study of strawberries' preservation. This licorice extract (LE) can be produced with high yield (c.a. 25%), and contains 85.71 ± 3.07 mg GAE/g, moderate antioxidant activity (102.52 ± 0.49 µmol Trolox/g) and good antimicrobial activity (IC_{50} values of 0.84 ± 0.04 and 1.43 ± 0.06 mg/mL for *E. coli* and *S. aureus*, respectively).

3.3. Rheology of the Edible Coating Solution

The rheological properties of coating solutions are of great importance because they affect the structure and apparent viscosity of the film matrix. The uniformity, the spreadability and the thickness of the coating could be greatly influenced by the flow characteristics of the coating solution [53]. Different chitosan-licorice coating solutions were prepared employing 1% (Ch + LE-1) and 5% (Ch + LE-5) of the licorice extract EWLe50 and pure chitosan (Ch) was used as control. In order to obtain a good knowledge of fluid behavior, the viscosity and viscoelastic properties of coating solutions were measured to assess the processability of the coating.

3.3.1. Steady Rheological Properties

Figure 3a shows the flow curve of different samples: the changes observed in shear stress are a consequence of the addition of licorice root extract to chitosan control, due to the interaction between chitosan and licorice phytochemicals and complexation. All coating solutions exhibited the behavior of non-Newtonian fluid and can be adjusted to the Power Law model (Equation (1)). The model parameters (K and n) are given in Table 4, and high fitting degree ($R^2 > 0.994$) was achieved. The K (consistency coefficient) values decrease significantly and n (flux index) values increase with the addition of LE. Chitosan solution presents a consistency coefficient of 0.469 ± 0.015 and flux index of 0.795 ± 0.005. Moreover, the samples Ch + LE-1 and Ch + LE-5 present K values of $3.72 \cdot 10^{-4}$ ± 0.001 and $1.76 \cdot 10^{-4}$ ± 0.001 and n values of 1.436 ± 0.018 and 1.503 ± 0.012, respectively. n values equal to 1 indicate the Newtonian fluids, lower than 1 indicate shear thinning fluids, while n values higher than 1 indicate a shear thickening behavior [54]. Then, the chitosan coating solution presents a shear thinning behavior due to alignment of molecules in the direction of flow, and a decrease of viscosity with the increase of shear rate. Similar results were obtain by Zhang [55] in a coating solution of chitosan/zein with the addition of alpha-tocopherol.

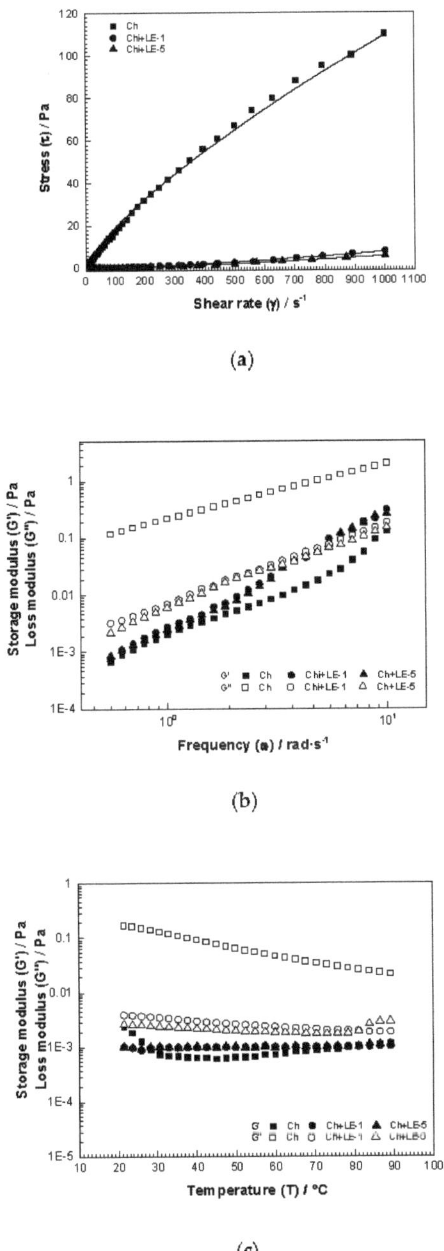

Figure 3. Rheological properties of edible coating solutions produced with chitosan and chitosan-licorice: (**a**) flow curve, (**b**) frequency sweep and (**c**) temperature sweep.

Table 4. Power Law parameters obtained for the edible coating solutions. Ch = Chitosan; Ch + LE-1 = Chitosan + 1% of licorice extract; Ch + LE-5 = Chitosan + 5% of licorice extract.

Treatment	K (Pa·s^n)	n	R^2	$\sigma\left(\dot{\gamma}=10\ s^{-1}\right)$ (Pa)
Ch	0.469 ± 0.015	0.795 ± 0.005	0.998	0.2
Ch + LE-1	3.72×10^{-4} ± 0.001	1.436 ± 0.018	0.994	0.0047
Ch + LE-5	1.76×10^{-4} ± 0.001	1.503 ± 0.012	0.998	0.0021

On the other hand, the addition of LE caused a decrease in the shear stresses and a non-Newtonian behavior. The shear thickening behavior (n > 1) observed in samples Ch + LE-1 and Ch + LE-5 are probably a consequence of particle hydro-clustering and an order-to-disorder transition. At large shear rates and stresses, convective and hydrodynamic force dominate over inter-particle force, and cause hydro-clustering of particles [56,57]. The rheological properties of coating solutions are consistently used among the literature to provide data on films' microstructure, thickness, coating spreadability and uniformity on substrate. In general, highly viscous solutions retained air bubbles in the casting process, and low viscous solutions facilitated its spreading on the plate where the films were formed [58].

3.3.2. Dynamic Rheological Properties

The study of dynamic viscoelastic properties of coating solutions led to obtain information about the molecular entanglement and molecular network formation during drying [59]. The LVR used in this study was the maximum stress value in the flat region of storage modulus (G′) and stress curve. Stress value applied for Ch, Ch + LE-1 and Ch + LE-5 was of 0.1 Pa.

Figure 3b shows the frequency dependence of the storage (G′) and loss modules (G′) of the different edible coating solutions studied, from 0.01 to 100 rad·s^{-1}. According to mechanical spectra (Figure 3b), increased G′ and G″ values were observed with the increase of angular frequency (ω), with G″ exhibiting a larger increase for all the samples in comparison with G′. Higher G″ values were obtained for chitosan-licorice solutions, indicating that these solutions will behave as liquids during the mixer process for lower processing times. Then, to deform the material, the supplied energy would be lower in the case of the chitosan-licorice coatings [60].

Furthermore, G″ > G′ for the pure chitosan coating solution in the entire frequency range applied, without any crossover point observed, exhibited the typical behavior of liquid-like solutions [61]. But, in the case of chitosan-licorice solutions, crossover points were reached, indicating a longest time required for chain disengagement of the polymer structures in the solution [62,63]. This observed frequency dependence of G′ and G″ indicates that chitosan-licorice solutions are a useful class of materials to be applied as films in food coating, since they can enhance adhesiveness and the hardness of the coating solution.

Figure 3c shows the effect of temperature on G′ and G″ values of the edible chitosan and chitosan-licorice coating solutions for a heating rate of 5 °C/min in the LVR and 0.1 Hz frequency. The effect of temperature on the variation of G′ and G″ of the coating solutions exposes the phase transitions and elasticity and allows the selection of appropriate temperature ranges for the formulation and the application of the coating solution on the product.

For all coatings, G″ was higher than G′ in the whole range of temperatures studied, denoting the liquid-like behavior. Chitosan coating solution presents higher G′ values than samples with 1% and 5% of LE, the ones which present similar values. Gelatinization does not occur because the high temperatures accelerate the mobility of chitosan molecules in the solution and reduce the intermolecular hydrogen bonding interactions, removing energized water molecules surrounding the chitosan chains [64].

3.3.3. Rheological Properties of Edible Film

The physicochemical and functional properties of edible films, such as mechanical, rheological and optical properties, are directly related to their microstructure and the interaction between film components and the drying conditions [65]. The rheological properties of edible films were evaluated in order to analyze the influence of combining chitosan with licorice extracts in the coating solutions, and to assess the particular characteristics and modifications in comparison with using only chitosan.

All the films were flexible and easily detachable from the sample holder, without evident defects in the form of cracks or pores. Figure 4a shows the frequency dependence of storage modulus (G′) and loss modulus (G″) of films of Ch, Ch + LE-1 and Ch + LE-5. For Ch films, G′ >> G″ over the entire frequency range studied, indicating that chitosan film exhibits dominant elastic behavior. G′>G″ in the case of Ch + LE-1, and the film presents a more robust network and exhibits a gel-like behavior. Finally, for Ch + LE-5 film, G″ > G′, exhibiting a viscous behavior. Then, the addition of licorice phytochemicals to chitosan increases the viscous properties of film, displaying a variation in G′. Tan δ (= G″/G′) increase with the addition of licorice extract to chitosan and values become Tan δ ≈ 1, corroborating that the viscous quality of the film is greatly affected by the addition of LE and suggesting that the films containing LE have higher flexibility compared to the chitosan film and thus, they are more suitable than pure chitosan for application as edible food coatings.

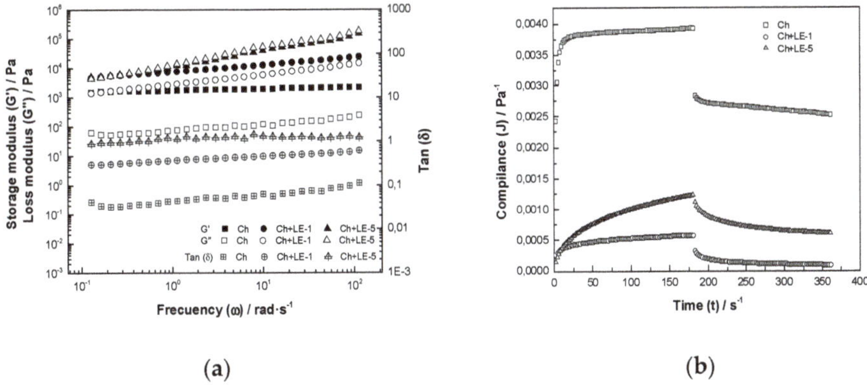

Figure 4. Rheological properties of chitosan and chitosan-licorice edible films: (**a**) frequency sweep and (**b**) creep and recovery.

The creep compliance data are shown in Figure 4b. The extensional creep curves of all films showed a typical behavior of viscoelastic material, with varying degrees of viscoelasticity. The compliance of films decreases with the addition of licorice extracts and with the percentage amount of extract added. It can be inferred that films became more flexible with the addition of LE to chitosan, indicating an interaction between LE phytochemicals with possible hydrogen bonds formation.

The creep results were simulated by applying the Burger model (Equation (2)) and a satisfactory description of the essential features of the film viscoelasticity was attained, with high correlation coefficients ($R^2 > 0.999$) (Table 5). The parameters of the Burger model reflect the structure of any system, and the decreasing/increasing of these parameters shows the weakening/strengthening of the structure. The instantaneous compliance (J_o) of chitosan was 3.793, and decreased to 0.482 and 0.801 for Ch + LE-1 and Ch + LE-5 respectively, indicating a reduction of rigidity or strength [66,67]. The retarded compliance (J_1) of chitosan was 1.408 mPa^{-1}, decreasing to 0.481 and 0.626 respectively, with the addition of 1% and 5% of LE, indicating a reduction of the gel cohesive force and also a reduction of the resistance to deformation caused by the three-dimensional network structure [66,67]. Newtonian viscosity (μ_o) increased significantly with the addition of LE to chitosan, revealing that mechanical strength and mobility of polymer chains increased with the addition of LE to chitosan in the coating formulation.

Table 5. Burger model parameters obtained for the films. Ch = Chitosan; Ch + LE-1 = Chitosan + 1% of licorice extract; Ch + LE-5 = Chitosan + 5% of licorice extract.

Table 1	J_o (mPa^{-1})	J_1 (mPa^{-1})	λ_{ret} (s)	μ_o (Pa·s)	R^2	Recovery (%)
Ch	3.793	1.408	4864.0	1.282×10^6	0.998	35.82
Ch + LE-1	0.482	0.481	906.3	1.878×10^6	0.999	83.78
Ch + LE-5	0.801	0.6258	340.8	4.252×10^5	0.997	51.09

Recovery percentages are an indirect measure of gel elasticity and the calculated values for the different coating films are given in Table 5. The recovery percentages of the films with LE extracts (83.78% and 51.09% for Ch + LE-1 and Ch + LE-5, respectively) were considerably higher than that of chitosan (35.82%). Then, the addition of licorice extract changed the viscous behavior and improved the elasticity of the films, with the Ch+LE-1 coating being the one which presents the highest recovery percentage (best gel elasticity).

3.4. *Effect of Chitosan-Licorice Edible Coatings on Strawberries' Quality*

Four sets of strawberries were prepared to carry out the study of the effect of chitosan-licorice edible coatings on the fruit preservation: uncoated (control) fruit (C), coated with chitosan (Ch) and coated with Ch + LE-1 and Ch + LE-5. The samples were stored at 4 °C for 10 days.

3.4.1. Physicochemical Effects

In order to determine the effect of licorice extracts on the coated fruits, photographs were picked up during storage, which are shown in Figure 5. Pictures corresponding to 14 days of storage were also included in the figure, because the effect of coatings became very obvious. The coated strawberries presented a brilliant appearance, which was better than the uncoated sample.

The coated forming on berries depends on the rheological parameters of the coating solutions, which present different shear stress and viscoelastic properties. The amount of coating solution adhered to strawberry during application strongly depends on the viscosity and the surface tension of the coating solutions. The roughness on the surface of strawberries require formulations with ingredients that can reduce the coating solution surface tension, in order to ensure coating uniformity and absence of void holes [4]. The total sugars present in fruit can act as a plasticizer, interacting with the polymer chains and generating "free" volumes within the chains, weakening the intermolecular forces and, consequently, reducing the resistance of the films [68]. Some authors determined that solutions with high viscosity and low surface tension promote a better film-forming surface [69]. In this study, the chitosan coating solution presented a viscosity of 170 mPa·s, while lower values were observed in the case of chitosan-licorice solutions (4.40 and 2.10 mPa·s for Ch + LE-1 and Ch + LE-5, respectively). Nevertheless, despite the observed viscosity reduction, the addition of LE improved the viscoelastic properties and allowed a better surface extensibility on berries, forming thinning coatings with higher uniformity and adherence, which can be related with the better physical appearance observed (Figure 5). The application of coatings did not affect the color of berries, maintaining good visual quality and no water migration on the surface was observed.

The presence of mold was visually observed in the control sample on day 10 (Figure 6), while no mold was observed in the coated berries. Ch + LE-5 coating exhibits the best preservation of strawberry appearance. The fungal decay of samples is depicted in Figure 6. The percentage of fungal decay was higher in control samples, followed by chitosan samples, while chitosan-licorice coatings did not present fungal decay during 10 days of storage. The licorice extract increased the antimicrobial properties of chitosan, according to the antibacterial activity assessed in this study and in other studies available in the literature [32].

Figure 5. Visual appearance of uncoated, chitosan- and chitosan-licorice-coated strawberries corresponding to 1, 3, 10 and 14 days of storage at 4 °C.

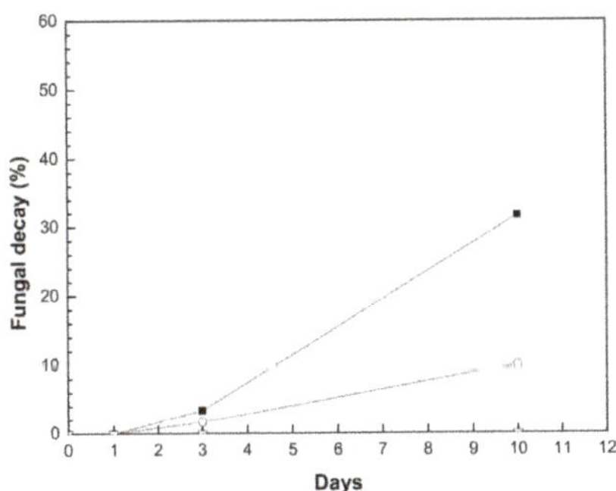

Figure 6. Fungal decay of uncoated and licorice-chitosan-coated strawberries stored for 10 days at 4 °C: (○) Ch (Chitosan); (■) C (uncoated).

Figure 7 shows the weight loss of uncoated and coated samples with the different films throughout the cold storage time. Postharvest mass loss of fruits and vegetables is mostly due to moisture loss through transpiration, and is recognized to accelerate the susceptibility of fruit and vegetables to physiological disorders [70,71]. After only three days of storing, the uncoated sample presented less weight loss than the samples treated with any of the coatings. Nevertheless, on day 10, the barrier effect exerted by the coatings could be observed, since all coatings reduced the loss of fruit weight in comparison with uncoated samples, with the coatings containing Ch + LE-1 and Ch + LE-5 being more efficient than the one containing solely chitosan. For these coatings, weight losses of samples are lower than 2% after 10 days of storing. Similar weight loss values were obtained by Martínez et al. [15] using chitosan coatings combined with *Thymus capitatus* essential oil. Furthermore, these loss weight values were lower than those observed by Petriccione et al. [72] when studying different species of strawberry coated with 2% chitosan (weight losses of 7–9%) or Ventura-Aguilar [73] using an edible film of chitosan combined with a *Roselle calyces* extract (weight loss of 6%).

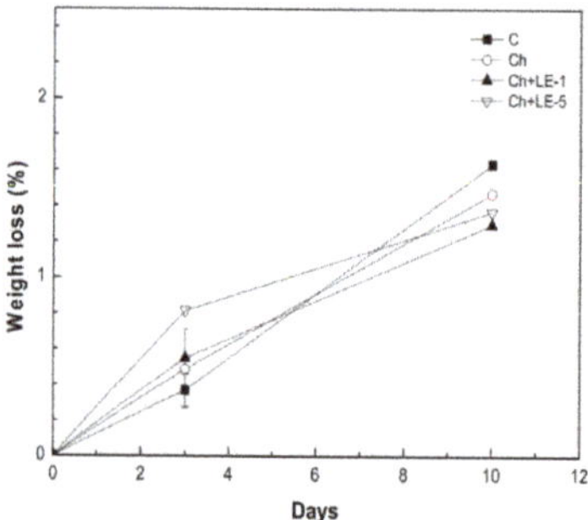

Figure 7. Effect of different licorice-chitosan coatings on the weight loss of strawberries stored for 10 days at 4 °C: (○) Ch (chitosan) (▲) Ch + LE-1 (Chitosan + 1% of licorice extract); (▽) Ch + LE- 5 (Chitosan + 5% of licorice extract; (■) C = uncoated.

The effect of the different coatings in comparison with uncoated samples on the pH and titratable acidity (TA) of strawberries were also examined, since these properties are a measure of ripening effect. Fruit maturation produces the enzymatic reaction of sugars and the amounts of acids tend to decrease and the pH increases. The initial pH values of coated strawberries were slightly higher than uncoated samples (Figure 8a). Although there were some differences on day 1, pH values of coated samples did not change significantly during storage and no significant differences were observed between coated samples on day 10 (mean pH value is 4.05 ± 0.10), while a significantly higher pH value was obtained for the uncoated fruit (4.5 ± 0.20).

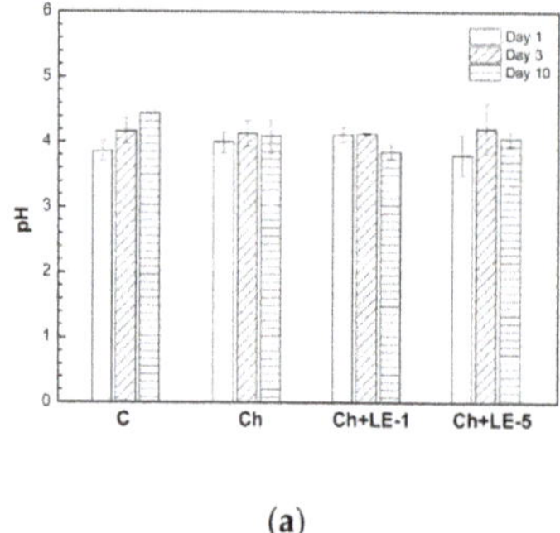

(a)

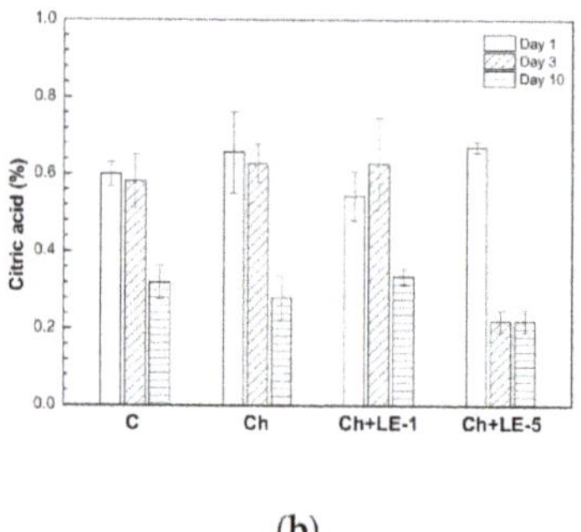

(b)

Figure 8. Effect of different coatings based on chitosan and licorice extracts on (**a**) pH and (**b**) titratable acidity (TA) expressed as citric acid percentage of strawberries stored for 10 days at 4 °C. C = uncoated; Ch = chitosan; Ch + LE-1 = Chitosan + 1% of licorice extract; Ch + LE-5 = Chitosan + 5% of licorice extract. Means values and intervals of Tukey's test at 95% according to the analysis of variance (ANOVA) test.

Furthermore, the TA values were significantly reduced ($p < 0.05$) both in coated and uncoated strawberries, but due to the nature of fruit organic acids, these usual decreases in fruit acidity during maturity did not induce a notable change in the pH value. As can be observed in Figure 8b, a significant decrease of TA values ($p < 0.05$) with storing time was observed in all cases, but differences between control and the different coatings were not significant. These results are in agreement with similar research conducted by Bhimrao et al. [74], who studied the extension of shelf life of strawberries using a coating formed with chitosan and whey protein isolate.

3.4.2. Total Phenolic Compounds Content and Microbiology Analysis

As stated before, the phenol content of a vegetal sample can be related with its antioxidant activity. Thus, the TPC content of the uncoated and coated strawberries was evaluated after 10 storing days, as an indicator of the coating antioxidant functional capacity (Table 6). The TPC values of uncoated strawberry on day 1 are in accordance with the values reported in the literature [75], which indicated that TPC values of 27 strawberry cultivars varied from 0.57 to 1.33 GAE mg/g of fresh weight.

Table 6. Total phenolic compounds (TPC) of strawberries treated with different coatings. Ch = Chitosan; Ch + LE-1 = Chitosan + 1% of licorice extract; Ch + LE-5 = Chitosan + 5% of licorice extract.

Treatment	TPC (mg GAE/g)	
	Day 1	Day 10
Uncoated	1.41 ± 0.12	0.97 ± 0.04
Ch	1.28 ± 0.06	0.99 ± 0.08
Ch + LE-1	1.52 ± 0.10	1.32 ± 0.05
Ch + LE-5	1.44 ± 0.02	1.38 ± 0.35

The results given in Table 6 indicate that, on day 10 and for all cases (uncoated and coated fruit), the TPC values were lower after the cold storage. But, the type of coating had a significant effect on the observed TPC decrease, as is depicted in Figure 9, and the control samples (uncoated fruit) presented the highest TPC loss percentage (30.93 mg GAE/g), followed by chitosan coating (22.65 mg GAE/g), chitosan + 1% LE (13.1 mg GAE/g) and chitosan + 5% LE (13.3 mg GAE/g). Then, LE extract contributed significantly to preserve the phenolic compounds of samples.

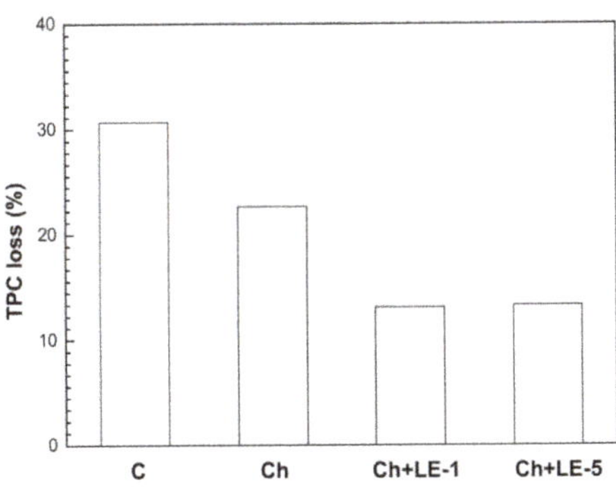

Figure 9. Effect of chitosan-licorice coatings on total phenolic compounds (TPC) of strawberries stored for 10 days at 4 °C: C = uncoated; Ch = chitosan; LE1 = Chitosan + 1% of licorice extract; LE5 = Chitosan + 5% of licorice extract.

Concerning the microbiological analysis, Table 7 shows the aerobic bacteria and yeast determined in coated and uncoated strawberries after 6 days of cold storage. No initial growth of bacteria or yeasts was observed in the uncoated and coated strawberries. But after 6 days of storage, viable count of bacteria and yeasts were determined in uncoated samples and in berries coated with chitosan, while in the samples coated with chitosan-licorice mixtures, no growth was observed. Then, the addition of phytochemicals of licorice in order to prepared edible coating improved the antimicrobial properties of chitosan.

Table 7. Effects of edible coatings prepared with chitosan and chitosan + licorice extract on the growth of mesophilic and psychrophilic bacteria (total aerobic psychrophilic bacteria) and yeasts during storage at 4 °C. Ch = Chitosan; Ch + LE-1 = Chitosan + 1% of licorice extract; Ch + LE-5 = Chitosan + 5% of licorice extract.

Bacteria and Yeasts	Treatment	Total Viable Count (Log CFU/g) on Day 6 of Cold Storage
Mesophilic	Uncoated	7.0 ± 2.0
	Ch	4.5 ± 0.7
	Ch + LE-1	n.g.
	Ch + LE-5	n.g.
Psychrophilic	Uncoated	4.0 ± 0.01
	Ch	4.0 ± 0.01
	Ch + LE-1	n.g.
	Ch + LE-5	n.g.
Yeasts	Uncoated	4.0 ± 0.01
	Ch	4.0 ± 0.01
	Ch + LE-1	n.g.
	Ch + LE-5	n.g.

n.g.: no growth observed. CFU: Colony forming units

4. Conclusions

In this study, the ultrasound-assisted extraction using solvents with different polarities and two temperatures (25 and 50 °C) was carried out, in order to obtain bioactive licorice extracts (LE) for the elaboration of edible coatings. The polarity of the solvent contributed to the extraction of specific compounds, and while temperature did not significantly affect the extraction yield, it influenced the TPC content of samples. A higher concentration of glabridin was obtained with ethanol solvent, showing a positive effect on the antioxidant activity of the extract. Nevertheless, ethanolic licorice root extracts did not present antimicrobial activity. On the other hand, the extraction with mixtures of ethanol or methanol with water (50:50 *v*/*v*) enhanced the extraction of polar biomolecules (such as glycyrrhizin), and the extracts presented good antimicrobial activity but less antioxidant capacity in comparison with the ethanolic extract.

The edible coating solutions elaborated with pure chitosan and with chitosan and licorice ethanol:water extract, presented a non-Newtonian flow behavior. Solutions of pure chitosan presented shear thinning behavior, and the addition of LE decreased the shear stress value and presented a shear thickening behavior, probably due to particle hydro-clustering and an order-to-disorder transition. The viscoelastic properties of the coating solution show that the restructuring ability of Ch + LE physical gels improve the properties of film forming. Then, the addition of licorice phytochemical to chitosan increased the viscous properties of film, displayed a variation in the dissipation energy (G″) and films became more flexible with the addition of LE. Furthermore, the addition of LE to chitosan decreased the shear stress values of samples, allowing better physical properties of the films applied on strawberries.

Concerning the preservation capacity, the addition of LE increased the antimicrobial properties of chitosan, according to the antibacterial activities studied in vitro. The values of weight loss of these coatings were significantly lower than those corresponding to pure chitosan. Additionally, the chitosan + LE edible coatings better preserved the content of phenolic compounds in the berries.

Then, the ultrasound-assisted extraction of bioactive compounds of licorice root using ethanol:water (50:50 *v*/*v*) is an alternative for producing bioactive licorice extracts with potential application in the elaboration of edible coatings with good rheological properties and preservative capacity.

Author Contributions: Conceptualization, M.R.G.-R.; L.A.G.-Z., and T.F.; methodology, S.E.Q.; M.R.G.-R.; L.A.G.-Z., and T.F.; software, O.L. and S.E.Q.; validation, T.F., formal analysis, S.E.Q., M.R.G.-R., L.A.G.-Z., and T.F.; investigation, S.E.Q., O.L., resources, S.E.Q., O.L., M.R.G.-R.; L.A.G.-Z. and T.F., writing—original draft preparation, S.E.Q. and T.F.; writing—review and editing, S.E.Q., O.L., L.A.G.-Z., M.R.G.-R. and T.F., supervision, T.F., M.R.G.-R. and L.A.G.-Z. and project administration, M.R.G.-R. and T.F.; funding acquisition, S.E.Q.; O.L., M.R.G.-R. and T.F. All authors have read and agreed to the published version of the manuscript.

Funding: This research was funded by Comunidad Autónoma de Madrid, grant number P2013/ABI27, project Bolívar Gana con Ciencia, MinCiencias Contract 368-2019 and Programa Nacional de Innovación Agraria—PNIA of Perú, Contract: No. 152-2018-INIA-PNIA-PASANTIA.

Acknowledgments: The authors gratefully acknowledge the financial support from Comunidad Autónoma de Madrid (ALIBIRD, project: P2013/ABI2728). Somaris E. Quintana is grateful for the funding provided by Gobernación de Bolivar and Fundación Ceiba, Colombia, in the project "Bolívar Gana con Ciencia" and MinCiencias Contract 368-2019. Olimpia Ollalla thanks the Programa Nacional de Innovación Agraria—PNIA of Perú (Contract: No. 152-2018-INIA-PNIA-PASANTIA).

Conflicts of Interest: The authors declare no conflict of interest.

References

1. Valdés, A.; Ramos, M.; Beltrán, A.; Jiménez, A.; Garrigós, M.C. State of the Art of Antimicrobial Edible Coatings for Food Packaging Applications. *Coatings* **2017**, *7*, 56. [CrossRef]
2. Elsabee, M.; Abdou, E.S. Chitosan based edible films and coatings: A review. *Mater. Sci. Eng. C* **2013**, *33*, 1819–1841. [CrossRef] [PubMed]
3. Dhall, R.K. Advances in Edible Coatings for Fresh Fruits and Vegetables: A Review. *Crit. Rev. Food Sci. Nutr.* **2013**, *53*, 435–450. [CrossRef] [PubMed]
4. García, M.A.; Pinotti, A.; Martino, M.N.; Zaritzky, N.E. Characterization of Starch and Composite Edible Films and Coatings. In *Edible Films and Coatings for Food Applications*; Springer: New York, NY, USA, 2009; pp. 169–209.
5. Silva-Weiss, A.; Bifani, V.; Ihl, M.; Sobral, P.J.A.; Guillén, M.C.G. Structural properties of films and rheology of film-forming solutions based on chitosan and chitosan-starch blend enriched with murta leaf extract. *Food Hydrocoll.* **2013**, *31*, 458–466. [CrossRef]
6. Aguirre-Joya, J.A.; De Leon-Zapata, M.A.; Alvarez-Perez, O.B.; Torres-León, C.; Nieto-Oropeza, D.E.; Ventura-Sobrevilla, J.M.; Aguilar, M.A.; Ruelas-Chacón, X.; Rojas, R.; Ramos-Aguiñaga, M.E.; et al. Basic and Applied Concepts of Edible Packaging for Foods. In *Food Packaging and Preservation*; Elsevier: Amsterdam, The Netherlands, 2018; pp. 1–61.
7. Ravi Kumar, M.N.V. A review of chitin and chitosan applications. *React. Funct. Polym.* **2000**, *46*, 1–27. [CrossRef]
8. Devlieghere, F.; Vermeulen, A.; Debevere, J. Chitosan: Antimicrobial activity, interactions with food components and applicability as a coating on fruit and vegetables. *Food Microbiol.* **2004**, *21*, 703–714. [CrossRef]
9. Duan, J.; Cherian, G.; Zhao, Y. Quality enhancement in fresh and frozen lingcod (Ophiodon elongates) fillets by employment of fish oil incorporated chitosan coatings. *Food Chem.* **2010**, *119*, 524–532. [CrossRef]
10. Odila Pereira, J.; Soares, J.; Sousa, S.; Madureira, A.R.; Gomes, A.; Pintado, M. Edible films as carrier for lactic acid bacteria. *LWT Food Sci. Technol.* **2016**, *73*, 543–550. [CrossRef]
11. Ghaouth, A.; Arul, J.; Ponnampalam, R.; Boulet, M. Chitosan Coating Effect on Storability and Quality of Fresh Strawberries. *J. Food Sci.* **1991**, *56*, 1618–1620. [CrossRef]
12. Zhang, D.; Quantick, P.C. Antifungal effects of chitosan coating on fresh strawberries and raspberries during storage. *J. Hortic. Sci. Biotechnol.* **1998**, *73*, 763–767. [CrossRef]
13. Duran, M.; Aday, M.S.; Zorba, N.N.D.; Temizkan, R.; Büyükcan, M.B.; Caner, C. Potential of antimicrobial active packaging 'containing natamycin, nisin, pomegranate and grape seed extract in chitosan coating' to extend shelf life of fresh strawberry. *Food Bioprod. Process.* **2016**, *98*, 354–363. [CrossRef]

14. Pinzon, M.I.; Sanchez, L.T.; Garcia, O.R.; Gutierrez, R.; Luna, J.C.; Villa, C.C. Increasing shelf life of strawberries (*Fragaria* ssp) by using a banana starch-chitosan-Aloe vera gel composite edible coating. *Int. J. Food Sci. Technol.* **2019**, *55*, 92–98. [CrossRef]
15. Martínez, K.; Ortiz, M.; Albis, A.; Castañeda, C.G.G.; Valencia, M.E.; Tovar, C.D.G. The Effect of Edible Chitosan Coatings Incorporated with Thymus capitatus Essential Oil on the Shelf-Life of Strawberry (*Fragaria x ananassa*) during Cold Storage. *Biomolecules* **2018**, *8*, 155. [CrossRef] [PubMed]
16. Shahbazi, Y. Application of carboxymethyl cellulose and chitosan coatings containing Mentha spicata essential oil in fresh strawberries. *Int. J. Biol. Macromol.* **2018**, *112*, 264–272. [CrossRef] [PubMed]
17. Apriyanti, D.; Rokhati, N.; Mawarni, N.; Khoiriyah, Z.; Istirokhatun, T. Edible Coating from Green Tea Extract and Chitosan to Preserve Strawberry (*Fragaria vesca* L.). In Proceedings of the The 24th Regional Symposium on Chemical Engineering (RSCE 2017), Semarang, Indonesia, 15–16 November 2017.
18. Almenar, E.; Del-Valle, V.; Hernández-Muñoz, P.; Lagarón, J.M.; Catalá, R.; Gavara, R. Equilibrium modified atmosphere packaging of wild strawberries. *J. Sci. Food Agric.* **2007**, *87*, 1931–1939. [CrossRef]
19. Van De Velde, F.; Tarola, A.M.; Güemes, D.; Pirovani, M.E. Bioactive Compounds and Antioxidant Capacity of Camarosa and Selva Strawberries (*Fragaria x ananassa* Duch.). *Foods* **2013**, *2*, 120–131. [CrossRef]
20. Vu, K.; Hollingsworth, R.G.; Leroux, E.; Salmieri, S.; Lacroix, M. Development of edible bioactive coating based on modified chitosan for increasing the shelf life of strawberries. *Food Res. Int.* **2011**, *44*, 198–203. [CrossRef]
21. Dhital, R.; Mora, N.B.; Watson, D.G.; Kohli, P.; Choudhary, R. Efficacy of limonene nano coatings on post-harvest shelf life of strawberries. *LWT* **2018**, *97*, 124–134. [CrossRef]
22. Oliveira, J.; Da Glória, E.M.; Da Silva, P.P.M.; Baggio, J.S.; Da Silva, P.P.M.; Ambrosio, C.M.; Spoto, M.H.F. Antifungal activity of essential oils associated with carboxymethylcellulose against Colletotrichum acutatum in strawberries. *Sci. Hortic.* **2019**, *243*, 261–267. [CrossRef]
23. Hosseinzadeh, H.; Nassiri-Asl, M. Pharmacological Effects of Glycyrrhiza spp. and Its Bioactive Constituents: Update and Review. *Phytother. Res.* **2015**, *29*, 1868–1886. [CrossRef]
24. Al-Ani, B.M.; Owaid, M.N.; Al-Saeedi, S.S.S. Fungal interaction between Trichoderma spp. and Pleurotus ostreatus on the enriched solid media with licorice Glycyrrhiza glabra root extract. *Acta Ecol. Sin.* **2018**, *38*, 268–273. [CrossRef]
25. Jiang, J.; Zhang, X.; True, A.D.; Zhou, L.; Xiong, Y.L. Inhibition of Lipid Oxidation and Rancidity in Precooked Pork Patties by Radical-Scavenging Licorice (*Glycyrrhiza glabra*) Extract. *J. Food Sci.* **2013**, *78*, C1686–C1694. [CrossRef] [PubMed]
26. Qiu, X.; Chen, S.; Liu, G.; Yang, Q. Quality enhancement in the Japanese sea bass (*Lateolabrax japonicas*) fillets stored at 4 °C by chitosan coating incorporated with citric acid or licorice extract. *Food Chem.* **2014**, *162*, 156–160. [CrossRef] [PubMed]
27. Madanipour, S.; Alimohammadi, M.; Rezaie, S.; Nabizadeh, R.; Khaniki, G.J.; Hadi, M.; Yousefi, M.; Bidgoli, S.M.; Yousefzadeh, S. Influence of postharvest application of chitosan combined with ethanolic extract of liquorice on shelflife of apple fruit. *J. Environ. Health Sci. Eng.* **2019**, *17*, 331–336. [CrossRef] [PubMed]
28. Samaram, S.; Mirhosseini, H.; Tan, C.P.; Ghazali, H.M.; Bordbar, S.; Serjouie, A. Optimisation of ultrasound-assisted extraction of oil from papaya seed by response surface methodology: Oil recovery, radical scavenging antioxidant activity, and oxidation stability. *Food Chem.* **2015**, *172*, 7–17. [CrossRef]
29. Dzah, C.S.; Duan, Y.; Zhang, H.; Wen, C.; Zhang, J.; Chen, G.; Ma, H. The effects of ultrasound assisted extraction on yield, antioxidant, anticancer and antimicrobial activity of polyphenol extracts: A review. *Food Biosci.* **2020**, *35*, 100547. [CrossRef]
30. Singleton, V.L.; Orthofer, R.; Lamuela-Raventós, R.M. Analysis of total phenols and other oxidations substrates and antioxidants by means of Folin–Ciocalteu reagent. *Polyphen. Flavonoids* **1974**, *25*, 152–178.
31. Brand-Williams, W.; Cuvelier, M.; Berset, C. Use of a free radical method to evaluate antioxidant activity. *LWT* **1995**, *28*, 25–30. [CrossRef]
32. Quintana, S.E.; Cueva, C.; Villanueva-Bermejo, D.; Moreno-Arribas, M.V.; Fornari, T.; García-Risco, M.R. Antioxidant and antimicrobial assessment of licorice supercritical extracts. *Ind. Crop. Prod.* **2019**, *139*, 111496. [CrossRef]

33. Wei, S.-S.; Yang, M.; Chen, X.; Wang, Q.-R.; Cui, Y.-J. Simultaneous determination and assignment of 13 major flavonoids and glycyrrhizic acid in licorices by HPLC-DAD and Orbirap mass spectrometry analyses. *Chin. J. Nat. Med.* **2015**, *13*, 232–240. [CrossRef]
34. Ali, A.; Muhammad, M.T.M.; Sijam, K.; Siddiqui, Y. Effect of chitosan coatings on the physicochemical characteristics of Eksotika II papaya (*Carica papaya* L.) fruit during cold storage. *Food Chem.* **2011**, *124*, 620–626. [CrossRef]
35. Torres-León, C.; Vicente, A.; Flores-López, M.L.; Rojas, R.; Serna-Cock, L.; Alvarez-Pérez, O.B.; Aguilar, C.N. Edible films and coatings based on mango (var. Ataulfo) by-products to improve gas transfer rate of peach. *LWT* **2018**, *97*, 624–631. [CrossRef]
36. Steffe, J.F. *Rheological Methods in Food Process Engineering*, 2nd ed.; Freeman Press: East Lansing, MI, USA, 1996.
37. AOAC (Association of Official Analytical Chemist). *Official Methods of Analysis*, 17th ed.; AOAC: Gaithersburg, MD, USA, 2000.
38. Hajji, S.; Younes, I.; Affes, S.; Boufi, S.; Nasri, M. Optimization of the formulation of chitosan edible coatings supplemented with carotenoproteins and their use for extending strawberries postharvest life. *Food Hydrocoll.* **2018**, *83*, 375–392. [CrossRef]
39. Nieto, J.A.; Santoyo, S.; Prodanov, M.; Reglero, G.; Jaime, L. Valorisation of Grape Stems as a Source of Phenolic Antioxidants by Using a Sustainable Extraction Methodology. *Foods* **2020**, *9*, 604. [CrossRef] [PubMed]
40. Arranz, E.; Villalva, M.; Guri, A.; Ortego-Hernández, E.; Jaime, L.; Reglero, G.; Santoyo, S.; Corredig, M. Protein matrices ensure safe and functional delivery of rosmarinic acid from marjoram (*Origanum majorana*) extracts. *J. Sci. Food Agric.* **2019**, *99*, 2629–2635. [CrossRef] [PubMed]
41. Kaderides, K.; Goula, A.M.; Adamopoulos, K.G. A process for turning pomegranate peels into a valuable food ingredient using ultrasound-assisted extraction and encapsulation. *Innov. Food Sci. Emerg. Technol.* **2015**, *31*, 204–215. [CrossRef]
42. Anwar, F.; Przybylski, R. Effect of solvents extraction on total phenolics and antioxidant activity of extracts from flaxseed (*Linum usitatissimum* L.). *Acta Sci. Pol. Technol. Aliment.* **2012**, *11*, 293–302.
43. Yehuda, I.; Madar, Z.; Leikin-Frenkel, A.; Tamir, S. Glabridin, an isoflavan from licorice root, downregulates iNOS expression and activity under high-glucose stress and inflammation. *Mol. Nutr. Food Res.* **2015**, *59*, 1041–1052. [CrossRef]
44. Nassiri, A.M.; Hosseinzadeh, H. Review of pharmacological effects of *Glycyrrhiza* sp. and its bioactive compounds. *Phyther. Res.* **2008**, *22*, 709–724.
45. Spigno, G.; Tramelli, L.; De Faveri, D.M. Effects of extraction time, temperature and solvent on concentration and antioxidant activity of grape marc phenolics. *J. Food Eng.* **2007**, *81*, 200–208. [CrossRef]
46. Tian, M.; Yan, H.; Row, K.H. Simultaneous extraction and separation of liquiritin, glycyrrhizic acid, and glabridin from licorice root with analytical and preparative chromatography. *Biotechnol. Bioprocess. Eng.* **2009**, *13*, 671–676. [CrossRef] [PubMed]
47. Xie, P.-J.; Huang, L.-X.; Zhang, C.-H.; You, F.; Zhang, Y.-L. Reduced pressure extraction of oleuropein from olive leaves (*Olea europaea* L.) with ultrasound assistance. *Food Bioprod. Process.* **2015**, *93*, 29–38. [CrossRef]
48. Khemakhem, I.; Ahmad-Qasem, M.H.; Barrajón-Catalán, E.; Micol, V.; García-Pérez, J.V.; Ayadi, M.A.; Bouaziz, M. Kinetic improvement of olive leaves' bioactive compounds extraction by using power ultrasound in a wide temperature range. *Ultrason. Sonochem.* **2017**, *34*, 466–473. [CrossRef]
49. Amado, I.R.; Franco, D.; Sánchez, M.; Zapata, C.; Vázquez, J. Optimisation of antioxidant extraction from Solanum tuberosum potato peel waste by surface response methodology. *Food Chem.* **2014**, *165*, 290–299. [CrossRef] [PubMed]
50. Villanueva-Bermejo, D.; Zahran, F.; Troconis, D.; Villalva, M.; Reglero, G.; Fornari, T. Selective precipitation of phenolic compounds from Achillea millefolium L. extracts by supercritical anti-solvent technique. *J. Supercrit. Fluids* **2017**, *120*, 52–58. [CrossRef]
51. Thakur, D.; Jain, A.; Ghoshal, G. Evaluation of phytochemical, antioxidant and antimicrobial properties of glycyrrhizin extracted from roots of Glycyrrhiza Glabra. *J. Sci. Ind. Res.* **2016**, *75*, 487–494.
52. Bartnick, D.D.; Mohler, C.M.; Houlihan, M. Methods for the Production of Food Grade Extracts. WO 2006/047404 A3 4 May 2006. Available online: https://patentscope2.wipo.int/search/en/detail.jsf?docId=WO2006047404 (accessed on 31 October 2020).

53. Chen, C.-H.; Kuo, W.-S.; Lai, L.-S. Rheological and physical characterization of film-forming solutions and edible films from tapioca starch/decolorized hsian-tsao leaf gum. *Food Hydrocoll.* **2009**, *23*, 2132–2140. [CrossRef]
54. Haddarah, A.; Bassal, A.; Ismail, A.; Gaiani, C.; Ioannou, I.; Charbonnel, C.; Hamieh, T.; Ghoul, M.; Gaiani, C. The structural characteristics and rheological properties of Lebanese locust bean gum. *J. Food Eng.* **2014**, *120*, 204–214. [CrossRef]
55. Zhang, L.; Liu, Z.; Sun, Y.; Wang, X.; Li, L. Effect of α-tocopherol antioxidant on rheological and physicochemical properties of chitosan/zein edible films. *LWT* **2020**, *118*, 108799. [CrossRef]
56. Maranzano, B.J.; Wagner, N.J. The effects of particle size on reversible shear thickening of concentrated colloidal dispersions. *J. Chem. Phys.* **2001**, *114*, 10514–10527. [CrossRef]
57. Maranzano, B.J.; Wagner, N.J. The effects of interparticle interactions and particle size on reversible shear thickening: Hard-sphere colloidal dispersions. *J. Rheol.* **2001**, *45*, 1205–1222. [CrossRef]
58. Peressini, D.; Bravin, B.; Lapasin, R.; Rizzotti, C.; Sensidoni, A. Starch–methylcellulose based edible films: Rheological properties of film-forming dispersions. *J. Food Eng.* **2003**, *59*, 25–32. [CrossRef]
59. Li, C.; Xiang, F.; Wu, K.; Jiang, F.; Ni, X. Changes in microstructure and rheological properties of konjac glucomannan/zein blend film-forming solution during drying. *Carbohydr. Polym.* **2020**, *250*, 116840. [CrossRef] [PubMed]
60. Powles, J.G.; Rickayzen, G.; Heyes, D.M. Purely viscous fluids. *Proc. R. Soc. A Math. Phys. Eng. Sci.* **1999**, *455*, 3725–3742. [CrossRef]
61. Calero, N.; Muñoz, J.; Ramírez, P.; Guerrero, A. Flow behaviour, linear viscoelasticity and surface properties of chitosan aqueous solutions. *Food Hydrocoll.* **2010**, *24*, 659–666. [CrossRef]
62. Lauten, R.A.; Nyström, B. Linear and nonlinear viscoelastic properties of aqueous solutions of cationic polyacrylamides. *Macromol. Chem. Phys.* **2000**, *201*, 677–684. [CrossRef]
63. Rweiab, S.-P.; Chen, Y.-M.; Lin, W.-Y.; Chiang, W.-Y. Synthesis and Rheological Characterization of Water-Soluble Glycidyltrimethylammonium-Chitosan. *Mar. Drugs* **2014**, *12*, 5547–5562. [CrossRef]
64. Tang, Y.-F.; Du, Y.; Hu, X.-W.; Shi, X.-W.; Kennedy, J.F. Rheological characterisation of a novel thermosensitive chitosan/poly(vinyl alcohol) blend hydrogel. *Carbohydr. Polym.* **2007**, *67*, 491–499. [CrossRef]
65. Fabra, M.J.; Jiménez, A.; Atarés, L.; Talens, P.; Chiralt, A. Effect of Fatty Acids and Beeswax Addition on Properties of Sodium Caseinate Dispersions and Films. *Biomacromolecules* **2009**, *10*, 1500–1507. [CrossRef]
66. Huang, J.; Zeng, S.; Xiong, S.; Huang, Q. Steady, dynamic, and creep-recovery rheological properties of myofibrillar protein from grass carp muscle. *Food Hydrocoll.* **2016**, *61*, 48–56. [CrossRef]
67. Toker, O.S.; Karaman, S.; Yuksel, F.; Dogan, M.; Kayacier, A.; Yilmaz, M.T. Temperature Dependency of Steady, Dynamic, and Creep-Recovery Rheological Properties of Ice Cream Mix. *Food Bioprocess. Technol.* **2013**, *6*, 2974–2985. [CrossRef]
68. Martelli, M.R.; Barros, T.T.; De Moura, M.R.; Mattoso, L.H.C.; Assis, O.B.G. Effect of Chitosan Nanoparticles and Pectin Content on Mechanical Properties and Water Vapor Permeability of Banana Puree Films. *J. Food Sci.* **2013**, *78*, N98–N104. [CrossRef] [PubMed]
69. Vargas, M.; Chiralt, A.; Albors, A.; González-Martínez, C. Effect of chitosan-based edible coatings applied by vacuum impregnation on quality preservation of fresh-cut carrot. *Postharvest Biol. Technol.* **2009**, *51*, 263–271. [CrossRef]
70. Alquezar, B.; Mesejo, C.; Alferez, F.; Agustí, M.; Zacarias, L. Morphological and ultrastructural changes in peel of 'Navelate' oranges in relation to variations in relative humidity during postharvest storage and development of peel pitting. *Postharvest Biol. Technol.* **2010**, *56*, 163–170. [CrossRef]
71. Alferez, F.; Alquezar, B.; Burns, J.K.; Zacarias, L. Variation in water, osmotic and turgor potential in peel of 'Marsh' grapefruit during development of postharvest peel pitting. *Postharvest Biol. Technol.* **2010**, *56*, 44–49. [CrossRef]
72. Petriccione, M.; Mastrobuoni, F.; Pasquariello, M.S.; Zampella, L.; Nobis, E.; Capriolo, G.; Scortichini, M. Effect of Chitosan Coating on the Postharvest Quality and Antioxidant Enzyme System Response of Strawberry Fruit during Cold Storage. *Foods* **2015**, *4*, 501–523. [CrossRef]
73. Ventura-Aguilar, R.I.; Bautista-Baños, S.; Flores-García, G.; Zavaleta-Avejar, L. Impact of chitosan based edible coatings functionalized with natural compounds on *Colletotrichum fragariae* development and the quality of strawberries. *Food Chem.* **2018**, *262*, 142–149. [CrossRef]

74. Muley, A.B.; Singhal, R.S. Extension of postharvest shelf life of strawberries (*Fragaria ananassa*) using a coating of chitosan-whey protein isolate conjugate. *Food Chem.* **2020**, *329*, 127213. [CrossRef]
75. Gomes, M.D.S.; Cardoso, M.D.G.; Guimarães, A.C.G.; Guerreiro, A.C.; Gago, C.M.L.; Boas, E.V.D.B.V.; Dias, C.M.B.; Manhita, A.C.C.; Faleiro, M.L.; Miguel, M.G.C.; et al. Effect of edible coatings with essential oils on the quality of red raspberries over shelf-life. *J. Sci. Food Agric.* **2017**, *97*, 929–938. [CrossRef]

Article

Mucin-Grafted Polyethylene Glycol Microparticles Enable Oral Insulin Delivery for Improving Diabetic Treatment

Momoh A. Mumuni [1,*], Ugwu E. Calister [2], Nafiu Aminu [3], Kenechukwu C. Franklin [1], Adedokun Musiliu Oluseun [4], Mohammed Usman [5], Barikisu Abdulmumuni [6], Oyeniyi Y. James [3], Kenneth C. Ofokansi [1], Attama A. Anthony [1], Emmanuel C. Ibezim [1] and David Díaz Díaz [7,8,9,*]

1 Drug Delivery Research Unit, Department of Pharmaceutics, University of Nigeria Nsukka, 410001 Enugu State, Nigeria; frankline.kenechukwu@unn.edu.ng (K.C.F.); kenneth.ofokansi@unn.edu.ng (K.C.O.); anthony.attama@unn.edu.ng (A.A.A.); emmanuel.ibezim@unn.edu.ng (E.C.I.)
2 Department of Pharmaceutical Technology and Industrial Pharmacy, University of Nigeria, Nsukka, 410001 Enugu State, Nigeria; calister.ugwu@unn.edu.ng
3 Department of Pharmaceutics and Pharmaceutical Microbiology, Usmanu Danfodiyo University, 840101 Sokoto, Nigeria; nabgus@yahoo.com (N.A.); iyeniyi.yinka@udusok.edu.ng (O.Y.J.)
4 Department of Pharmaceutics and Pharmaceutical Technology, University of Uyo, 520271 Akwa-Ibom State, Nigeria; mo_adedokun@yahoo.com
5 Department of Chemistry, Usmanu Danfodiyo University, 840 Sokoto, Nigeria; hammeduss@yahoo.com
6 Department of Geology, University of Nigeria Nsukka, 410001 Enugu State, Nigeria; barikisu.abdulmumuni@unn.edu.ng
7 Departamento de Química Orgánica, Universidad de La Laguna, Avda. Astrofísico Francisco Sánchez 3, La Laguna, 38206 Tenerife, Spain
8 Instituto de Bio-Orgánica Antonio González, Universidad de La Laguna, Avda. Astrofísico Francisco Sánchez 2, La Laguna, 38206 Tenerife, Spain
9 Institut für Organische Chemie, Universität Regensburg, Universitätsstr. 31, 93053 Regensburg, Germany
* Correspondence: audu.momoh@unn.edu.ng or momohmumuni123@gmail.com (M.A.M.); ddiazdiaz@chemie.uni-regensburg.de or ddiazdiaz@ull.edu.es (D.D.D.); Tel.: +234-8037784357 or +34-934006145 (M.A.M.); +34-922318584 (D.D.D.)

Received: 20 March 2020; Accepted: 7 April 2020; Published: 11 April 2020

Abstract: In this study, different ratios of mucin-grafted polyethylene-glycol-based microparticles were prepared and evaluated both in vitro and in vivo as carriers for the oral delivery of insulin. Characterization measurements showed that the insulin-loaded microparticles display irregular porosity and shape. The encapsulation efficiency and loading capacity of insulin were >82% and 18%, respectively. The release of insulin varied between 68% and 92% depending on the microparticle formulation. In particular, orally administered insulin-loaded microparticles resulted in a significant fall of blood glucose levels, as compared to insulin solution. Subcutaneous administration showed a faster, albeit not sustained, glucose fall within a short time as compared to the polymeric microparticle-based formulations. These results indicate the possible oral delivery of insulin using this combination of polymers.

Keywords: insulin; mucin; polyethylene glycol; microparticles; toxicology

1. Introduction

Globally, diabetes is one of the most common metabolic diseases, affecting more than 345 million people worldwide [1]. It is estimated that in 2030, this number will increase to about 552 million, which is outrageous and a serious threat to public health [2]. Diabetes is broadly classified into two major groups, namely, Type 1, also known as insulin-dependent diabetes, which occurs when the body cannot produce enough insulin or when the pancreas is unable to produce sufficient insulin, causing hyperglycemia; and Type 2, a more common type of diabetes that results from the combination of inadequate insulin secretion and insulin resistance. In Type 1, the patient requires routine administration of insulin, being nowadays considered the most effective treatment with high specificity and activity [3]. Currently, the conventional route of insulin administration is subcutaneous, which faces many hurdles such as the difficulty of achieving a normal pattern of nutrient-related and basal insulin, significant tissue trauma and pain, particularly with multiple dosing, and hypoglycemia episodes [4].

In order to abrogate these challenges, it is of utmost importance to develop adequate oral insulin delivery systems for effective diabetes therapy. Indeed, oral ingestion remains the most desirable route for the application of pharmaceuticals, as it does not require a skilled health care professional and it therefore conveniently allows self-administration of the drug [5]. Oral ingestion of insulin would deliver the drug directly to the liver through portal circulation, mimicking the fate of endogenously secreted insulin [6]. However, oral delivery of many peptidic drugs faces important challenges. The high molecular weight and hydrophilic nature of many peptides limit their passive diffusion across the cell membrane, which forces them to pass through the gaps in the paracellular space (1–5 nm) [7]. Moreover, in the gastrointestinal tract (GIT), insulin absorption after oral administration is hampered by some pharmacokinetic challenges due to acidic gastric pH and the presence of digestive enzymes [3]. Insulin oral delivery has been investigated using some strategies to address these physicochemical problems by the use of colloidal systems such as microparticles (MPs), liposomes, nanoparticles, and microemulsions, just to mention a few [8]. Among these carriers, MPs have been proposed to overcome the pharmacokinetic problems associated with oral insulin delivery since they prevent peptide degradation in the acidic environment, prolong residence time, increase drug absorption, and promote controlled release, resulting in a better therapeutic response and patient compliance. MP delivery systems are usually formulated by encapsulation of the desired drug using a suitable inert polymeric carrier [9].

Within this context, a suitable biopolymer with mucoadhesive properties can be used as an enclosing carrier in MP preparation in order to enhance stability and achieve controlled release of, especially, some drugs that metabolize quickly. Mucoadhesiveness helps enhance closeness and prolong contact between drugs in polymeric carriers and their mucous surfaces, which maximizes the rate of drug absorption [10]. Such delivery systems confer significant advantages, particularly for drugs that are unstable in the GIT. In this context, we recently developed a polyethylene glycol (PEG)/mucin system for oral delivery of metformin hydrochloride, where PEG acts as an adhesion enhancer [11]. Regarding the other component of this system, mucins are a family of high-molecular-weight glycosylated proteins produced by epithelial tissues in most animals, being the major structure-forming component of the viscoelastic mucus [12,13]. These glycoconjugates have gained increasing attention from researchers working on drug delivery as a biocompatible modifier for a variety of enzymes and proteins, enhancing their mucoadhesiveness [14,15]. Additionally, improved in vivo retention time and a reduction in toxicity, antigenicity, and immunogenicity are also advantages provided by the use of mucins in drug delivery systems.

Thus, based on the above-mentioned reports, we hypothesized that a polymeric drug carrier based on blends of mucin and PEG could be advantageous for insulin delivery due to the biocompatible, biodegradable, and protective nature of these polymers. As compared to previous reports on MP-based formulations [16], PEG/mucin-based MPs offer a competitive advantage for insulin protection in an acidic medium due to the PEG component. Moreover, the mucin is resistant to proteolytic enzymes and improves mucoadhesion to the mucosal wall, which leads to an increase in the residence time in the GIT [14,17]. Consequently, the main objective of this study was to develop MPs based on snail mucin and PEG-4000 coated with Eudragit RS100, a pH-sensitive polymer, to facilitate the encapsulation of insulin in the core of the MPs. Additionally, the ability of the MPs to enhance the intestinal bioavailability of insulin was also evaluated in vivo in diabetic rats.

2. Materials and Methods

2.1. Materials

Human insulin (Humulin) was obtained from Elly Pharm. Ltd., India; Eudragit® RS 100 from Evonik, Darmstadt, Germany; and acetone from BDH. Purified and deionized water was obtained from the Department of Pharmaceutics, University of Nigeria, Nsukka. Mucin was extracted using acetone as described by Adikwu and co-workers [13]. Wistar rats were obtained from the Department of Veterinary Medicine, University of Nigeria, Nsukka. All other chemicals and solvents used in this study were of analytical grade and used without further purification.

2.2. Preparation of Empty MPs

Polymeric MPs were prepared at five different ratios of mucin to polyethylene glycol (1:1, 1:2, 1:3, 1:4, and 1:5), as described in Table 1. Approximately 10 g of mucin and 10 g of PEG-4000 were dispersed separately in 40 mL of distilled water to obtain a homogenous dispersion and were allowed to stand for 24 h. The solution of PEG-4000 was gradually added to the mucin dispersion and dispersed using a Silverson L4R mixer (Silverson Machines Ltd, Waterside, England), affording a homogeneous mixture. Both dispersions were added gradually into a 250 mL container containing 20 mL of soft liquid paraffin. The resulting mixture was subsequently subjected to homogenization using an ultrasonic probe sonicator (Model: ATP 500, Mumbai, India) at 50 Watt for 10 min in an ice bath. The so-formed MPs were then recovered through filtration using a millipore filter. Thereafter, the MPs were washed several times with cold acetone to remove any trace of liquid paraffin. The recovered MPs were dried in a desiccator and stored in airtight containers for further studies. The same procedure was repeated for all other batches and coded A0–A5 (Table 1).

Table 1. Formulations based on different ratios of mucin and PEG-4000 used for preparing insulin-containing microparticle-based carriers.

Formulation Batch	Mucin (g)	PEG-4000 (g)	Insulin (mL)
A0	1	1	0.0
A1	1	1	15.0
A2	1	2	15.0
A3	1	3	15.0
A4	1	4	15.0
A5	1	5	15.0

2.3. Coating of the MPs

The coating of the MPs was carried out following the procedure reported in the literature [14,18–20] with slight modifications. In brief, a 2.0% solution of Eudragit® RS-100 in dichloromethane was freshly prepared, and MPs (5.0 g) were gently dispersed and shaken for about 3 min in the corresponding coating solution. The so-formed coated MPs were recovered by filtration using a filter tube (Advantec Toyo Kaisha, Ltd., Tokyo, Japan) and dried by flushing cold air. This procedure was repeated five times; then, the coated MPs were recovered, air dried, and stored in an airtight bottle for further studies.

2.4. Loading of Insulin in MPs

A diffusion method was employed for loading insulin into the MPs and to prevent the disruption of the insulin structure during the mixing and homogenization steps. Specifically, coated MPs (5.0 g) were placed in a beaker, and insulin solution (15 mL of a 100 IU/mL solution) was added. The MPs were allowed to hydrate by interaction with the insulin solution for 20 min. The hydrated MPs were then recovered and subsequently freeze-dried for 12 h. Thereafter, the resultant insulin-loaded MPs were stored in airtight screw-capped bottles and stored at 5 °C for further use. The same procedures were employed for batches A1–A5, while batch A0 was prepared in the absence of insulin and served as a negative control.

2.5. Characterization of Insulin-Loaded MPs

2.5.1. Thermal Analysis

The melting transitions and changes in heat capacity of the batches of insulin-loaded MPs were determined using a Netzsch differential scanning calorimeter (DSC), model DSC 204 F1 (Erich NETZSCH GmbH & Co. Holding KG, Selb, Germany). Each sample (ca. 3–5 mg) was weighed into a hermetically sealed aluminum pan, and the thermal properties were determined within the range 20–400 °C at a heating rate of 10 K/min under a 20 mL/min nitrogen flux. Baselines were determined using an empty pan, and all the thermograms were baseline-corrected.

2.5.2. Recovery Values of Insulin-Loaded MPs

The amount of MPs recovered from the formulation was calculated using Equation (1), while the recovery rate was estimated using Equation (2):

$$\% \text{ Recovery} = \frac{W_1}{W_2 + W_3} \times 100 \quad (1)$$

where W_1 is the weight of the MPs (g), W_2 is the amount of insulin (g), and W_3 is the amount of carrier and additives (g).

$$\text{Recovery rate } (\%) = \frac{\text{Amount recovered}}{\text{Original amount}} \times 100 \quad (2)$$

2.5.3. Encapsulation Efficiency (EE) and Drug Loading (DL) of Insulin-Loaded MPs

Each batch of insulin-loaded MPs (10 mg) was dispersed in phosphate buffer (10 mL, pH 7.4) and was placed into a microconcentrator (5000 MWCO Vivascience, Hanover, Germany). The mixtures were centrifuged (TDL-4 B. Bran Scientific and Instru. Co., London, England), for 120 min at 1500 rpm, and the supernatants were assayed using a spectrophotometric method. The percentage of insulin encapsulated in the MPs was obtained with reference to a standard Beer's plot of pure insulin and drug loading using Equations (3) and (4), respectively.

$$\text{Encapsulation efficiency} = \frac{\text{Actual drug content}}{\text{Theoretical drug content}} \times 100 \quad (3)$$

$$\text{Drug loading} = \frac{\text{Amount of drug encapsulated}}{\text{Weight of the formulation}} \times 100 \quad (4)$$

2.5.4. Morphology and Particle Size Evaluations

The morphology of the MPs was determined using scanning electron microscopy (SEM) (JEOL-6500F, Tokyo, Japan) under an accelerated voltage of 4 KV and a working distance of 6 mm. Briefly, a drop of sample dispersion was spread onto a metal slab, and the excess droplets were removed using filter paper. The samples were then coated in a cathode evaporator with a fine carbon layer and observed by SEM.

The mean particle size and distribution of the insulin-loaded MPs were analyzed using a microscopic imaging analysis principle, employing a polarized light microscope (Leica Microsystems GmbH, Wetzlar, Germany) with a Motic image analyzer attachment (Moticam, Fujian, China). All tests were carried out in triplicate and the results were averaged.

2.6. In Vitro Release Study

The in vitro release profiles of the insulin were evaluated using a dialysis membrane in acidic pH of 1.2 (HCl solution) or in pH phosphate buffer solution as a release medium to simulate the gastric juice and intestinal juice of the gastrointestinal environment, respectively. Herein, insulin-loaded MPs (50 mg) were enclosed in a dialysis membrane tubing (80–100 kDa, Spectrum Laboratories Inc., Lorzweiler, Germany) end-to-end secured with a thread to avoid leakage of the content. Thereafter, the membrane containing insulin MPs was suspended in 200 mL of HCl solution (pH 1.2) or phosphate buffer solution (pH 7.4) in a beaker mounted on a magnetic stirrer set at 50 rpm and maintained at 37 ± 1 °C. At different time intervals, aliquots of release medium (5 mL) were withdrawn and replaced with the same volume of fresh medium to maintain sink conditions. The withdrawn samples were filtered and analyzed by UV–visible spectrophotometry at 271 nm. The amount of insulin released was calculated using the standard calibration curve for insulin. The experiments were done in triplicate and the results were averaged.

2.7. Antidiabetic Study

Wistar albino rats with an average weight of 160 g were purchased from the animal house of the Department of Pharmacology and Toxicology, University of Nigeria, Nsukka, and were allowed to acclimatize in laboratory conditions for 7 days before the experimental procedures. Prior to animal studies, the ethical animal procedures were obtained from the institutional ethics department and the reference number DOR/UNN/17/00014 was assigned. The ethical principle is in line with the approved standard [21,22]. Before starting the experiments, the animals were fasted for 12 h but allowed free access to water.

2.7.1. Induction of Diabetes

Albino male rats with average weight 160 g were fasted for 12 h before the initiation of diabetes. Diabetes was elicited with intraperitoneal administration of 0.5 mL alloxan monohydrate prepared with normal saline (0.9% NaCl) at a dose of 200 mg/kg. The blood glucose levels were monitored for 72 h using an Accu-Check digital glucose meter (F. Hoffmann-La Roche AG, Basel, Switzerland) for all the rats until the induction and stabilization of a diabetic state. Parameters such as urinary urgency and weight loss were monitored for 5–7 days post-administration. Rats with elevated blood glucose levels of ≥100 mg/dL along with other signs of diabetes were enlisted in this investigation.

2.7.2. Antidiabetic Efficacy of Insulin-Loaded MPs

Batch A5 (mucin/PEG-4000 ratio of 1:5) was selected for this evaluation due to the maximum in vitro insulin release and high encapsulation efficiency observed for this batch. Twenty-five diabetic rats were randomly divided into five groups (five rats per group) and caged separately. The groups were treated as follows:

- Group 1: received orally 50 IU /kg of batch A5;
- Group 2: received orally 50 IU/kg of pure insulin solution;
- Group 3: received subcutaneously 5 IU/kg of insulin;
- Group 4: received orally 5 mL/kg of water;
- Group 5: received 50 mg/kg of batch A0 (blank).

Groups 3 and 4 served as positive and negative controls, respectively. All the oral administrations were done using a gastric nasal tube. At predetermined time intervals, blood samples were collected from the marginal tail of the rats for 24 h post-administration and the blood glucose levels were checked using a glucometer (Accu-Check, Switzerland).

The serum enzymes were determined by using commercially available kits according to the manufacturers' instructions.

2.8. Statistical Analysis

Data were analyzed using SPSS Version 16.0m (SPSS Inc. Chicago, IL, USA). All values are expressed as mean ± SD. Differences between means were assessed using one-way ANOVA and Student's *t*-test. $p < 0.05$ was considered statistically significant.

3. Results and Discussion

3.1. Thermal Analysis

DSC measurements were carried out to determine the thermal properties of the MPs. According to the van't Hoff theory, peak integration and melting temperature values are used to compare the consistency of formulations from different batches and to determine impurities that will change the melting profiles of the materials. Melting temperature is a strong indication of drug purity and therefore allows not only for a quick screening of the melting process, but also for the identification of drug populations that may be in a different conformation or interacting with an excipient, resulting in shoulder regions in the thermograms. The thermograms of insulin, mucin, and PEG-4000 showed endothermic peaks at 94.88, 95.0, and 62.4 °C, respectively (Figure S1), while batches A1–A5 showed peaks at 94.9 °C (Figure 1). In comparison, insulin and mucin presented broader peaks, while PEG-4000 showed a very sharp endothermic peak. The peaks observed for the MPs are consistent with a thermal-induced transition. PEG-4000 is a nonionic amphiphilic polymer with thermotropic phase behavior, including melting temperature and enthalpy changes, that is highly affected by the presence of mucin due to the interactions between both polymers. All insulin-loaded MPs showed endothermic peaks around the melting temperature of mucin and insulin (≈95 °C), independently of the PEG-4000 concentration. A transition peak was observed for each batch due to the solubilization of PEG into the mucin. Therefore, the thermal behavior of the matrixes depends on the interactions between the constituents, as well as their concentrations. The sharpness and symmetry of the endothermic peaks of the insulin-loaded MPs, corresponding to the mucin, are indicative of matrixes with high purity (Figure 1). Furthermore, batches A5, A4, and A3 showed sharper peaks, while batches A1 and A2 displayed slightly broader peaks, suggesting an amorphous nature for the latter. This amorphous feature was observed at lower enthalpy and less crystallinity of the matrixes, which could lead to retention of the entrapped drug over time. The reduced crystallinity may indicate imperfect lattices leading to a special pocket that can accommodate a large number of drug molecules [23]. Hence, an increase in the quantity of PEG-4000 was observed to increase the crystallinity of the formulations, the order being as follows: A5 > A4 > A3 > A2 > A1. The thermogram of batch A0 (blank) clearly showed the expected glass transition (Figure 1).

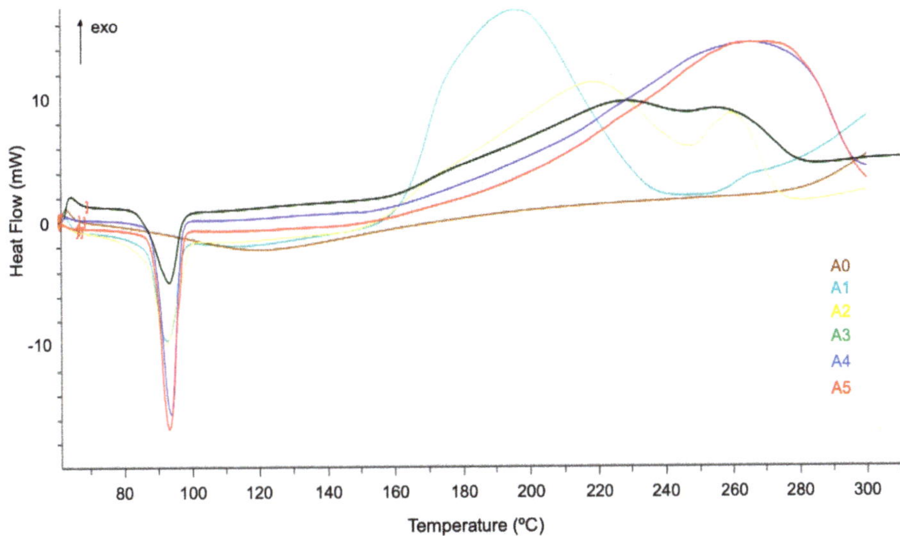

Figure 1. DSC thermograms of insulin-loaded PEG/mucin matrixes in superposition. Batch description: A0 = mucin/PEG/insulin (1:1:0), A1 = mucin/PEG/insulin (1:1:15), A2 = mucin/PEG/insulin (1:2:15), A3 = mucin/PEG/insulin (1:3:15), A4 = mucin/PEG/insulin (1:4:15), A5 = mucin/PEG/insulin (1:5:15). Note: mucin and PEG are measured in grams, while insulin is measured in milliliters.

3.2. *Morphology and Particle Size of Insulin-Loaded MPs*

The morphologies of the MPs were studied by scanning electron microscope (SEM). Several studies have demonstrated that the morphology and pore size of matrixes intended for drug delivery play an important role in the control of the release kinetics [11]. SEM images of batches A1 and A2 revealed small nonspherical particles that clumped together and mixed with little needle-like crystals (Figure 2). The morphology of batch A3 was characterized by large lumpy crystals, while mixtures of lumpy and small crystals were identified in batches A4 and A5 (Figure 2). In general, the MPs showed rough porous surfaces and irregular shapes. Batches A1 and A2 had almost identical morphologies with agglomerated particles, while batches with higher concentration of PEG-4000 (i.e., batches A3–A5) showed a clearer distribution of small and large MPs.

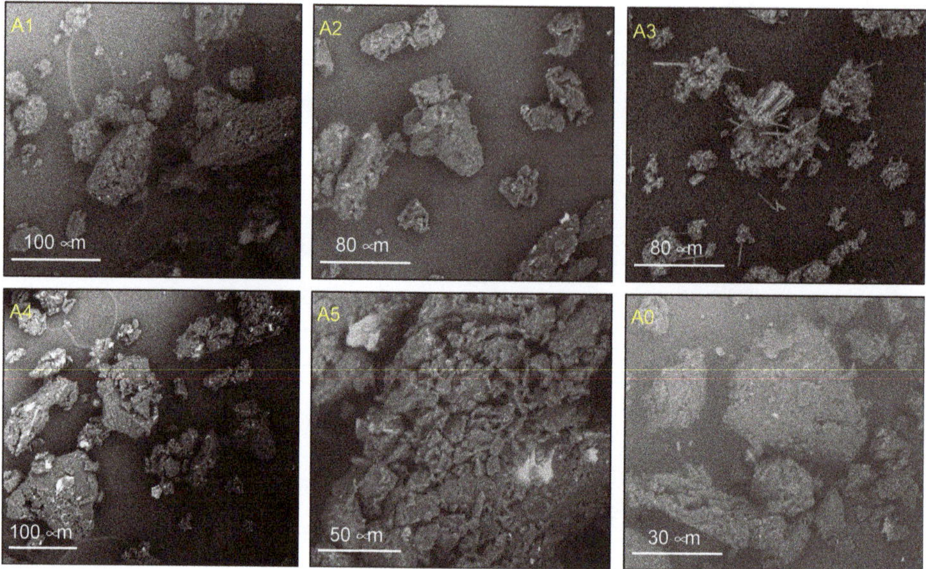

Figure 2. SEM of representative batches of insulin-loaded microparticles (MPs) containing different ratios of mucin to PEG. Formulations: A1 = mucin/PEG/insulin (1:1:15), A2 = mucin/PEG/insulin (1:2:15), A3 = mucin/PEG/insulin (1:3:15), A4 = mucin/PEG/insulin (1:4:15), A5 = mucin/PEG/insulin (1:5:15). Control A0 corresponds to the formulation containing no insulin (mucin/PEG, 1:1). Each formulation is indicated in the top left corner of the corresponding microscopy image.

Moreover, particle size may be affected by one or more factors, including, among others, the degree of homogenization, homogenization speed, composition ratio, rate of particle size growth, and crystal habit of the particle [24]. During Ostward ripening, the particles tend to move towards one another, which leads to the formation of aggregates due to the kinetic advantage gain by larger crystals as a result of an increase in the concentration [23,24]. In this study, the particle size was found to increase with increasing PEG/mucin ratio. Therefore, batch A5, having the highest PEG-4000/mucin ratio, displayed a significantly ($p < 0.05$) higher particle size (398 ± 0.21 μm) as compared to batch A1 with a particle size of 311 ± 0.22 μm, while unloaded MPs (batch A0) showed a particle size of 276 ± 0.03 μm (Table 2). Thus, the particle size tends to increase after drug loading due to the intensification of interparticle attractive interactions (van der Waals forces) leading to particle aggregation.

Table 2. Physicochemical parameters of insulin-loaded MPs (n = 5) [1].

Batch [2]	EE (%)	DL (%)	PS (μm)	RV (%)
A0	–	–	276 ± 0.03	66 ± 0.14
A1	82 ± 0.12	18 ± 0.01	311 ± 0.22	75 ± 0.13
A2	81 ±0.11	26 ± 0.24	321 ± 0.13	80 ± 0.14
A3	83 ±0.23	28 ± 0.04	328 ± 0.11	83 ± 0.17
A4	84 ±0.16	35 ± 0.17	346 ± 0.05	86 ± 0.13
A5	92 ± 0.12	39 ± 0.12	398 ± 0.21	89 ± 0.12

[1] Abbreviations: EE = encapsulation efficiency; DL = drug loading; RV = recovery values; PS = particle size. [2] Formulation A0 = unloaded MPs; formulations A1–A5: A1 = mucin/PEG/insulin (1:1:15), A2 = mucin/PEG/insulin (1:2:15), A3 = mucin/PEG/insulin (1:3:15), A4 = mucin/PEG/insulin (1:4:15), A5 = mucin/PEG/insulin (1:5:15). Data represent mean values ± SD. Note: mucin and PEG are measured in grams, while insulin is measured in milliliters.

3.3. Encapsulation Efficiency, Drug Loading, and Recovery Yield of MPs

Encapsulation efficiency (EE) is defined as the ratio between the weight of entrapped drug and the total original weight of the drug, while drug loading (DL) is defined as the ratio between the amount of entrapped drug and the total weight of the matrix [25]. The nature and concentration of the carrier, the formulation technique, and the nature of the active pharmaceutical ingredient are some of the factors that influence the EE and DL of MPs. The insulin-loaded MPs showed EE, DL, and recovery values (RV) that ranged over 81%–92%, 18%–39%, and 75%–89%, respectively (Table 2). These encouraging values were attributed to the method used for the inclusion of insulin; the insulin was not incorporated during the homogenization step because this could lead to degradation, as has been observed for protein drugs subjected to high homogenization forces [7]. The observed high EE% is ascribed to a high adsorption capacity for insulin in these polymeric matrixes. Additionally, an increase in PEG-4000 concentration caused increases in the EE, DL, and RV (e.g., see batch A5 vs. A1) (Table 2). The recovery values (RV) of batches loaded with insulin MPs ranged from 75% to 89%, being much higher than that of the unloaded batch A0 (66.5%). This could be attributed to good adsorption of insulin into the polymeric MPs. Furthermore, batch A5 showed the maximum RV value of 89%, while batch A1 and unloaded batch A0 displayed RV values of 75% and 66%, respectively. Hence, the recovery values increased with drug loading, which is in good agreement with the observations made in previous studies [26].

3.4. In Vitro Release of Insulin-Loaded MPs

Figure 3 shows the cumulative release of insulin from the MPs at pH 1.2 and pH 7.4. All formulations showed a gradual release of insulin. Interestingly, there was a minor release of insulin (>12%) at pH 1.2 (Figure 3a), indicating that the MPs coated with Eudragit RS-100 resisted dissolution in the acidic medium, thereby preventing the expulsion of the entrapped insulin. The slight amount of insulin observed in this case could be the result of insulin molecules physically adhered to the surface of the MPs.

Interestingly, higher release of insulin was observed at pH 7.4 (Figure 3b), as compared to the release in acidic medium. Maximum and minimum release were observed in batch A5 (92%) and batch A1 (68%), respectively. The results showed a strong dependence between the insulin release and the PEG concentration used in the formulation. For instance, batch A1 (mucin/PEG = 1:1) showed low release compared to A5 (mucin/PEG = 1:5). This can be correlated to the viscoelastic behavior of the formulation provided by the mucin component, which would create better adherence to the gastrointestinal mucosal wall, resulting in increased drug residence time, site-targeted drug release, and improved drug bioavailability. Furthermore, the mucous gel layer has been shown to have a mean turnover time that varies between 47 and 270 min; this is an important parameter in designing mucoadhesive drug delivery systems [27,28]. It has been noted that the mucous aqueous gel layer blocks drug diffusion and adsorption through the epithelium. The insulin release kinetics followed the order A1 > A2 > A3 > A4 > A5. The maximum drug release observed for the MPs with the highest PEG concentration was ascribed to the capability of PEG chains to disseminate across the mucus and facilitate drug solubilization. This resulted in faster release at the absorption site compared to that of mucin, which is good agreement with previous reports [29]. From the extrapolation of the graph, the T_{40} (i.e., the time taken to release 40% of the drug) values of batch A1, A2, A3, A4, and A5 MPs were 8, 4, 5, 4, and 4 h, respectively. This was in agreement with the significantly delayed release and higher retention time of batch A1 compared to the other batches.

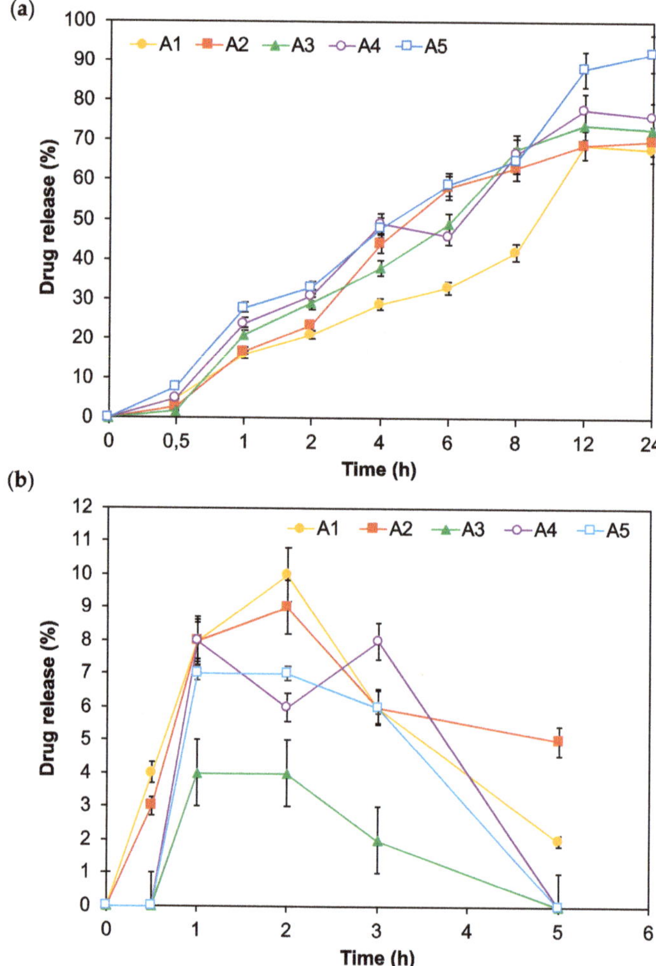

Figure 3. Insulin release profiles at pH 7.4 (**a**) and at pH 1.2 (**b**). Formulations A1–A5: A1 = mucin/PEG/insulin (1:1:15), A2 = mucin/PEG/insulin (1:2:15), A3 = mucin/PEG/insulin (1:3:15), A4 = mucin/PEG/insulin (1:4:15), A5 = mucin/PEG/insulin (1:5:15). Data represent mean values ± SD ($n = 3$; $p < 0.05$). Note: mucin and PEG are measured in grams, while insulin is measured in milliliters.

3.5. *In Vivo Antidiabetic and Toxicological Studies*

Alloxan was used to induce a hyperglycemic condition in the male rats. Alloxan is a diabetogenic chemical substance that inactivates the β cells of the pancreases, eliciting insulin deficiency without affecting other islet types [30]. The results of the blood glucose reduction test, presented in Figure 4, indicate that insulin-loaded MPs (Group 1) displayed a more pronounced and significant ($p < 0.05$) blood glucose decrease, from 100 mg/dL to 48 mg/dL post-administration, compared to other samples. This is an indication that the MPs were able to provide protection to the insulin and effectively deliver it to the absorption site in vivo. A dose of 50 IU/Kg has been reported to have a similar effect to a higher dose (100 IU/Kg) with a minimum blood glucose level of 55%, and is thus taken as an effective therapeutic dose to saturate the insulin absorption sites [31]. Group 2, which received pure insulin by oral administration, showed minor hypoglycemia when compared to the group receiving insulin-loaded MPs, indicating that no significant absorption occurred in this group. This small hypoglycemia effect may be due to a small fraction of insulin being directly absorbed through the intestinal wall [32]. Such direct insulin absorption occurs at specific insulin receptors located in intestinal enterocytes [33] and undergoes rapid internalization by the epithelial cells through the interstitial space, ending up in the blood stream. The upper intestinal area has been reported to be the most active area of insulin absorption during fasting conditions [30]. Furthermore, no apparent blood glucose reduction was observed in the groups that received water (negative control, Group 4) or placebo (Group 5). In contrast, Group 3, which received a subcutaneous insulin injection, showed the highest glucose reduction level 2 h after administration. After this time, the glucose level increased again beyond normal levels, unlike in Group 1, where the level was maintained after reaching the highest glucose reduction. This result is an indication that subcutaneous insulin will require multiple doses to stabilize the patient within 12 h.

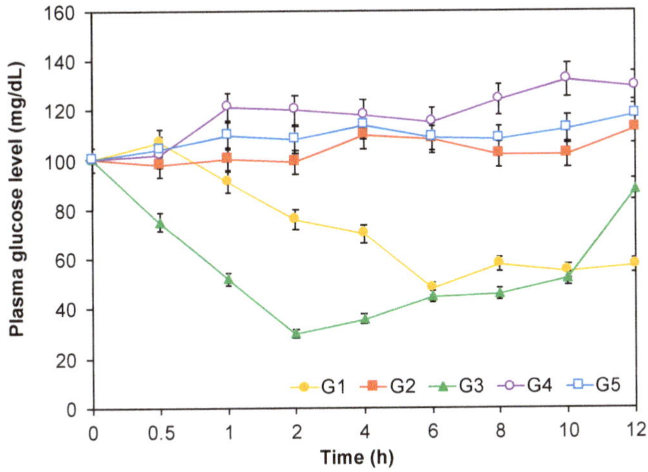

Figure 4. Percentage decrease of the blood glucose level in rats treated with insulin-loaded MP formulations ($n = 5$). Groups G1–G5: G1 = received insulin-loaded MPs (formulation A5), G2 = received pure insulin solution, G3 = received subcutaneous insulin, G4 = received water, G5 = received unloaded MPs (batch A0). Data represent mean values ± SD ($n = 3$).

Toxicological evaluations of the formulations were based on enzymes such as aspartate transaminase (AST) or serum glutamic oxaloacetic transaminase (SGOT), alanine transaminase (ALT) or serum glutamic pyruvic transaminase (SGPT), and alkaline phosphatase (ALP). These enzymes are found mainly in the liver, red blood cells, heart, kidneys, and pancreas, being useful biomarkers for the diagnosis of injured body tissues, particularly in organs such as heart and liver [34,35]. It has been demonstrated that the more AST and ALT are released into the bloodstream, the more tissues are damaged [36]. The results obtained for our formulations showed that the quantities of these enzyme markers (i.e., SGPT, SGOT, ALP) in the serum were within the established reference values (Table 3). In addition, no significant variation from the control batch (water) was observed. Overall, these values suggest that the constituents of the insulin-loaded MPs are safe with no hepatotoxic effects.

Table 3. Toxicology study via liver function test (LFT) [1].

Batch [2]	SGPT (IU/L)	SGOT (IU/L)	ALP (IU/L)
A1	38.0 ± 0.16	66 ± 0.01	113 ± 0.11
A2	39.5 ± 0.21	65 ± 0.33	118 ± 0.23
A3	36.5 ± 0.15	67 ± 0.13	120 ± 0.21
A4	37.0 ± 0.11	66 ± 0.04	115 ± 0.22
A5	38.5 ± 0.13	68 ± 0.34	116 ± 0.52
Control (water)	36.0 ± 0.30	64 ± 0.12	115 ± 0.11
Reference value	10-40	50-150	30-130

[1] Abbreviations: SGOT = serum glutamic oxaloacetic transaminase; SGPT = serum glutamic pyruvic transaminase; ALP = alkaline phosphatase. [2] Batch description: A1 = mucin/PEG/insulin (1:1:15), A2 = mucin/PEG/insulin (1:2:15), A3 = mucin/PEG/insulin (1:3:15), A4 = mucin/PEG/insulin (1:4:15), A5 = mucin/PEG/insulin (1:5:15). Data represent mean values ± SD ($n = 3$). Note: mucin and PEG are measured in grams, while insulin is measured in milliliters.

4. Conclusions

In conclusion, despite great progress in the knowledge on insulin oral delivery using polymeric systems, there are still some limitations with respect to the stability and toxicity of the formulations. In this work, we demonstrated that insulin-loaded PEG-4000/mucin MPs can help to overcome some of these challenges. Such MPs display a smooth nonspherical morphology and can be prepared with more than 81% encapsulation efficiency via an emulsification–coacervation method followed by drug diffusion. The developed formulations exhibited good physicochemical performance and bioactivity. In vitro release studies at intestinal pH revealed significant insulin release, while maximum insulin was found to be retained within the MPs at stomach acid pH. Specifically, insulin-loaded MPs prepared with a 1:1 mucin/PEG ratio exhibited the highest prolonged drug release, while MPs containing higher PEG concentrations displayed faster drug release. Thus, insulin release was tuned within a range of 68%–92%. Furthermore, insulin-loaded MPs significantly ($p < 0.05$) reduced the blood glucose level to 62 mg/dL after oral treatment. The blood glucose reduction effect was higher in animals that received orally administered insulin-loaded MPs than in those that were treated with insulin solution administered orally as a negative control. Interestingly, insulin-loaded MPs exhibited maximum glucose level reduction after 6 h, which was equivalent to that observed in the case of conventional subcutaneously administered insulin. Finally, no hepatotoxicity was observed after oral administration of insulin-loaded MPs. These results indicate that the use of insulin-loaded MPs based on the combination of PEG-4000 and mucin, without the use of organic crosslinkers, may be a promising alternative approach for diabetes treatment. Further pharmacokinetic studies to fully evaluate the potential of these polymeric carriers are underway in our laboratories, and the results will be published in due course.

Supplementary Materials: The following are available online at http://www.mdpi.com/2076-3417/10/8/2649/s1, Figure S1: DSC thermograms of mucin (*top*), insulin (*middle*), and PEG-4000 (*bottom*).

Author Contributions: Conceptualization, M.A.M.; Formal analysis, D.D.D.; Funding acquisition, M.A.M. and D.D.D.; Investigation, M.A.M., U.E.C., N.A., K.C.F., A.O.M., M.U., B.A., O.Y.J., K.C.O., A.A.A. and E.C.I.; Supervision, M.A.M.; Writing—original draft, M.A.M., K.C.O., A.A.A. and E.C.I.; Writing—review and editing, D.D.D. All authors have read and agreed to the published version of the manuscript.

Funding: This study was funded by the Tertiary Education Trust Fund (TETFUND)–National Research Fund (NFR) of Nigeria with grant number TETFUND/DESS/NRF/STI/13/VOL.1.

Acknowledgments: Momoh is very grateful to the Classical and Biomedical Laboratory (Nsukka), InterCEDD Centre (Nsukka), and Projex Biomedical Laboratory (Nsukka). We are very thankful to Isa Yakubu, from the Department of Chemical Engineering at the Ahmadu Bello University Zaria, for assisting with thermal characterization of the materials. David Díaz Díaz thanks the Deutsche Forschungsgemeinschaft (DFG) for the Heisenberg Professorship Award and the Spanish Ministry of Science, Innovation and Universities for the Senior Beatriz Galindo Award (Distinguished Researcher).

Conflicts of Interest: The authors declare no conflict of interest.

References

1. Sah, S.P.; Singh, B.; Choudhary, S.; Kumar, A. Animal models of insulin resistance: A review. *Pharmacol. Rep.* **2016**, *68*, 1165–1177. [CrossRef] [PubMed]
2. Whiting, D.R.; Guariguata, L.; Weil, C.; Shaw, J. IDF Diabetes atlas: Global estimates of the prevalence of diabetes for 2011 and 2030. *Diabetes Res. Clin. Pract.* **2011**, *94*, 311–321. [CrossRef] [PubMed]
3. Fonte, P.; Araújo, F.; Silva, C.; Pereira, C.; Reis, S.; Santos, H.A. Polymer-based nanoparticles for oral insulin delivery: Revisited approaches. *Biotechnol. Adv.* **2014**, *33*, 1342–1354. [CrossRef] [PubMed]
4. Rolla, A.R.; Rakel, R.E. Practical approaches to insulin therapy for type 2 diabetes mellitus with premixed insulin analogues. *Clin. Ther.* **2005**, *27*, 1113–1125. [CrossRef]
5. Sastry, S.V.; Nyshadham, J.V.; Fix, J.A. Recent technological advances in oral drug delivery—A review. *Pharmaceut. Sci. Tech. Today* **2000**, *3*, 138–143. [CrossRef]
6. Yaturu, S. Insulin therapies: Current and future trends at dawn. *World J. Diabetes* **2013**, *4*, 1–7. [CrossRef]
7. Builders, P.F.; Kunle, O.O.; Adikwu, M.U. Preparation and characterization of mucinated agarose: A mucin-agarose physical crosslink. *Int. J. Pharm.* **2008**, *356*, 174–180. [CrossRef]
8. Elsayed, A.; Remawi, M.A.; Qinna, N.; Farouk, A.; Badwan, A. Formulation and characterization of an oily-based system for oral delivery of insulin. *Eur. J. Pharm. Biopharm.* **2009**, *73*, 269–279. [CrossRef]
9. Huang, Y.J.; Wang, C.H. Pulmonary delivery of insulin by liposomal carriers. *J. Control Release* **2006**, *113*, 9–14. [CrossRef]
10. Pothal, R.K.; Sahoo, S.K.; Chatterjee, S.; Sahoo, D.; Barik, B.B. Preparation and evaluation of mucoadhesive microcapsules of theophylline. *Ind. Pharm.* **2004**, *3*, 74–79.
11. Momoh, M.A.; Adedokun, M.O.; Adikwu, M.U.; Kenechukwu, F.C.; Ibezim, E.C.; Ugwoke, E.E. Design, characterization and evaluation of PEGylated-mucin for oral delivery of metformin hydrochloride. *Afr. J. Pharm. Pharmacol.* **2013**, *7*, 347–355.
12. Mortazavi, S.A.; Carpenter, B.G.; Smart, J.D. Comparative study on the role played by mucus glycoprotein in the rheological behaviors of the mucoadhesive/mucosal interaction. *Int. J. Pharm.* **1992**, *94*, 195–201. [CrossRef]
13. Adikwu, M.U.; Aneke, K.O.; Builders, P.F. Biophysical properties of mucin and its use as a mucoadhesive agent in drug delivery: Current development and future concepts. *Nigerian J. Pharm. Res.* **2005**, *4*, 60–69.
14. Builders, P.F.; Kunle, O.O.; Okpaku, L.C.; Builders, M.I.; Attama, A.A.; Adikwu, M.U. Preparation and evaluation of mucinated sodium alginate microparticles for oral delivery of insulin. *Eur. J. Pharm. Biopharm.* **2008**, *70*, 777–783. [CrossRef]
15. Momoh, M.A.; Kenechukwu, F.C.; Ernest, O.C.; Oluseun, A.; Abdulmumin, B.; Youngson, D.C.; Kenneth, O.C.; Anthony, A.A. Surface-modified mucoadhesive microparticles as a controlled release system for oral delivery of insulin. *Heliyon* **2019**, *5*, e02366.
16. Li, L.; Yang, L.; Li, M.; Zhang, L. A cell-penetrating peptide mediated chitosan nanocarriers for improving intestinal insulin delivery. *Carbohydr. Polym.* **2017**, *174*, 182–189. [CrossRef]
17. Momoh, M.A.; Emmanuel, O.C.; Onyeto, A.C.; Darlington, Y.; Kenechukwu, F.C.; Ofokansi, K.C.; Attama, A.A. Preparation of snail cyst and PEG-4000 composite carriers via PEGylation for oral delivery of insulin: An in vitro and in vivo evaluation. *Trop. J. Pharm. Res.* **2019**, *18*, 919–926.
18. Builders, P.F.; Ibekwe, N.; Okpaku, L.C.; Attama, A.A.; Kunle, O.O. Preparation and characterization of mucinated cellulose microparticles for therapeutic and drug delivery purposes. *Eur. J. Pharm. Biopharm.* **2009**, *72*, 34–41. [CrossRef]
19. Chawla, A.; Sharma, P.; Pawar, P. Eudragit S-100 coated sodium alginate microspheres of naproxen sodium: Formulation, optimization and in vitro evaluation. *Acta Pharm.* **2012**, *62*, 529–545. [CrossRef]
20. Ramasamy, T.; Ruttala, H.B.; Shanmugam, S.; Umadevi, S.K. Eudragit-coated aceclofenac-loaded pectin microspheres in chronopharmacological treatment of rheumatoid arthritis. *Drug Deliv.* **2013**, *20*, 65–77. [CrossRef]
21. Bin, X.; Guohua, J.; Weijiang, Y.; Depeng, L.; Yongkun, L.; Xiangdong, K.; Juming, Y. Preparation of poly(lactic-co-glycolic acid) and chitosan composite nanocarriers via electrostatic self assembly for oral delivery of insulin. *Mater. Sci. Eng. C* **2017**, *78*, 420–428.
22. European Community, Council Directive on the Ethics of Experiments Involving Laboratory Animals (86/609/EEC). 2003. Available online: http://data.europa.eu/eli/dir/1986/609/oj (accessed on 9 April 2020).

23. Kenechukwu, F.C.; Umeyor, C.E.; Momoh, M.A.; Ogbonna, J.D.N.; Chime, S.A.; Nnamani, P.O.; Attama, A.A. Evaluation of gentamicin-entrapped solid lipid microparticles formulated with a biodegradable homolipid from Capra hircus trop. *J. Pharm. Res.* **2014**, *13*, 1999-1205.
24. Attama, A.A.; Okafor, C.E.; Builders, P.F.; Okorie, O. Formulation and in vitro evaluation of a PEGylated microscopic lipospheres delivery system for ceftriaxone sodium. *Drug Deliv.* **2009**, *16*, 448–457. [CrossRef]
25. Potta, S.G.; Minemi, S.; Nukala, R.K.; Peinado, C.; Lamprou, D.A.; Urquhart, U. Development of solid lipid nanoparticles for enhanced solubility of poorly soluble drugs. *J. Biomed. Nanotech.* **2010**, *6*, 634–640. [CrossRef] [PubMed]
26. Momoh, M.A.; Ossai, E.C.; Omeje, C.E.; Omenigbo, O.P.; Kenechukwu, F.C.; Ofokansi, K.C.; Attama, A.A.; Olobayo, K.O. A new lipid-based oral delivery system of erythromycin for prolong sustain release activity. *Mater. Sci. Eng. C* **2019**, *97*, 245–253. [CrossRef] [PubMed]
27. Des Rieux, A.; Fievez, V.; Garinot, M.; Schneider, Y.J. Nanoparticles as potential oral delivery systems of proteins and vaccines: a mechanistic approach. *J. Control. Release* **2006**, *116*, 1–27. [CrossRef] [PubMed]
28. Raffaele, F.; Monica, L.; Claudio, C.; Massimo, M.; Maurizio, D.; Davide, M. Investigation of size, surface charge, PEGylation degree and concentration on the cellular uptake of polymer nanoparticles. *Colloid Surf. B* **2014**, *123*, 639–647.
29. Kenneth, O.; Gerhard, W.; Gert, F.; Conrad, C. Matrix-loaded biodegradable gelatin nanoparticles as new approach to improve drug loading and delivery. *Eur. J. Pharm. Biopharm.* **2010**, *76*, 1–9.
30. Abdallah, M.; Yuichi, T.; Hirofumi, T. Design and evaluation of novel pH-sensitive chitosan nanoparticles for oral insulin delivery. *Eur. J. Pharm. Sci.* **2011**, *42*, 445–451.
31. Sarmento, B.; Veiga, F.; Ferreira, D. Development and characterization of new insulin containing polysaccharide nanoparticles. *Colloids Surf. B* **2006**, *53*, 193–202. [CrossRef]
32. Ziv, E.; Bendayan, M. Intestinal absorption of peptides through the enterocytes. *Microsc. Res. Tech.* **2000**, *49*, 346–352. [CrossRef]
33. Salatin, S.; Maleki, D.S.; Yari, K.A. Effect of the surface modification, size, and shape on cellular uptake of nanoparticles. *Cell Biol. Int.* **2015**, *39*, 881–890. [CrossRef] [PubMed]
34. Ma, Z.; Lim, T.M.; Lim, L.Y. Pharmacological activity of peroral chitosan-insulin nanoparticles in diabetic rats. *Int. J. Pharm.* **2005**, *293*, 271–280.
35. Simon-Giavarotti, K.A.; Giavarotti, L.; Gomes, L.F.; Lima, A.F.; Veridiano, A.M.; Garcia, E.A.; Mora, O.A.; Fernández, V.; Viodela, L.A.; Junqueira, V.B. Enhancement of lindane-induced liver oxidative stress and hepatotoxicityby thyroid hormone is reduced by gadolinium chloride. *Free Radic. Res.* **2000**, *36*, 1033–1039. [CrossRef] [PubMed]
36. Kasarala, G.; Tillmann, H.L. Standard liver tests. *Clin. Liver Dis.* **2016**, *8*, 13–18. [CrossRef] [PubMed]

Article

Hydrothermal Synthesis of ZnO–doped Ceria Nanorods: Effect of ZnO Content on the Redox Properties and the CO Oxidation Performance

Sofia Stefa [1], Maria Lykaki [1], Vasillios Binas [2], Pavlos K. Pandis [3], Vassilis N. Stathopoulos [3,*] and Michalis Konsolakis [1,*]

1 Industrial, Energy and Environmental Systems Lab (IEESL), School of Production Engineering and Management, Technical University of Crete, GR-73100 Chania, Crete, Greece; sstefa@isc.tuc.gr (S.S.); mlykaki@isc.tuc.gr (M.L.)

2 Institute of Electronic Structure and Laser, Foundation for Research and Technology-Hellas (FORTH–IESL), P.O. Box 1527, Vasilika Vouton, GR-71110 Heraklion, Greece; binasbill@iesl.forth.gr

3 Laboratory of Chemistry and Materials Technology, General (Core) Department, National and Kapodistrian University of Athens, GR-34400 Psachna Campus, Evia, Greece; ppandis@teiste.gr

* Correspondence: vasta@uoa.gr (V.N.S.); mkonsol@pem.tuc.gr (M.K.); Tel.: +30-22280-99688 (V.N.S.); +30-28210-37682 (M.K.)

Received: 30 September 2020; Accepted: 26 October 2020; Published: 28 October 2020

Abstract: The rational design of highly efficient, noble metal-free metal oxides is one of the main research priorities in the area of catalysis. To this end, the fine tuning of ceria-based mixed oxides by means of aliovalent metal doping has currently received particular attention due to the peculiar metal-ceria synergistic interactions. Herein, we report on the synthesis, characterization and catalytic evaluation of ZnO–doped ceria nanorods (NR). In particular, a series of bare CeO_2 and ZnO oxides along with CeO_2/ZnO mixed oxides of different Zn/Ce atomic ratios (0.2, 0.4, 0.6) were prepared by the hydrothermal method. All prepared samples were characterized by X-ray diffraction (XRD), N_2 physisorption, temperature-programmed reduction (TPR), scanning electron microscopy with energy dispersive X-ray spectroscopy (SEM-EDS) and transmission electron microscopy (TEM). The CO oxidation reaction was employed as a probe reaction to gain insight into structure-property relationships. The results clearly showed the superiority of mixed oxides as compared to bare ones, which could be ascribed to a synergistic ZnO–CeO_2 interaction towards an improved reducibility and oxygen mobility. A close correlation between the catalytic activity and oxygen storage capacity (OSC) was disclosed. Comparison with relevant literature studies verifies the role of OSC as a key activity descriptor for reactions following a redox-type mechanism.

Keywords: CeO_2/ZnO mixed oxides; zinc oxide; ceria nanorods; oxygen storage capacity; CO oxidation

1. Introduction

Cerium oxide or ceria (CeO_2) is considered one of the most promising metal oxides for numerous catalytic applications, such as catalytic oxidation of CO [1–3], NO reduction [4–6], water–gas shift reaction [7–10], reforming reactions [11,12] and soot combustion [13–15]. In particular, CeO_2 has attracted extensive interest due to its abundant oxygen vacancies, interchangeability between Ce^{3+} and Ce^{4+} oxidation states and high oxygen storage capacity (OSC) [16–18]. During the past few decades, nanotechnology has made remarkable progress toward the development of ceria nanocomposites with distinct physical and chemical properties, greatly expanding their potential applications in energy and environmental catalysis [19,20]. In this perspective, we recently showed that among ceria nanocomposites of different shape, ceria nanorods possess enhanced reducibility that is linked to their high population in intrinsic defects and oxygen vacancies. The latter is considered responsible for their superior CO oxidation performance [21].

More importantly, ceria-based mixed oxides have recently gained increasing attention in the fields of materials science and heterogeneous catalysis, due to their unique surface, structural and electronic properties, which are completely different to those of parent counterparts. The combination of various transition metals with CeO_2 nanoparticles leads to enhanced surface and redox properties due to the "synergistic" interactions between the oxide phases. Accordingly, various CeO_2–based transition metals (M/CeO_2, M = Mn, Fe, Co, Ni, Cu) have been widely employed in heterogeneous catalysis [22,23].

Among the different transition metal oxides, zinc oxide (ZnO) is a wide and direct band gap semiconductor, which has been used in many fields, due to its low cost and environmental sustainability [24–26]. Zinc is not a critical raw material, thus no direct impact on the environmental resources is considered. Moreover, it is abundant in the steel industry as one of the main recovered products in the metal scrap recycling process [27]. In view of this fact, the potential of a low-cost metal, able to further enhance the catalytic performance of CeO_2, is of particular importance both from an environmental and economic point of view.

Although CeO_2–ZnO composites have been extensively used in photocatalysis [28–32], only a few studies have been devoted to their catalytic applications. Xie et al. [33] reported on the enhanced CO oxidation activity of CeO_2–ZnO composites, which is ascribed to the synergistic interaction between ZnO hollow microspheres and commercial CeO_2 powders. In a similar manner, the synergistic interaction between ZnO and CeO_2 is considered responsible for the enhanced CO oxidation performance of three-dimensional ordered macroporous CeO_2–ZnO [34].

In light of the above aspects, the present work aims for a first time at investigating the impact of ZnO as modifier for ceria nanorods (NR), in an attempt to further adjusting their surface/redox properties. To this end, a series of pure CeO_2 and ZnO along with CeO_2/ZnO mixed oxides of different Zn:Ce atomic ratios (0.2, 0.4, 0.6) were prepared by the hydrothermal method. The as-prepared materials were thoroughly characterized by X-ray diffraction (XRD), N_2 adsorption at −196 °C (Brunauer–Emmett–Teller (BET) method), scanning electron microscopy-energy dispersive X-ray spectroscopy (SEM–EDS), transmission electron microscopy (TEM) and H_2 temperature-programmed reduction (H_2–TPR). The oxidation of CO was used as a probe reaction to disclose structure-activity relationships.

2. Materials and Methods

2.1. Materials Synthesis

All of the chemicals used in this work were of analytical reagent grade. $Ce(NO_3)_3 \cdot 6H_2O$ (purity ≥99.0%, Fluka, Bucharest, Romania), $Zn(CH_3COO)_2 \cdot 2H_2O$ (purity ≥99%, Sigma-Aldrich, St. Louis, MO, USA) were used as precursors for the preparation of ceria-zinc materials. HO_2CCO_2H (purity ≥99%, Sigma-Aldrich, St. Louis, MO, USA), NaOH (purity ≥98%, Honeywell Fluka, Seelze, Germany) and absolute ethanol (purity 99.8%, ACROS Organics, Geel, Belgium) were also employed during the synthesis procedure.

Bare ceria nanorods were initially prepared by the hydrothermal method, as described in detail in our previous work [21]. CeO_2/ZnO mixed oxides of different Zn:Ce ratios (0.2, 0.4, 0.6) were prepared by a modified hydrothermal method. In particular, 0.38 g HO_2CCO_2H, 0.64 g $Zn(CH_3COO)_2 \cdot 2H_2O$ and a certain quantity of CeO_2 nanorods (2.42, 1.21, 0.81 g for Zn/Ce = 0.2, 0.4 and 0.6, respectively), were dispersed in 40 mL of double deionized water under stirring for 20 min. Then, the suspension obtained was transferred to a plastic bottle and aged at 70 °C for 1 h. The precipitate was recovered by centrifugation, washed with double deionized water and ethanol, dried at room temperature overnight, and calcined at 500 °C for 2 h under air flow (heating ramp 5 °C/min). For comparison purposes, a pure zinc oxide sample was also prepared by the same method. The as-prepared materials are denoted as CeO_2/ZnO–x, where x refers to Zn:Ce atomic ratio.

For comparison purposes, a mechanical mixture (CeO_2+ZnO–0.4) was synthesized in agate by hand, by physically mixing ceria nanorods and ZnO, with the same composition as the CeO_2/ZnO–0.4 sample.

2.2. Materials Characterization

The textural characteristics of the investigated samples were assessed by N_2 adsorption-desorption isotherms at −196 °C (Nova 2200e Quantachrome flow apparatus, Boynton Beach, FL, USA). The structural properties were determined by X-ray diffraction (XRD) in a Rigaku diffractometer (model RINT 2000, Tokyo, Japan). Morphological/surface analysis was performed by scanning electron microscopy (SEM, JEOL JSM-6390LV, JEOL Ltd., Akishima, Tokyo, Japan) operating at 20 keV, equipped with an energy dispersive X-ray spectrometry (EDS) system and transmission electron microscopy (TEM) on a JEM-2100 instrument (JEOL, Tokyo, Japan). Redox properties were determined by means of H_2 temperature-programmed reduction (H_2–TPR). More details about the aforementioned characterization studies and the corresponding apparatus can be found elsewhere [2,35].

2.3. Catalytic Evaluation Studies

Catalytic experiments were carried out in a fixed-bed reactor (12.95 mm i.d.) loaded with 100 mg of catalyst. Feed composition was 0.2 vol.% of CO and 1 vol.% O_2 balanced with He. The total feed stream was 80 cm^3 min^{-1}, corresponding to a gas hourly space velocity (GHSV) of 40,000 h^{-1}.

All samples were pretreated with 10 °C min^{-1} up to 490 °C (30 min) under 20 vol.% O_2 in He. Then, the temperature was leveled off to 25 °C, followed by He purging to remove any physisorbed species. Catalytic activity measurements were carried out up to 500 °C. Both reactants and products were analyzed by gas chromatography equipped with thermal conductivity detectors (TCD) and two capillary columns (Molecular Sieve 5X and PoraPlot Q).

CO conversion (X_{CO}) was calculated according to the equation:

$$X_{CO}(\%) = \frac{[CO]_{in} - [CO]_{out}}{[CO]_{in}} \times 100 \quad (1)$$

where $[CO]_{in}$ and $[CO]_{out}$ are the CO concentrations in the inlet and outlet gas streams, respectively.

Specific reaction rates of CO consumption, in terms of the catalyst's mass (mol g^{-1} s^{-1}) or surface area (mol m^{-2} s^{-1}) were estimated under differential reaction conditions (W/F = 0.075 g s cm^{-3}, $X_{CO} < 15\%$, T = 200 °C):

$$r\left(\frac{mol}{g{\cdot}s}\right) = \frac{X_{CO} \times [CO]_{in} \times F\left(\frac{cm^3}{min}\right)}{100 \times 60\left(\frac{s}{min}\right) \times V_m\left(\frac{cm^3}{mol}\right) \times m_{cat}(g)} \quad (2)$$

$$r\left(\frac{mol}{m^2{\cdot}s}\right) = \frac{X_{CO} \times [CO]_{in} \times F\left(\frac{cm^3}{min}\right)}{100 \times 60\left(\frac{s}{min}\right) \times V_m\left(\frac{cm^3}{mol}\right) \times m_{cat}(g) \times S_{BET}\left(\frac{m^2}{g}\right)} \quad (3)$$

where V_m and F are gas molar volume and total flow rate, respectively, at 298 K and 1 bar, m_{cat} is the mass of catalyst and S_{BET} is the surface area.

3. Results and Discussion

3.1. Textural and Structural Characterization

The determination of the textural properties (surface area, pore volume, pore size) of CeO_2/ZnO samples was carried out by means of nitrogen adsorption-desorption (BET) analysis.

The results obtained are summarized in Table 1. The lowest value in BET surface area is demonstrated by ZnO (7.05 m^2/g). CeO_2 nanorods exhibit the highest surface area (79.31 m^2/g) followed by CeO_2/ZnO–0.2 (76.22 m^2/g), CeO_2/ZnO–0.4 (62.21 m^2/g), CeO_2/ZnO–0.6 (56.12 m^2/g) and ZnO (7.05 m^2/g). In general, ZnO addition progressively decreases the BET surface area. This is also in agreement with the gradual increase of pore size with the increase of ZnO content, as discussed below.

Table 1. Textural characteristics of bare CeO_2, ZnO and the CeO_2/ZnO samples.

Sample	BET Analysis		
	Surface Area (m^2/g)	Pore Volume (cm^3/g)	Average Pore Size (nm)
CeO_2	79.31 ± 0.15	0.48	24.2
CeO_2/ZnO–0.2	76.22 ± 0.12	0.67	34.4
CeO_2/ZnO–0.4	62.21 ± 0.11	0.65	40.6
CeO_2/ZnO–0.6	56.12 ± 0.09	0.91	55.9
ZnO	7.05 ± 0.01	0.06	23.0

The adsorption-desorption isotherms and the Barrett–Joyner–Halenda (BJH) desorption pore size distribution (PSD) of as-prepared samples are shown in Figure 1a,b, respectively. The pore size distribution in all samples is well within the mesopores region, which is further corroborated by the existence of type IV isotherms (Figure 1a). As presented in Table 1 and Figure 1b, the addition of ZnO leads to an increase in pore volume and average pore size, with the CeO_2/ZnO–0.6 sample exhibiting the highest pore size (55.9 nm) followed by CeO_2/ZnO–0.4 (40.6 nm), CeO_2/ZnO–0.2 (34.4 nm), CeO_2 (24.2 nm) and ZnO (23.0 nm).

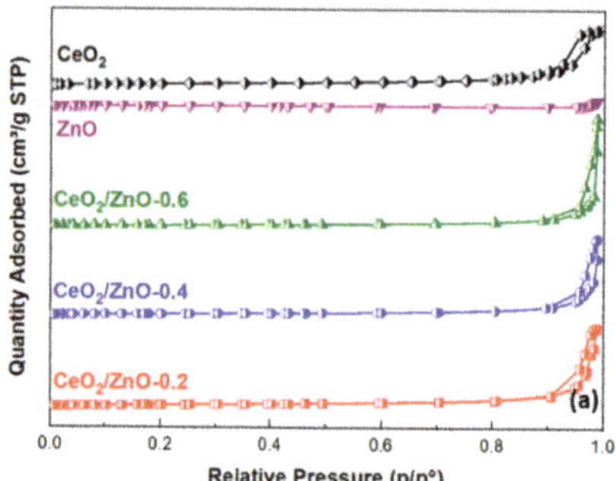

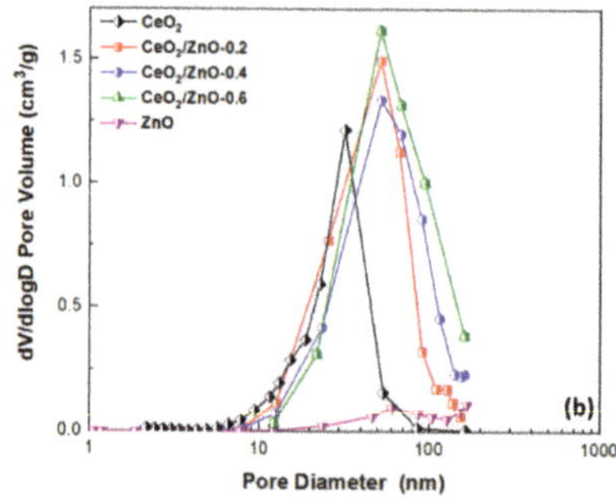

Figure 1. (**a**) N_2 adsorption-desorption isotherms and (**b**) pore size distribution of bare CeO_2, ZnO and the CeO_2/ZnO samples.

Figure 2 depicts the XRD patterns of CeO_2, ZnO and CeO_2/ZnO (Zn:Ce = 0.2, 0.4, 0.6) mixed oxide phases. The main diffraction peaks for bare ceria at $2\theta = 28.5°, 33.1°, 47.5°$ and $56.3°$ correspond to (111), (200), (220) and (311) planes, respectively. The diffraction peaks are attributed to ceria face-centered cubic fluorite structure (Fm3m symmetry, no. 225) (JCPDS card: 01-081-0792). Similarly, the XRD patterns of pure ZnO exhibit the typical hexagonal wurtzite structures (P63mc symmetry, no. 186) (JCPDS card: 01-079-0208). The strong peaks at $2\theta = 31.7°, 34.4°$ and $36.2°$ can be attributed to (100), (002), (101) lattice planes, respectively. For the CeO_2/ZnO samples, reflection planes perfectly matched to both indexed CeO_2 cubic and ZnO hexagonal structures. The XRD patterns indicate the formation of mixed oxides with finely dispersed phases of parent oxides. When ZnO content is increased, the intensity of ZnO reflections at (100), (002) and (101) is also increased, while the XRD profile of CeO_2 remains unaffected. The sharp diffraction peaks in the XRD patterns indicate that the synthesized catalysts are well crystallized. By applying the Scherrer equation, the average crystallite size of the as-prepared samples was calculated and the results are presented in Table 2. The lattice parameters are also included in Table 2. Bare CeO_2 exhibits a crystallite size of 13.0 nm, while pure ZnO exhibits a much larger crystallite size (34.5 nm). For the CeO_2/ZnO samples, ceria crystallite size displays a small decrease with the addition of ZnO, while the crystallite size of ZnO is increased. In addition, the lattice parameters of mixed oxides remain practically unaffected, as compared to the pure oxides, revealing well-dispersed phases of the different constituent oxides.

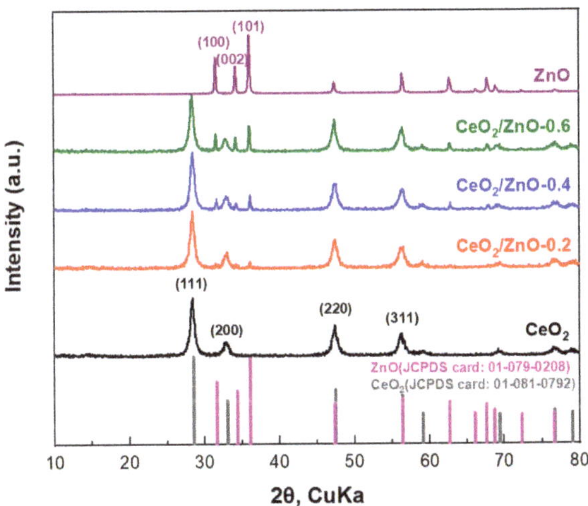

Figure 2. XRD patterns of CeO_2, ZnO and the CeO_2/ZnO samples.

Table 2. Structural characteristics of CeO_2, ZnO and the CeO_2/ZnO samples.

Samples	XRD Analysis			
	Phase Detected	Average Crystallite Size, D_{XRD} (nm)		Lattice Parameters (nm)
		CeO_2	ZnO	
CeO_2	Cerium (IV) oxide	12.99 ± 0.01	-	a = b = c = 0.5430 ± 0.0001
CeO_2/ZnO-0.2	Cerium (IV) oxide Zincite	12.14 ± 0.01	44.41 ± 0.05	a = b = c = 0.5439 ± 0.0001 a = b = 0.3180 ± 0.0001; c = 0.5232 ± 0.0001
CeO_2/ZnO-0.4	Cerium (IV) oxide Zincite	11.92 ± 0.01	44.65 ± 0.05	a = b = c = 0.5430 ± 0.0001 a = b = 0.3262 ± 0.0001; c = 0.5225 ± 0.0001
CeO_2/ZnO-0.6	Cerium (IV) oxide Zincite	11.56 ± 0.01	39.32 ± 0.05	a = b = c = 0.5439 ± 0.0001 a = b = 0.3267 ± 0.0001; c = 0.523 ± 0.0001
ZnO	Zincite	-	34.50 ± 0.05	a = b = 0.3272 ± 0.0001; c = 0.5233 ± 0.0001

3.2. Morphological Characterization (TEM, SEM–EDS)

The morphological features of CeO_2, ZnO and CeO_2/ZnO mixed oxides were characterized by transmission electron microscopy analysis (TEM). Pure ZnO (Figure 3a) displays an irregular shape (50–100 nm), while bare CeO_2 (Figure 3b) exhibits a distinct rod-like morphology (50–200 nm in length). In Figure 3c–e, the CeO_2/ZnO mixed oxides are presented. It is evident that the mixed oxides exhibit the rod-like morphology of CeO_2 nanorods, while separated ZnO particles of irregular morphology are also detected. These findings are in accordance with XRD results, implying the formation of distinct oxide phases.

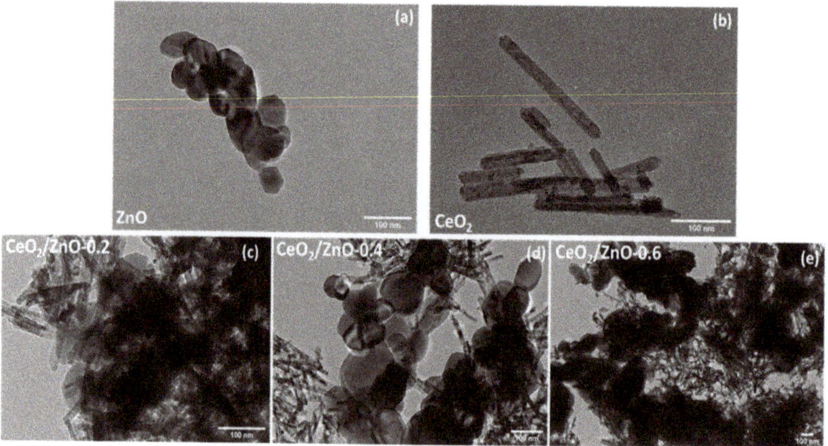

Figure 3. TEM images of (**a**) ZnO, (**b**) CeO_2, (**c**) CeO_2/ZnO–0.2, (**d**) CeO_2/ZnO–0.4 and (**e**) CeO_2/ZnO–0.6.

Elemental analysis was indicatively carried out by scanning electron microscopy analysis along with energy dispersive X-ray spectrometry (SEM–EDS) over the CeO_2/ZnO–0.2 sample. The SEM image along with the corresponding elemental analysis is shown in Figure 4. SEM–EDS analysis implies a Zn:Ce atomic ratio of ca. 0.19, being in good agreement with the nominal Zn/Ce ratio. The latter implies a uniform distribution of CeO_2 and ZnO phases on the entire material.

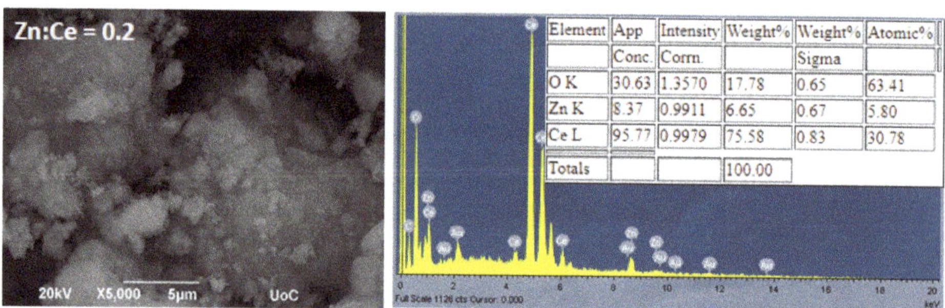

Element	App Conc.	Intensity Corrn.	Weight%	Weight% Sigma	Atomic%
O K	30.63	1.3570	17.78	0.65	63.41
Zn K	8.37	0.9911	6.65	0.67	5.80
Ce L	95.77	0.9979	75.58	0.83	30.78
Totals			100.00		

Figure 4. SEM-EDS analysis of the CeO_2/ZnO–0.2 sample.

3.3. Redox Properties (H_2–TPR)

By means of H_2–TPR experiments, the quality and quantity of the active oxygen sites of the catalysts can be determined. Such information is important for reactions following a redox-type mechanism, such as CO oxidation. Figure 5 shows the reduction profiles of bare and mixed oxides in the temperature range of 100–800 °C. The CeO_2 sample shows a peak at 500–700 °C (peak C), which is ascribed to the loosely bound surface species, contrary to bulk oxygen, which is reduced at temperatures higher than 700 °C [1,36–38]. For the pure ZnO sample, two broad peaks are found located at 150–300 °C (peak A) and 410–530 °C (peak B) regions, which can be attributed to the reduction of the hydroxyl species related to ZnO and the surface oxygen reduction from ZnO, respectively [39]. Additionally, in the pristine ZnO, the Zn^{2+} reduction to Zn^0 has been reported at 465 °C [40]. The TPR profiles of CeO_2/ZnO samples consist of the distinct peaks of both CeO_2 and ZnO phases. It is worth noticing that ZnO addition to CeO_2 does not lead to a shift of TPR peaks, as recently found for CeO_2/TiO_2 [35] and Fe_2O_3/CeO_2 [2] mixed oxides. This implies the structurally independent nature of ZnO and CeO_2 in the mixed oxides. The latter is in agreement with XRD and TEM results showing distinct ZnO nanoparticles in the vicinity of CeO_2 nanorods without incorporation of Zn in the nanostructure of CeO_2 nanorods.

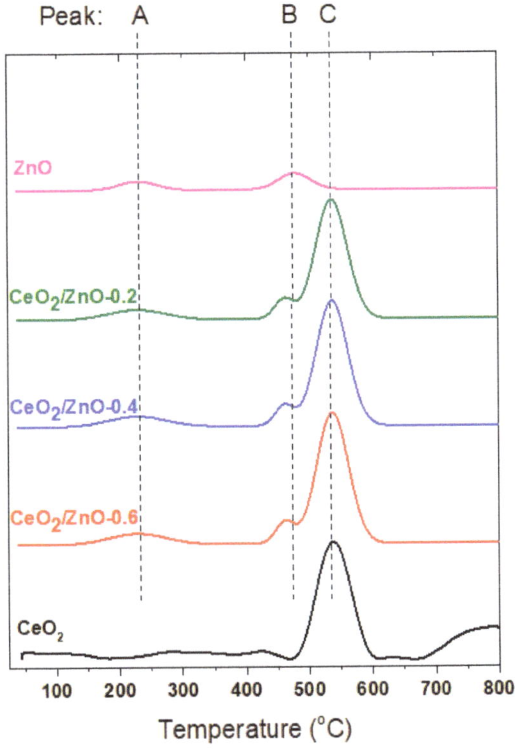

Figure 5. H_2-TPR profiles of CeO_2 and the CeO_2/ZnO samples.

To gain greater insight into the influence of ZnO on the reducibility of CeO_2/ZnO oxides, the H_2 uptake in the temperature range of 50–700 °C was estimated by the quantification of the TPR peaks (Table 3). Interestingly, ZnO addition results in an increase of H_2 uptake in the CeO_2/ZnO samples, implying a facilitation of surface oxygen species detachment. It is also of worth noticing the increase of H_2 uptake related to ceria reduction (peak C), despite the progressive decrease of ceria content. The latter denotes the facile reduction of ceria capping oxygen upon the increase of ZnO content. In particular, all mixed oxides exhibit a total H_2 consumption of ca. 1.0 mmol H_2/g, as compared to 0.59 and 0.31 mmol H_2/g of bare CeO_2 and ZnO, respectively. In terms of oxygen storage capacity (OSC), the following trend is obtained: CeO_2/ZnO–0.6 (0.52 mmol O_2/g) > CeO_2/ZnO–0.4 (0.50 mmol O_2/g) > CeO_2/ZnO–0.2 (0.48 mmol O_2/g) > CeO_2 (0.29 mmol O_2/g) > ZnO (0.15 mmol O_2/g). This order coincides relatively well with the catalytic activity (see below), revealing the key role of reducibility.

The abundance of reducible oxygen species is expected to affect the oxygen mobility and in turn the CO oxidation process, as will be further discussed below.

Table 3. Redox features of CeO_2 and CeO_2/ZnO samples.

Sample	H_2 Uptake (mmol H_2/g) and OSC (mmol O_2/g)				
	Peak A	Peak B	Peak C	H_2 uptake	OSC
CeO_2	-	-	0.59	0.59	0.29
CeO_2/ZnO–0.2	0.15	0.21	0.61	0.97	0.48
CeO_2/ZnO–0.4	0.15	0.23	0.63	1.01	0.50
CeO_2/ZnO–0.6	0.16	0.25	0.63	1.04	0.52
ZnO	0.14	0.17	-	0.31	0.15

3.4. Catalytic Evaluation Studies

In Figure 6, the conversion profiles of CO over temperature for bare ZnO, CeO_2 and CeO_2/ZnO mixed oxides are shown. It is evident that all mixed oxides demonstrate superior performance as compared to bare ones. A similar performance was obtained between mixed oxides, with the CeO_2/ZnO–0.4 being slightly better. To gain insight into the role of preparation method and the synergistic interaction between the different counterparts, the catalytic performance of a mechanical mixture (CeO_2+ZnO–0.4) of exactly the same composition to the optimum sample (CeO_2/ZnO–0.4) was explored in parallel. The bare ceria nanorods (CeO_2) exhibit a catalytic profile shifted by ca. 30 °C to higher temperature than the optimum CeO_2/ZnO–0.4 sample. ZnO is much less active with a profile located ca. 170 degrees higher. More importantly, the conversion profile of the mechanical mixture CeO_2+ZnO–0.4 is shifted by ca. 100 °C to higher temperatures as compared to that of CeO_2/ZnO–0.4, revealing the beneficial interaction between CeO_2 and ZnO, which is induced by the preparation method (see experimental Section 2.1).

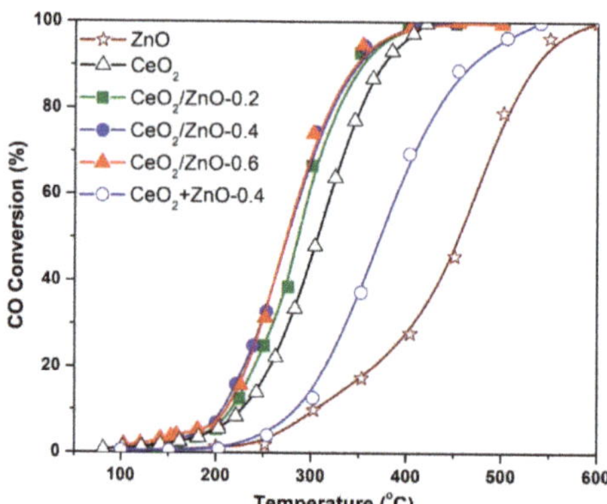

Figure 6. CO conversion profiles of CeO_2, ZnO and CeO_2/ZnO samples.

To obtain a better understanding in relation to the intrinsic activity of investigated samples, the specific activity, both in terms of catalyst mass (μmol g^{-1} s^{-1}) and surface area (μmol m^{-2} s^{-1}), was also estimated (Table 4). It is evident that the CeO_2/ZnO–0.4 sample offers the best performance, in terms of both conversion and mass-normalized specific activity, revealing its superior reactivity. It should be noted, however, that in terms of area-normalized activity, both bare ceria and mixed oxides exhibited an inferior activity as compared to bare ZnO. The latter can be attributed to the low surface area of ZnO (7.05 m^2/g), which is about one order of magnitude lower compared to bare CeO_2 and CeO_2/ZnO mixed oxides. Therefore, on the basis of the present results, the enhanced catalytic performance of CeO_2/ZnO mixed oxides can be attributed to a compromise between redox and surface properties. Moreover, the optimum CeO_2/ZnO–0.4 sample exhibits an apparent activation energy (E_a) of 32.1 kJ/mol, much lower than that of bare ceria (44.2 kJ/mol), ZnO (42.1 kJ/mol) and CeO_2+ZnO–0.4 mechanical mixture (43.2 kJ/mol). These findings unveil the lower energy barrier for CO oxidation over the hydrothermally prepared mixed oxides, as compared to single oxides and mechanical mixture, demonstrating the beneficial synergistic interactions induced by synthesis procedure.

Table 4. Conversion of CO and specific rates of CeO_2 and CeO_2/ZnO samples at 200 °C. Reaction conditions: 0.2 vol.% CO and 1 vol.% O_2 in He.

Sample	CO Conversion (%)	Specific Rate	
		r (μmol g^{-1} s^{-1})	r (×100) (μmol m^{-2} s^{-1})
CeO_2	5.1	0.056	0.070
CeO_2/ZnO–0.2	5.4	0.059	0.077
CeO_2/ZnO–0.4	6.9	0.075	0.121
CeO_2/ZnO–0.6	6.2	0.068	0.121
ZnO	1.3	0.014	0.201

The CO oxidation process investigated here can be corroborated on the basis of a redox-type (Mars-van Krevelen) mechanism. This particular mechanism involves CO chemisorption on $Ce^{\delta+}$ active sites, followed by oxygen activation on oxygen vacancies [3,21,41]. CO oxidation is taking place between the $Ce^{\delta+}$–CO and adjacent oxygen species, followed by the regeneration of active sites and the reoccupation of oxygen vacancies through gas phase oxygen. The proposed reaction sequence is described in detail in our previous work [21].

In view of the above mechanistic aspects, the key role of the catalyst's redox properties is clearly manifested. In detail, the high oxygen storage capacity (Table 3), linked to enhanced reducibility and oxygen exchange kinetics, can be accounted for the improved oxidation performance. Any structural or compositional modification in ceria that can affect the ceria-oxygen or ceria-oxygen-metal bond can eventually facilitate O_2 activation and in turn CO oxidation. In this regard, we recently showed that the enhanced reducibility of CeO_2 nanorods, linked to their abundance in oxygen vacancies, is the pivotal factor for their superior catalytic performance [21]. Furthermore, we demonstrated that when a transition metal element is incorporated in the structure of ceria nanorods, even if catalytically non-active (i.e., Ti), it can affect oxygen species coordination environment and in turn the OSC and the catalytic performance [35].

These arguments in relation to the crucial role of reducibility are clearly supported by the direct relationship between the redox properties (OSC, mmol g^{-1}) and the normalized reaction rate (μmol g^{-1} s^{-1}), as shown in Figure 7. Interestingly, the key role of OSC as an activity descriptor is further demonstrated in the present work by including in Figure 7 relevant literature data, previously obtained with bare CeO_2 and CeO_2/TiO_2 oxides [2,35]. The latter is of particular importance towards the development of cost-effective and highly active metal oxides by appropriately adjusting their redox features.

Thus, regarding the present findings, although bare ZnO is catalytically far less active compared to CeO_2 (Figure 6), their combination results in a synergistic effect towards the formation of CeO_2/ZnO mixed oxides of improved activity. Similar synergistic interactions were previously revealed over TiO_2–doped CeO_2 [35], where, however, the doping element was incorporated in the nanostructure, resulting in the formation of Ce–O–Ti active sites. Notably, in the present work, ZnO is not incorporated in the CeO_2 nanorod structure. As clearly shown by TEM images (Figure 3), ZnO nanoparticles are in close interaction with nanorods but clearly as separate particles. This is also verified by the XRD results (Figure 2), which indicate distinct oxide phases without the formation of solid solution. Hence, the improved catalytic activity of CeO_2/ZnO mixed oxides, which is related to the population of active oxygen species and to the facile reduction of surface oxygen, can be attributed to the facilitation of ceria capping oxygen reduction by adjacent ZnO nanoparticles, which in turn leads to higher OSC (Figure 5, Table 3). Thus, although ZnO is not incorporated in the ceria nanorods structure, it notably contributes to surface oxygen reduction, most likely through the interfacial ZnO–CeO_2 sites. The latter in combination with the abundance and lower cost of zinc compared to cerium oxide is a very interesting aspect in terms of catalyst design.

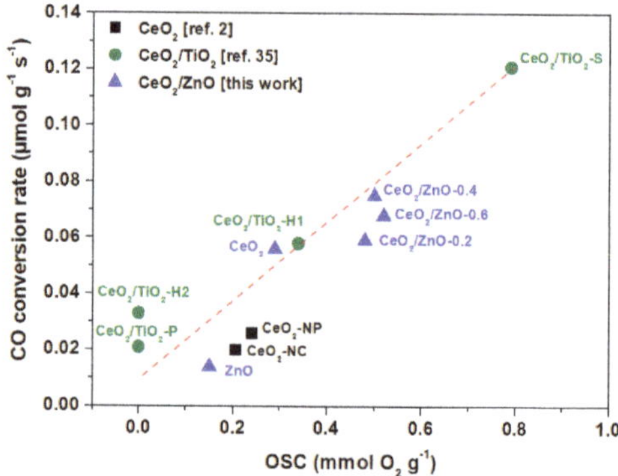

Figure 7. Correlation of specific activity with oxygen storage capacity (OSC). Literature data are also included. Specific rates were obtained at 200 °C under the reaction conditions: 0.2 vol.% CO and 1 vol.% O_2 in He. The dotted line simply represents the general trend of the data. The designation of the samples inside the figure refers to CeO_2 nanopolyhedra (CeO_2–NP), CeO_2 nanocubes (CeO_2–NC), CeO_2–TiO_2 nanoparticles prepared by precipitation (CeO_2/TiO_2–P), hydrothermal method in one and two steps (CeO_2/TiO_2–H1 and CeO_2/TiO_2–H2, respectively) and Stöber method (CeO_2/TiO_2–S).

4. Conclusions

In the present work, CeO_2/ZnO mixed oxides were synthesized by a two-step hydrothermal method. A thorough characterization study was carried out, revealing their textural, structural, morphological and redox features. CO oxidation was employed as a probe reaction to gain insight into the structure-activity relationships. The following order, both in terms of CO conversion and specific activity, was obtained: CeO_2/ZnO–0.4 > CeO_2/ZnO–0.6 > CeO_2/ZnO–0.2 > CeO_2 > ZnO. Despite the distinct appearance of ZnO and CeO_2 phases in mixed oxides and the low reactivity of ZnO, mixed oxides exhibit improved catalytic performance as compared to bare ones. The latter was attributed to the synergistic CeO_2–ZnO interactions towards an improved oxygen mobility and reducibility. Interestingly, a close relationship between the catalytic activity and the oxygen storage capacity was disclosed. Moreover, comparison with relevant literature studies verifies the role of OSC as a key activity descriptor for reactions following a redox-type mechanism. The present work reveals the significance of the rational design of noble metal-free mixed oxides towards the development of highly active materials that can be used in a variety of environmental and energy-related processes.

Author Contributions: S.S., M.L., V.B. and P.K.P. contributed to materials synthesis and characterization, results interpretation and paper–writing; V.N.S. and M.K. contributed to the conception, design, results interpretation and writing of the paper. M.K. reviewed, edited and submitted the manuscript in the final form. All authors have read and agreed to the published version of the manuscript.

Funding: This research has been co-financed by the European Union and Greek national funds through the Operational Program Competitiveness, Entrepreneurship and Innovation, under the calls RESEARCH—CREATE—INNOVATE (project code: T1EDK-00094).

Conflicts of Interest: The authors declare no conflict of interest.

References

1. Lykaki, M.; Pachatouridou, E.; Iliopoulou, E.; Carabineiro, S.A.C.; Konsolakis, M. Impact of the synthesis parameters on the solid state properties and the CO oxidation performance of ceria nanoparticles. *RSC Adv.* **2017**, *7*, 6160–6169. [CrossRef]
2. Lykaki, M.; Stefa, S.; Carabineiro, S.A.C.; Pandis, P.K.; Stathopoulos, V.N.; Konsolakis, M. Facet-Dependent Reactivity of Fe_2O_3/CeO_2 Nanocomposites: Effect of Ceria Morphology on CO Oxidation. *Catalysts* **2019**, *9*, 371. [CrossRef]
3. Wu, Z.; Li, M.; Overbury, S.H. On the structure dependence of CO oxidation over CeO_2 nanocrystals with well-defined surface planes. *J. Catal.* **2012**, *285*, 61–73. [CrossRef]
4. Zhang, S.; Li, Y.; Huang, J.; Lee, J.; Kim, D.H.; Frenkel, A.I.; Kim, T. Effects of Molecular and Electronic Structures in CoO_x/CeO_2 Catalysts on NO Reduction by CO. *J. Phys. Chem. C* **2019**, *123*, 7166–7177. [CrossRef]
5. Zhang, S.; Lee, J.; Kim, D.H.; Kim, T. NO reduction by CO over CoO_x/CeO_2 catalysts: Effect of support calcination temperature on activity. *Mol. Catal.* **2020**, *482*, 110703. [CrossRef]
6. Akter, N.; Zhang, S.; Lee, J.; Kim, D.H.; Boscoboinik, J.A.; Kim, T. Selective catalytic reduction of NO by ammonia and NO oxidation Over CoO_x/CeO_2 catalysts. *Mol. Catal.* **2020**, *482*, 110664. [CrossRef]
7. Cámara, A.L.; Corberán, V.C.; Martínez-Arias, A.; Barrio, L.; Si, R.; Hanson, J.C.; Rodriguez, J.A. Novel manganese-promoted inverse CeO_2/CuO catalyst: In situ characterization and activity for the water-gas shift reaction. *Catal. Today* **2020**, *339*, 24–31. [CrossRef]
8. Chen, C.; Zhan, Y.; Zhou, J.; Li, D.; Zhang, Y.; Lin, X.; Jiang, L.; Zheng, Q. Cu/CeO_2 Catalyst for Water-Gas Shift Reaction: Effect of CeO_2 Pretreatment. *ChemPhysChem* **2018**, *19*, 1448–1455. [CrossRef] [PubMed]
9. Ronda-Lloret, M.; Rico-Francés, S.; Sepúlveda-Escribano, A.; Ramos-Fernandez, E.V. CuO_x/CeO_2 catalyst derived from metal organic framework for reverse water-gas shift reaction. *Appl. Catal. A Gen.* **2018**, *562*, 28–36. [CrossRef]
10. Damaskinos, C.M.; Vasiliades, M.A.; Stathopoulos, V.N.; Efstathiou, A.M. The Effect of CeO_2 Preparation Method on the Carbon Pathways in the Dry Reforming of Methane on Ni/CeO_2 Studied by Transient Techniques. *Catalysts* **2019**, *9*, 621. [CrossRef]
11. Da Silva, A.M.; De Souza, K.R.; Jacobs, G.; Graham, U.M.; Davis, B.H.; Mattos, L.V.; Noronha, F.B. Steam and CO_2 reforming of ethanol over Rh/CeO_2 catalyst. *Appl. Catal. B Environ.* **2011**, *102*, 94–109. [CrossRef]
12. Hou, T.; Yu, B.; Zhang, S.; Xu, T.; Wang, D.; Cai, W. Hydrogen production from ethanol steam reforming over Rh/CeO_2 catalyst. *Catal. Commun.* **2015**, *58*, 137–140. [CrossRef]
13. Kumar, P.A.; Tanwar, M.D.; Russo, N.; Pirone, R.; Fino, D. Synthesis and catalytic properties of CeO_2 and Co/CeO_2 nanofibres for diesel soot combustion. *Catal. Today* **2012**, *184*, 279–287. [CrossRef]
14. Muroyama, H.; Hano, S.; Matsui, T.; Eguchi, K. Catalytic soot combustion over CeO_2-based oxides. *Catal. Today* **2010**, *153*, 133–135. [CrossRef]
15. Aneggi, E.; Wiater, D.; de Leitenburg, C.; Llorca, J.; Trovarelli, A. Shape-Dependent Activity of Ceria in Soot Combustion. *ACS Catal.* **2014**, *4*, 172–181. [CrossRef]
16. Konsolakis, M.; Lykaki, M. Recent Advances on the Rational Design of Non-Precious Metal Oxide Catalysts Exemplified by CuO_x/CeO_2 Binary System: Implications of Size, Shape and Electronic Effects on Intrinsic Reactivity and Metal-Support Interactions. *Catalysts* **2020**, *10*, 160. [CrossRef]
17. Montini, T.; Melchionna, M.; Monai, M.; Fornasiero, P. Fundamentals and Catalytic Applications of CeO_2 -Based Materials. *Chem. Rev.* **2016**, *116*, 5987–6041. [CrossRef]
18. Paier, J.; Penschke, C.; Sauer, J. Oxygen Defects and Surface Chemistry of Ceria: Quantum Chemical Studies Compared to Experiment. *Chem. Rev.* **2013**, *113*, 3949–3985. [CrossRef]
19. Cargnello, M.; Doan-Nguyen, V.V.T.; Gordon, T.R.; Diaz, R.E.; Stach, E.A.; Gorte, R.J.; Fornasiero, P.; Murray, C.B. Control of Metal Nanocrystal Size Reveals Metal-Support Interface Role for Ceria Catalysts. *Science* **2013**, *341*, 771–773. [CrossRef]
20. Tang, W.-X.; Gao, P.-X. Nanostructured cerium oxide: Preparation, characterization, and application in energy and environmental catalysis. *MRS Commun.* **2016**, *6*, 311–329. [CrossRef]
21. Lykaki, M.; Pachatouridou, E.; Carabineiro, S.A.C.; Iliopoulou, E.; Andriopoulou, C.; Kallithrakas-Kontos, N.; Boghosian, S.; Konsolakis, M. Ceria nanoparticles shape effects on the structural defects and surface chemistry: Implications in CO oxidation by Cu/CeO_2 catalysts. *Appl. Catal. B Environ.* **2018**, *230*, 18–28. [CrossRef]
22. Elias, J.S.; Risch, M.; Giordano, L.; Mansour, A.N.; Shao-Horn, Y. Structure, Bonding, and Catalytic Activity of Monodisperse, Transition-Metal-Substituted CeO_2 Nanoparticles. *J. Am. Chem. Soc.* **2014**, *136*, 17193–17200. [CrossRef] [PubMed]

23. Bion, N.; Epron, F.; Moreno, M.; Mariño, F.; Duprez, D. Preferential Oxidation of Carbon Monoxide in the Presence of Hydrogen (PROX) over Noble Metals and Transition Metal Oxides: Advantages and Drawbacks. *Top. Catal.* **2008**, *51*, 76–88. [CrossRef]
24. Wang, J.; Chen, R.; Xiang, L.; Komarneni, S. Synthesis, properties and applications of ZnO nanomaterials with oxygen vacancies: A review. *Ceram. Int.* **2018**, *44*, 7357–7377. [CrossRef]
25. Sajjad, M.; Ullah, I.; Khan, M.I.; Khan, J.; Khan, M.Y.; Qureshi, M.T. Structural and optical properties of pure and copper doped zinc oxide nanoparticles. *Results Phys.* **2018**, *9*, 1301–1309. [CrossRef]
26. Lee, C.-G.; Na, K.-H.; Kim, W.-T.; Park, D.-C.; Yang, W.-H.; Choi, W.-Y. TiO_2/ZnO Nanofibers Prepared by Electrospinning and Their Photocatalytic Degradation of Methylene Blue Compared with TiO_2 Nanofibers. *Appl. Sci.* **2019**, *9*, 3404. [CrossRef]
27. Stathopoulos, V.N.; Papandreou, A.; Kanellopoulou, D.; Stournaras, C.J. Structural ceramics containing electric arc furnace dust. *J. Hazard. Mater.* **2013**, *262*, 91–99. [CrossRef]
28. Rodwihok, C.; Wongratanaphisan, D.; Van Tam, T.; Choi, W.M.; Hur, S.H.; Chung, J.S. Cerium-Oxide-Nanoparticle-Decorated Zinc Oxide with Enhanced Photocatalytic Degradation of Methyl Orange. *Appl. Sci.* **2020**, *10*, 1697. [CrossRef]
29. Ma, T.-Y.; Yuan, Z.-Y.; Cao, J.-L. Hydrangea-Like Meso-/Macroporous ZnO-CeO_2 Binary Oxide Materials: Synthesis, Photocatalysis and CO Oxidation. *Eur. J. Inorg. Chem.* **2010**, *2010*, 716–724. [CrossRef]
30. Lv, Z.; Zhong, Q.; Ou, M. Utilizing peroxide as precursor for the synthesis of CeO_2/ZnO composite oxide with enhanced photocatalytic activity. *Appl. Surf. Sci.* **2016**, *376*, 91–96. [CrossRef]
31. De Lima, J.F.; Martins, R.F.; Neri, C.R.; Serra, O.A. ZnO:CeO_2-based nanopowders with low catalytic activity as UV absorbers. *Appl. Surf. Sci.* **2009**, *255*, 9006–9009. [CrossRef]
32. Liu, I.-T.; Hon, M.-H.; Teoh, L.G. The preparation, characterization and photocatalytic activity of radical-shaped CeO_2/ZnO microstructures. *Ceram. Int.* **2014**, *40*, 4019–4024. [CrossRef]
33. Xie, Q.; Zhao, Y.; Guo, H.; Lu, A.; Zhang, X.; Wang, L.; Chen, M.-S.; Peng, D.-L. Facile Preparation of Well-Dispersed CeO_2–ZnO Composite Hollow Microspheres with Enhanced Catalytic Activity for CO Oxidation. *ACS Appl. Mater. Interfaces* **2014**, *6*, 421–428. [CrossRef] [PubMed]
34. Mu, G.; Liu, C.; Wei, Q.; Huang, Y. Three dimensionally ordered macroporous CeO_2-ZnO catalysts for enhanced CO oxidation. *Mater. Lett.* **2016**, *181*, 161–164. [CrossRef]
35. Stefa, S.; Lykaki, M.; Fragkoulis, D.; Binas, V.; Pandis, P.K.; Stathopoulos, V.N.; Konsolakis, M. Effect of the Preparation Method on the Physicochemical Properties and the CO Oxidation Performance of Nanostructured CeO_2/TiO_2 Oxides. *Processes* **2020**, *8*, 847. [CrossRef]
36. Luo, J.-Y.; Meng, M.; Li, X.; Li, X.-G.; Zha, Y.-Q.; Hu, T.-D.; Xie, Y.-N.; Zhang, J. Mesoporous Co_3O_4–CeO_2 and Pd/Co_3O_4–CeO_2 catalysts: Synthesis, characterization and mechanistic study of their catalytic properties for low-temperature CO oxidation. *J. Catal.* **2008**, *254*, 310–324. [CrossRef]
37. Liu, J.; Zhao, Z.; Wang, J.; Xu, C.; Duan, A.; Jiang, G.; Yang, Q. The highly active catalysts of nanometric CeO_2-supported cobalt oxides for soot combustion. *Appl. Catal. B Environ.* **2008**, *84*, 185–195. [CrossRef]
38. Yu, S.-W.; Huang, H.-H.; Tang, C.-W.; Wang, C.-B. The effect of accessible oxygen over Co_3O_4–CeO_2 catalysts on the steam reforming of ethanol. *Int. J. Hydrogen Energy* **2014**, *39*, 20700–20711. [CrossRef]
39. Subramanian, V.; Potdar, H.S.; Jeong, D.-W.; Shim, J.-O.; Jang, W.-J.; Roh, H.-S.; Jung, U.H.; Yoon, W.L. Synthesis of a Novel Nano-Sized Pt/ZnO Catalyst for Water Gas Shift Reaction in Medium Temperature Application. *Catal. Lett.* **2012**, *142*, 1075–1081. [CrossRef]
40. Khan, A.; Smirniotis, P.G. Relationship between temperature-programmed reduction profile and activity of modified ferrite-based catalysts for WGS reaction. *J. Mol. Catal. A Chem.* **2008**, *280*, 43–51. [CrossRef]
41. Mukherjee, D.; Rao, B.G.; Reddy, B.M. CO and soot oxidation activity of doped ceria: Influence of dopants. *Appl. Catal. B Environ.* **2016**, *197*, 105–115. [CrossRef]

Publisher's Note: MDPI stays neutral with regard to jurisdictional claims in published maps and institutional affiliations.

Article

Electrical Monitoring as a Novel Route to Understanding the Aging Mechanisms of Carbon Nanotube-Doped Adhesive Film Joints

Xoan F. Sánchez-Romate [1,2,*], Alberto Jiménez-Suárez [2], María Sánchez [2], Silvia G. Prolongo [2], Alfredo Güemes [1] and Alejandro Ureña [2]

1 Department of Aerospace Materials and Processes, Escuela Técnica Superior de Ingeniería Aeronáutica y del Espacio, Universidad Politécnica de Madrid, Plaza del Cardenal Cisneros 3, 28040 Madrid, Spain; alfredo.guemes@upm.es

2 Materials Science and Engineering Area, Escuela Superior de Ciencias Experimentales y Tecnología, Universidad Rey Juan Carlos, Calle Tulipán s/n, 28933 Móstoles (Madrid), Spain; alberto.jimenez.suarez@urjc.es (A.J.-S.); maria.sanchez@urjc.es (M.S.); silvia.gonzalez@urjc.es (S.G.P.); alejandro.urena@urjc.es (A.U.)

* Correspondence: xoan.fernandez.sanchezromate@urjc.es; Tel.: +34914884771

Received: 19 March 2020; Accepted: 3 April 2020; Published: 8 April 2020

Featured Application: Structural Health Monitoring of novel carbon nanotube doped adhesive joints under aging conditions.

Abstract: Carbon fiber-reinforced plastic bonded joints with novel carbon nanotube (CNT) adhesive films were manufactured and tested under different aging conditions by varying the surfactant content added to enhance CNT dispersion. Single lap shear (SLS) tests were conducted in their initial state and after 1 and 2 months immersed in distilled water at 60 °C. In addition, their electrical response was measured in terms of the electrical resistance change through thickness. The lap shear strength showed an initial decrease due to plasticization of weak hydrogen bonds, and then a partial recovery due to secondary crosslinking. This plasticization effect was confirmed by differential scanning calorimetry analysis with a decrease in the glass transition temperature. The electrical response varied with aging conditions, showing a higher plasticity region in the 1-month SLS joints, and a sharper increase in the case of the non-aged and 2-month-aged samples; these changes were more prevalent with increasing surfactant content. By adjusting the measured electrical data to simple theoretical calculations, it was possible to establish the first estimation of damage accumulation, which was higher in the case of non-aged and 2-month-aged samples, due to the presence of more prevalent brittle mechanisms for the CNT-doped joints.

Keywords: carbon nanotubes; aging; structural health monitoring; water uptake; adhesive film; surfactant

---

1. Introduction

The increasing requirements of industry in terms of structural components make the development of novel materials necessary. In this context, composite structures present many advantages over conventional metallic alloys due to their exceptional specific properties that lead to energy efficiency and weight savings.

Therefore, the assembly of several composite parts is a challenging subject as the complexity of these structures is continuously increasing. For these purposes, bonded joints have some advantages over bolted ones as they are lightweight and avoid stress concentrations around the holes [1]. However, the inspection of adhesive joints sometimes is not always straightforward, since it involves many complex techniques, such as fiber Bragg grating sensors or Lamb waves, which often do not give a complete overview of the quality of the bonded joint [2–4]. Therefore, it is necessary to develop novel procedures that do not involve complex data analysis techniques and are not detrimental to the physical properties of the joint.

In this regard, carbon nanotubes (CNTs) seem to be a very promising solution. Their exceptional properties [5–7] and the enhancement of the electrical conductivity that they induce when added to an insulator resin [8–11] makes them very useful for multifunctional applications [12,13]. In fact, their use in structural health monitoring (SHM) applications is now of interest because of their piezoresistive and tunneling properties that lead to high sensitivities [14–17].

The aim of this work is to exploit the superb physical properties of CNTs in developing novel multifunctional adhesives with an inherent self-sensing capability. To date, most research into reinforced bonded joints has been focused on paste adhesives [18–21]. They exhibit excellent sensing properties and are capable of properly monitoring strain and debonding [22–24]. These paste adhesives can be treated as nanoreinforced composites, with the CNT dispersion procedure representing a challenging subject that often involves complex and expensive techniques, such as three-roll milling [25–27]. Therefore, this work is focused on the effect that CNT addition has on adhesive films, which allows for better thickness control and is used for structural applications in the aircraft industry.

In previous studies, CNT reinforced adhesive films have demonstrated high sensing properties and a good capability to properly monitor crack evolution [28–30]. The dispersion procedure has also been optimized in order to achieve a degree of good homogenization without any substantial detriment to the mechanical properties [31]. This is achieved by means of ultrasonication of a CNT dispersion in an aqueous solution, which is assisted by the addition of a surfactant, namely, sodium dodecyl sulfate (SDS). The addition of SDS improves the mechanical dispersion of the CNTs in the aqueous solution [32–35].

The effect of CNT dispersion on the mechanical and electrical properties of carbon fiber reinforced plastic (CFRP) bonded joints in their initial state has been characterized in previous works [31]. It has been concluded that these CNT adhesive films do not induce a detrimental effect on mechanical performance and they have proved to have excellent monitoring capabilities by means of electrical measurements [28]. This work takes a further step by analyzing the potential and applicability of these proposed bonded joints under aging conditions.

The amphiphilic behavior of SDS [36,37] plays an important role in the aging properties of adhesive joints. For this reason, immersion tests have been carried out in CNT-doped adhesive films once cured by varying the amount of SDS. In addition to this, single lap shear (SLS) tests have also been conducted in standard coupons in order to see the effect of water and temperature aging. The main application of these reinforced joints is the SHM. The electrical response has been also monitored during these tests so that the electrical properties can be better characterized in order to obtain a deeper knowledge of the aging mechanisms.

2. Materials and Methods

2.1. Materials

The multi-wall CNTs used for this study were NC7000 supplied by Nanocyl, with an average diameter of 10 nm and a length of up to 2 µm.

The adhesive was a FM300K adhesive film, supplied by Cytec. This is an epoxy-based adhesive with a knit tricot carrier, which allows enhanced bondline thickness control. It has a high elongation and toughness, together with an ultimate shear strength of 36.8 MPa. It is suitable for bonding metal-to-metal and CFRP-to-CFRP systems.

CNT dispersion takes place by means of ultrasonication by using a previously optimized dispersion procedure [31]. It consists of a 20-min ultrasonication of a CNT aqueous solution at 0.1 wt%. The disaggregation of larger agglomerates is enhanced by the addition of a SDS surfactant. To study the influence of this surfactant on the aging properties, the amount of SDS was fixed at 0.00, 0.25 wt% and 1.00 wt%.

After the dispersion procedure, the CNT suspension is sprayed over the adhesive surface prior to curing at a pressure of 1 bar at 40 cm for 0.5 s in order to achieve good homogenization of the CNTs over the surface.

In order to see the effects of aging, two types of specimens were prepared. One was the adhesive without substrate once cured—named in-bulk adhesive—in order to see the water uptake without any influence of the CFRP substrates. The second was the SLS specimens, which were made by secondary bonding of unidirectional CFRP substrates. The curing cycle was set for both the cured adhesive and the SLS joints in a hot press, as shown in Table 1. To improve the interfacial adhesion, the substrate surfaces were brushed.

Table 1. Cure-cycle parameters of secondary bonding.

Parameter	First Stage	Second Stage
Pressure	Ramp from 0 to 0.6 MPa for 15 min	0.6 MPa for 90 min
Temperature	Ramp from 25 to 175 °C for 45 min	175 °C for 60 min

2.2. Aging Tests

Cured adhesive and SLS specimens were subjected to aging conditions by immersion in distilled water at 60 °C, similar to some found in the literature [38]. Prior to immersion, the samples were dried in an oven at 50 °C for three days until weight loss was not observed between one and the next measurement. The aging time was set at 2 weeks (14 days) for the in-bulk adhesive, and up to 2 months (60 days) for SLS specimens. The reasons for these different immersion times were due to differences in the nature of each material and the exposed area subjected to water uptake. Adhesives tend to reach the water uptake saturation before the composite substrates [39] and in the case of the in-bulk adhesive, the exposed area of the adhesive is higher than in the CFRP joints. Therefore, the process of water uptake is accelerated [40].

Water absorption was measured in the in-bulk samples in their initial state and 1, 2, 3, 4, 7, 10 and 14 days after immersion. The water uptake was calculated by comparing the measured weight after immersion and the initial one in which is supposed that the samples were totally dry.

2.3. Electromechanical Tests

As commented before, bonded joints were subjected to SLS tests in order to study the aging effect on the electromechanical properties. They were conducted in three specimens for each condition (neat adhesive without CNTs and with 0.1 wt% CNTs with 0 wt%, 0.25 wt% and 1 wt% SDS). The tests were made according to standard ASTM D 5868-95 issue 01 using substrates of 100 × 25.4 × 2.5 mm with an overlapping area of 25.4 × 25.4 mm at a test rate of 13 mm/min. They were performed in a universal tensile Zwick machine.

Simultaneously, the electrical response was also monitored. Electrodes were made of copper wire sealed with silver ink in order to ensure a good electrical contact with the substrate surface. To protect the electrodes from environmental influences during testing, an adhesive layer was used. The measurements were carried out by an Agilent 34401A hardware and they were correlated to the mechanical response given by the tensile machine.

2.4. Characterization

Differential scanning calorimetry (DSC) measurements were conducted in a Metter Toledo mod 821 apparatus for the in-bulk adhesive. Two scans were carried out according to the standard ISO 11357-2:13, at 10 °C/min from ambient temperature to 250 °C. The glass transition temperature (T_g) was determined as the turning point of the heat capacity change. Two specimens of the non-aged- and

14-day-aged in-bulk specimens were measured in order to see the water uptake effect in the physical properties of the neat and CNT-doped adhesive.

3. Results

This section presents an analysis of the physical and mechanical evolution of SLS joints under aging conditions. First of all, the water uptake measurements for the in-bulk specimens are shown. Then, the mechanical properties of the SLS joints are discussed and finally, their electromechanical behavior is characterized.

3.1. Water uptake Measurements

Figure 1 shows the water uptake in terms of percentage of the initial weight for the in-bulk cured samples at each condition. The graph is in good agreement with the typical behavior of water uptake for this kind of samples, previously stated in other studies [38,41,42]. It is observed that water uptake is more prevalent in the initial stages and then the weight gain is going less significant until the water saturation is reached at 2 weeks of aging.

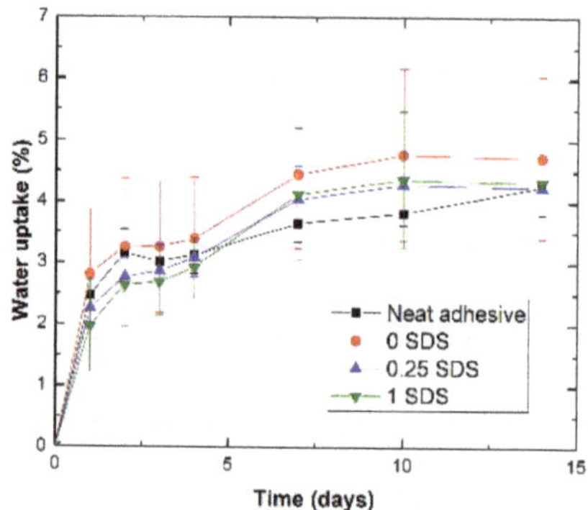

Figure 1. Water uptake graph for in-bulk specimens.

A similar water uptake behavior is found at every condition although some slight differences can be noticed. In this context, two opposite effects can play an important role. The first one is the hydrophobic behavior of carbon nanofillers, which can be introduced into the free volume of the polymer improving the barrier properties and leading to a reduction in the water uptake [43,44]. The second one is the amphiphilic behavior of the SDS that remains attached to the CNT surface [45,46], which can lead to an increase of the water absorption induced by the hydrophilic head groups [47]. In addition, CNT dispersion also plays a significant role. A poor dispersion can induce the presence of larger agglomerates, higher heterogeneity and higher distributed porosity. This irregular distribution leads, thus, to an irregular effect of the barrier properties of CNTs, which, in combination with the higher porosity, promotes a higher water uptake. However, a better dispersion of nanofillers improves the barrier properties leading, thus, to a lower water uptake.

The combination of these effects, as shown in the schematics of Figure 2, thus, explains the slight differences observed for each condition. In the case of the CNT reinforced adhesive without surfactant, a poor CNT dispersion is achieved, as stated in previous studies [31], so that the hydrophobic effect of CNTs is not so prevalent. Alternatively, the samples with 0.25 wt% and 1 wt% SDS show a similar trend, with a slightly lower water uptake than the sample without surfactant. In this case, the effect of the better CNT dispersion achieved was slightly prevalent over the amphiphilic effect of SDS surfactant. In the case of the neat adhesive, the water uptake was given directly by the physical behavior of the epoxy matrix.

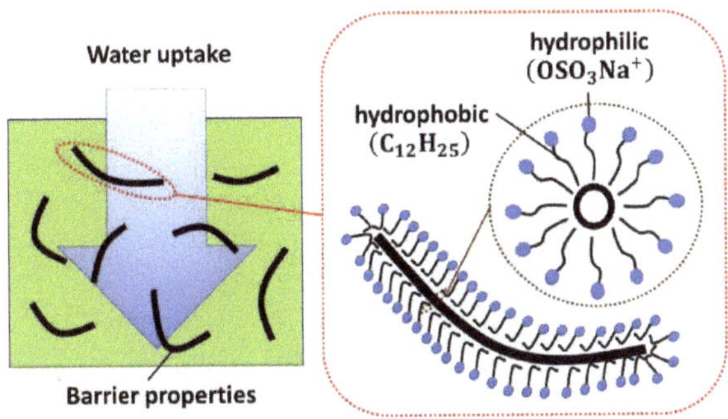

Figure 2. Schematics of combined effect of the water uptake (**left**) and amphiphilic behavior of sodium dodecyl sulfate (SDS) (**right**).

3.2. DSC Measurements

Table 2 shows the T_g values at different testing conditions for the in-bulk specimens. In the initial state, a drastic reduction of T_g was observed when comparing the non-doped with the doped adhesive. This means that CNTs accelerated the curing process, leading to the maximum conversion point of the system (T_g ~ 150 °C). This affirmation was supported by the measurements of the second scanning, where the T_g of all the samples were close to 150 °C, indicating the point of the maximum conversion of the resin. In addition, by observing the T_g of the aged samples, a significant decrease in comparison to the non-aged specimens was observed when adding CNTs, resulting in a similar glass transition temperature than for the neat adhesive. However, by observing the T_g obtained in a second scanning, it reaches the point of the maximum conversion in every case. This indicates that there was a plasticization effect caused by the water absorption in the case of CNT-doped samples. In the case of the neat adhesive, no significant differences were found when comparing aged and non-aged samples, so the plasticization effect was similar for the non-aged and aged samples. This affirmation was given by the fact that the network of the neat adhesive was not initially totally cured, with the plasticization effect less prevalent due to water absorption. It is important to note that the selected curing cycle was the same as that given by the supplier.

Table 2. Glass transition temperature for different in-bulk conditions.

Condition	Non-Aged T_g (°C)		2-Week-Aged T_g (°C)	
	1st Scanning	2nd Scanning	1st Scanning	2nd Scanning
Neat adhesive	117.0	146.0	115.5	146.0
0.00 SDS	144.0	148.0	119.0	146.0
0.25 SDS	148.5	149.5	118.0	148.5
1.00 SDS	146.0	147.5	109.0	149.5

Alternatively, when comparing the CNT-doped samples, it was observed that the addition of surfactant results in a more drastic reduction of the T_g, implying, thus, a higher plasticization effect. In order to better explain the possible effects that can take place in the material, it was necessary to focus on the mechanical testing of the SLS joints.

3.3. Single Lap Shear Tests

Figure 3 shows the lap shear strength (LSS) of the SLS specimens for each condition. It was observed that the LSS strength was significantly affected by the aging conditions. In every condition, a significant decrease of the LSS was observed after 1 month of aging while, for most of the cases, a slight recovery of the LSS was noticed by increasing the aging time. This different behavior can be explained by attending the water uptake results and also by the different chemical interactions inside the adhesive joint.

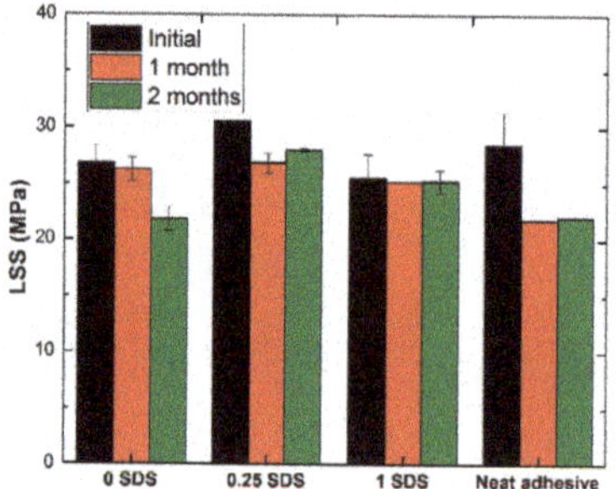

Figure 3. LSS of adhesive joints for each testing condition.

First of all, it is necessary to understand the role of aging mechanisms due to water absorption. The water uptake generally induces the creation of weak hydrogen bonds, leading to a swelling of the polymer chains and causing, thus, a drastic reduction of the physical properties. This effect is called plasticization and is detrimental to the mechanical properties [41]. However, after this initial stage, a slight recovery is generally observed. This has already been shown in other studies [43,48,49] and can be explained by a change in the mechanism of water absorption. After reaching the water uptake saturation, longer aging times can induce a transformation of the weak hydrogen bonds into multiple chemical connections between the water molecules and the polymer chains, promoting an increase in secondary crosslinking, thus leading to a stiffening of the material and also an embrittlement, as observed in the examples given in Figure 4a. This cannot be confirmed by the T_g as there was only one measurement. Therefore, it is necessary to focus on the mechanical response of the adhesive joints, particularly on the displacement at failure.

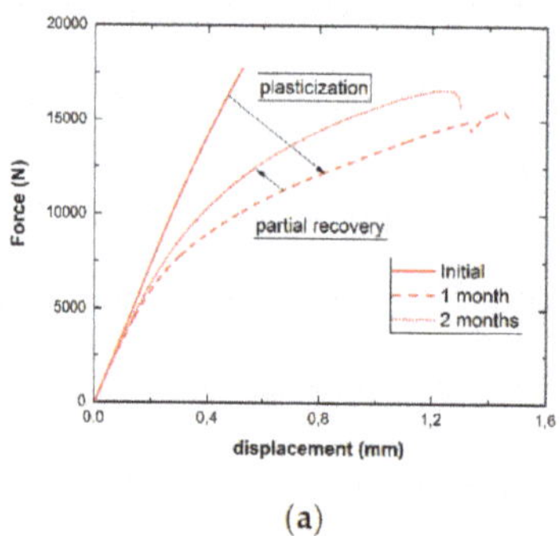

(a)

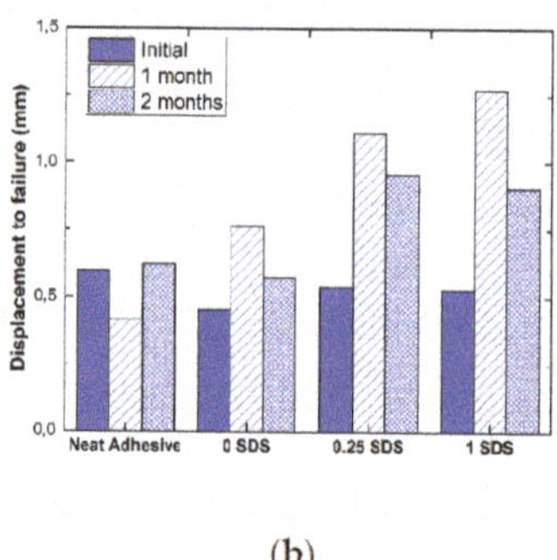

(b)

Figure 4. Mechanical effect of water uptake showing (**a**) a plasticization stage and (**b**) the displacement at failure.

In the particular case of the neat adhesive, there was only a partial conversion of the epoxy matrix at the initial state, as the T_g was significantly below the maximum value (~150 °C). Therefore, the water uptake does not affect significantly the plastic properties of the epoxy matrix, as it presents a more prevalent plasticization effect at the initial state. This was confirmed by a slight variation of the displacement at failure with aging time, as shown in Figure 4b.

However, in the case of CNT-doped adhesive joints, there are some different mechanisms. Here, the CNT addition itself and the dispersion state, dominated by the surfactant addition, play an important role in the aging mechanisms. As commented previously, a good dispersion implies a high homogeneity of CNT distribution in the matrix, improving the barrier properties. This acts in an opposite way to the amphiphilic behavior of SDS. In addition, the SDS also induces variations in the chemical interactions between the epoxy matrix and the CNTs, leading to a more drastic reduction of the T_g, as shown in Table 2. This was also observed by the displacement at failure shown in Figure 4b. In this case, the higher the SDS content, the higher the displacement at failure was. Moreover, it can be noted that the displacement at failure decreases after 2 months of aging. This was explained by the stiffening effect induced by the secondary crosslinking, which also leads to an embrittlement of the material, as previously stated. Therefore, the combination of the two effects explains the initial reduction of the LSS after 1 month and the general recovery for longer aging times.

In the case of the CNT sample without surfactant, a slight reduction of the LSS was observed after 1 month (~2%), while the effect on the mechanical properties was much more prevalent after 2 months, leading to a LSS reduction of more than 18%. In this case, the role of CNT dispersion was even more critical, as the absence of any surfactant leads to a much lower homogeneity of CNT distribution, inducing some areas with very high CNT content, acting as stress concentrators. Here, the embrittlement effect was more prevalent than the stiffening due to secondary crosslinking. This was confirmed by a higher relative reduction in the displacement at failure after 2 months of aging, similar to that initially obtained.

The higher displacement at failure observed in aging conditions for the samples with a higher amount of SDS can be explained accordingly to the presence of polar heads. They could interact with the matrix and the plasticizing effect is explained due to the unwinding of the chains around the nanotubes, as reported in other studies [50,51].

3.4. Electrical Monitoring

The aforementioned results give an initial idea of how aging conditions can affect the mechanical properties of CNT-doped adhesive film joints. In order to have a deeper understanding of aging effects on CNT-doped adhesive joints, it was necessary to analyze their electromechanical behavior. Figure 5 shows an example of electrical monitoring of a SLS specimen for different aging times. In every case, it was observed that the electrical resistance increases with displacement. This increase follows an approximately exponential behavior until failure, with the changes being more prevalent in the last stages of SLS testing. As stated in a previous study proving the monitoring capabilities of these CNT reinforced joints [28], the changes in the electrical resistance are due to the combination of two effects. The first effect was the increase of the tunneling distance between adjacent particles due to strain, leading to an increase of the tunneling resistance [52,53]. The second was the sudden crack propagation in the last stages of the tests, causing a prevalent breakage of electrical pathways through the joint. However, some important differences between the aged and non-aged specimens can be found.

By deeply analyzing the curves for the sample with 1 wt% SDS, the electrical behavior as a function of the applied strain changes was observed from the initial state to the 1-month-aged sample. In the aged sample, softer behavior was observed, due to the plasticization effect. This effect causes a steadier response of the material, with no abrupt changes in the electrical monitoring, as adhesive deformation and crack propagation take place in a softer way. By increasing the aging time, as discussed before, a secondary crosslinking was induced so the effect of plasticization was reduced, showing a mixed behavior between the initial state and the 1-month-aged sample, as noticed in the right graph of Figure 5c, where an abrupt change of the electrical behavior was observed.

The sample without surfactant shows similar behavior. At the initial state, some abrupt changes in the electrical resistance were observed, while the effect of plasticization was clearly shown after 1 month of aging (left graph of Figure 5b). By increasing the aging time, the stiffening effect of the secondary crosslinking was also observed by abrupt changes in the electrical behavior. In this case, as noticed before in the mechanical response, the behavior of the 2-month-aged sample was more similar to the non-aged one.

These initial results can give a good qualitative approximation of how aged and non-aged samples behave and prove the capability of CNT reinforced joints to properly monitor their mechanical behavior by means of electrical measurements. However, from the electrical response, it was possible to obtain estimations regarding damage evolution. To achieve this purpose, a simple analytical model, based on the tunneling effect of CNT reinforced polymers is proposed.

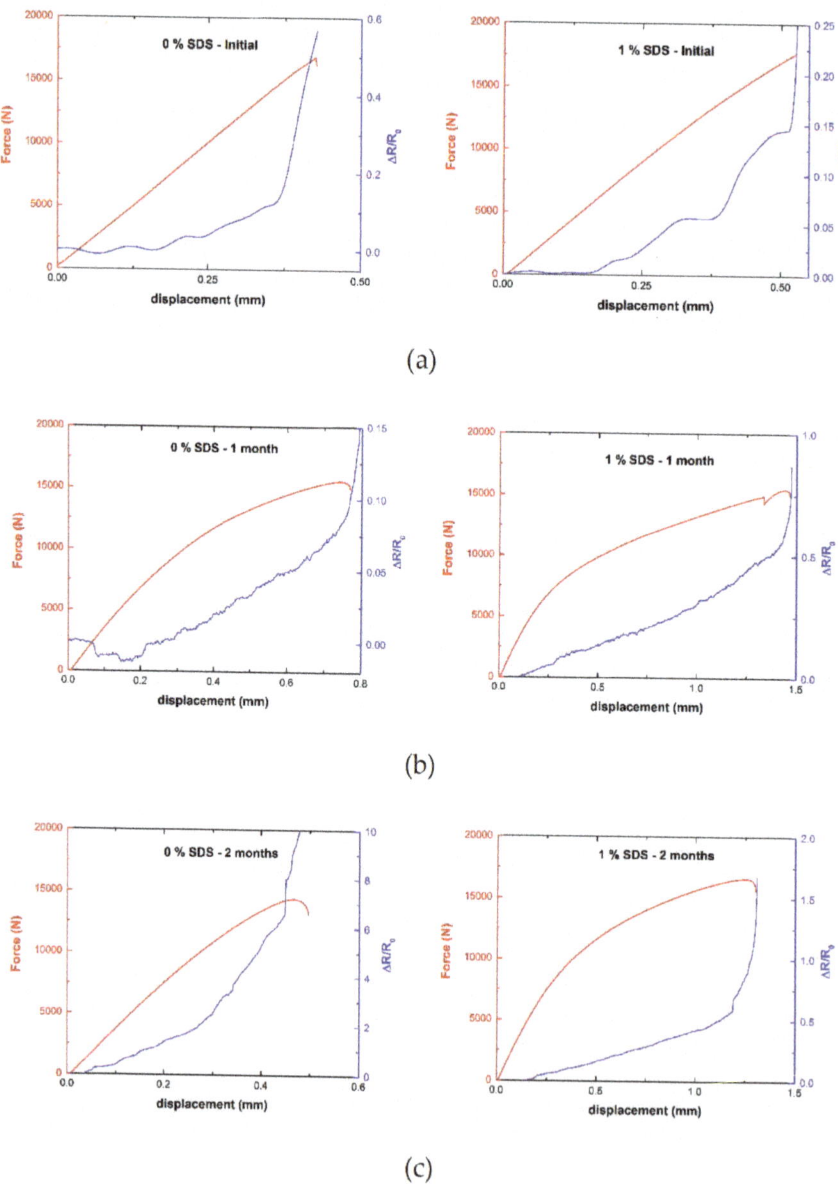

Figure 5. Electromechanical curves for single lap shear (SLS) specimens with (**left**) no surfactant and (**right**) 1 wt% SDS at (**a**) initial state and (**b**) after 1 and (**c**) 2 months of water immersion.

3.5. Theoretical Approach

The CNT-reinforced adhesive film was modeled as a nanocomposite with an homogeneous CNT distribution. The electrical mechanisms inside a nanoparticle network are given by the intrinsic electrical resistance of the CNTs themselves, the contact resistance between nanotubes and the tunneling resistance between near CNTs. In this particular case, the variation of electrical resistance due to applied strain are mainly dominated by the tunneling effect as intrinsic and contact resistance are assumed to be invariable with applied strain as stated in other studies [54].

Therefore, the changes in the electrical resistance can be divided into two terms, the first one, correlated to the variations due to the applied strain, which depends on the changes due to the tunneling effect between adjacent nanoparticles and the second one, which is correlated to the breakage of electrical pathways due to the effect of damage accumulation, as shown in the schematics of Figure 6, leading to the following expression:

$$\Delta R_{total} = \Delta R_{tunnel} + \Delta R_{damage} \tag{1}$$

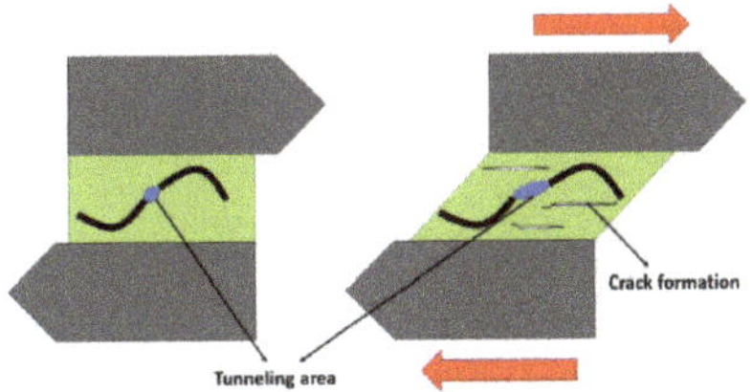

Figure 6. Schematics of electromechanical behavior in a SLS test showing the increase of tunneling distance and crack nucleation inside the adhesive.

The tunneling effect can be calculated by using the well-known Simmons formula for the tunneling resistance [55]. It has been proved to be an effective way to predict the electromechanical response of a strained CNT network as it is the most dominant electrical transport mechanism. It takes several aspects such as the polymer type and the contact area between adjacent CNTs:

$$R_{tunnel} = At{\cdot}e^{bt} \tag{2}$$

where A and b are two constants depending on the CNT geometry and matrix barrier characteristics and t is the tunneling distance, which changes with the applied strain.

The changes due to damage accumulation are not easy to model. There are many studies investigating this effect by proposing different damage evolution laws [56,57], but the particularities of the tested systems make the damage calculation very difficult. Therefore, damage accumulation is estimated by comparing the measured changes in the electrical resistance and the known tunneling effect.

$$\Delta R_{damage} = \Delta R_{measured} - \Delta R_{tunnel} \tag{3}$$

For this purpose, the initial tunneling distance is calculated as the distance that best fits the initial changes of the electrical resistance, where no damage is supposed.

Figure 7 shows the comparison between the theoretical line, taking only the tunneling effect and the experimental measurements for aged and non-aged samples with 1 wt% SDS into account. The pattern areas indicate the differences between the theoretical and the experimental ones being, thus, the damage accumulation during the SLS test.

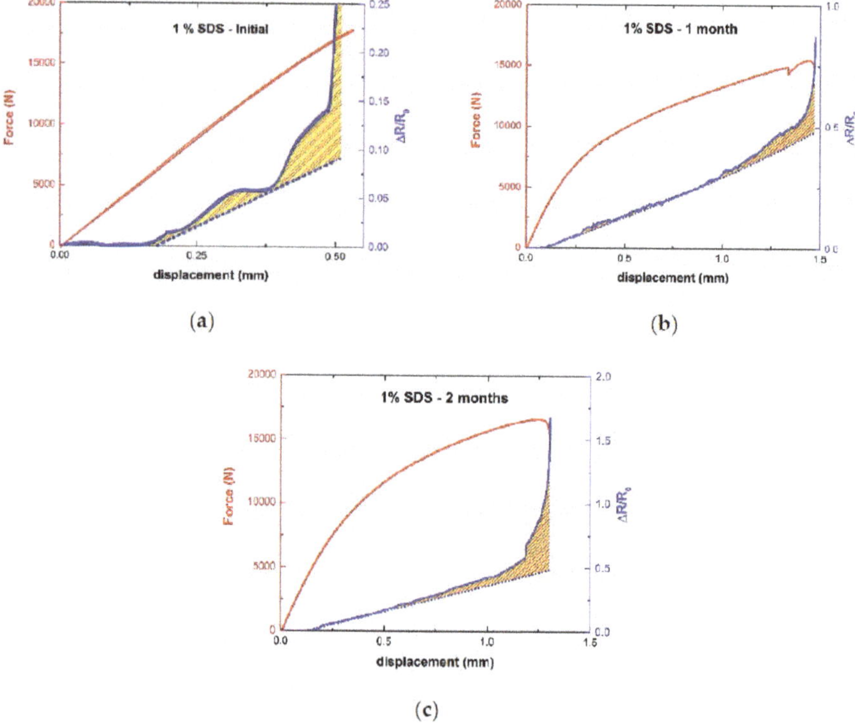

Figure 7. Damage accumulation evolution (red pattern area) by comparing experimental measurements (solid lines) and theoretical predictions (dashed lines) for 1 wt% SDS samples at (**a**) initial state and (**b**) after 1 and (**c**) 2 months of aging.

The non-aged sample shows significant irregular behavior. The threshold for damage accumulation was observed at ~0.25 mm displacement. After this point, the evolution of damage accumulation was very irregular. This can be explained due to the brittle mechanisms dominating the mechanical behavior of the adhesive. Secondary cracks start to nucleate and then they coalesce around the main crack [58], inducing a higher breakage of electrical paths in a similar way than that observed in other studies for fatigue testing [59]. This nucleation was not uniform so the unstable damage accumulation was explained. In the case of 1-month-aged samples, this damage accumulation starts to take place at 1 mm of displacement, that is, much later than in the non-aged specimen. This is in good agreement with the stated conclusions from water uptake and LSS measurements, as the induced plasticity in the first stages of water uptake avoids the early crack nucleation inside the adhesive joint. After that, damage accumulation takes places in a sudden way, that is, the cracks start to nucleate and then immediately coalesce. The 2-month-aged sample has a damage threshold of 0.5 mm, lower than in the 1-month-aged sample, due to the stiffening effect of the change in the water absorption mechanisms discussed above. Then, a softer evolution of damage accumulation is observed and finally, in the last stages of SLS tests a rapid coalescence takes place until final failure.

The previously described behavior was similar in the case of the 0.25 wt% SDS samples. However, in the case of the samples without surfactant, the electromechanical behavior shows some slight differences regarding the 1 wt% SDS samples, especially, concerning the sensitivity of the electrical response. Figure 8 presents the comparison between the theoretical and the experimental lines at different aging times. At the initial state, similar behavior for the 1 wt% SDS samples was observed with abrupt changes in the electrical resistance, inducing a high damage accumulation rate due to the rapid nucleation and coalescence of micro-cracks inside the material. The 1-month-aged specimen shows a much softer behavior, as expected due to the plasticization effect of the water uptake process. The threshold of damage accumulation was observed at a 0.7 mm displacement, that is, much later than for non-aged samples. However, unexpected behavior was observed for the 2 month-aged sample. In this case, the threshold for damage accumulation was observed nearly at the beginning of the SLS tests, that is, earlier than in the case of non-aged specimens. In addition, the damage accumulation was very high also in comparison to the other specimens as they show a much higher sensitivity.

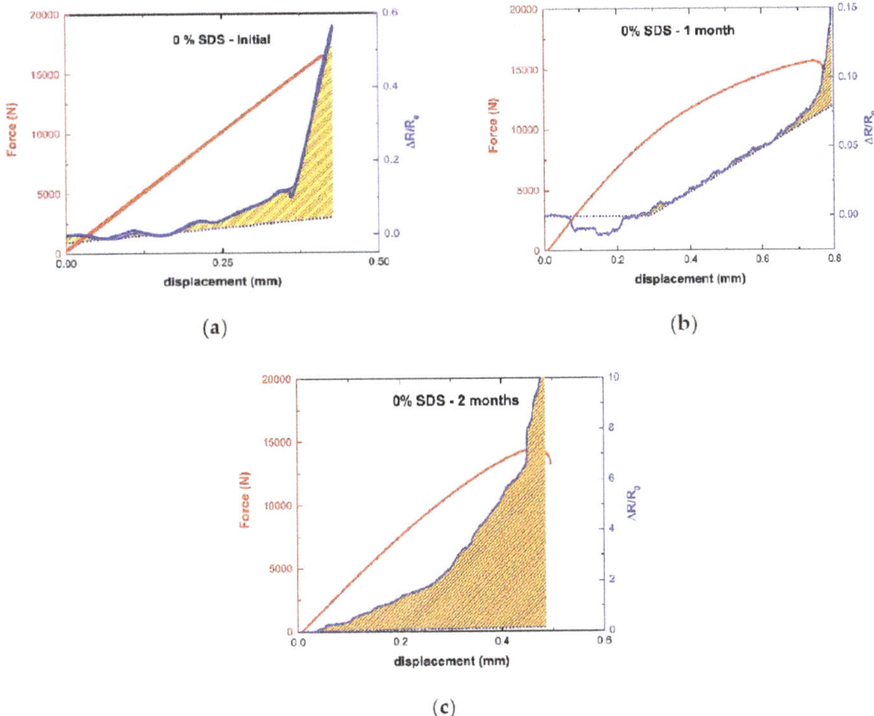

Figure 8. Damage accumulation evolution (red pattern area) by comparing experimental measurements (solid lines) and theoretical predictions (dashed lines) for non-surfactant samples at (**a**) initial state and (**b**) after 1 and (**c**) 2 months of aging.

This behavior can be explained by the interaction of two effects. One of them was due to the fact that the highly heterogeneous CNT distribution can induce high stress concentrations, leading to highly weak points in the matrix. This effect was also present in the non-aged and aged specimens, but in the case of the 2-month-aged one, the stiffening effect induced by the change in the mechanism of water absorption previously described can lead also to much higher embrittlement of the adhesive. This can also be stated in the reduction of the displacement at failure, so that the nucleation and coalescence of microcavities takes place more rapidly. The second one was correlated to the different interactions between the larger agglomerates of CNTs (which were much more prevalent in the non-surfactant samples) and the water molecules. This promotes a different electromechanical behavior than for the other samples.

4. Conclusions

SLS joints with CNT-doped adhesive films have been tested under aging conditions while their electromechanical properties have been monitored.

The water uptake measurements for the cured adhesive without substrate show that the behavior does not change significantly with the addition of both CNTs and SDS surfactant. This is explained by the combined effect of the amphiphilic behavior of the SDS and the barrier properties of CNT dispersion, acting in an opposite way.

The LSS of the bonded joints shows a general decrease after 1 month of aging because of the plasticizer effect of the water, which promotes the creation of weak hydrogen bonds. This statement was also confirmed by an increase of the displacement at failure. After 2 months of aging, there was a slight increase of LSS and a general reduction on the displacement at failure, which was explained by the secondary crosslinking that takes place due to water uptake saturation. In the case of the sample without surfactant, this behavior was slightly different because of the poor CNT distribution that can induce higher embrittlement, leading to a sudden decrease of LSS even after 2 months of aging.

Finally, the analysis of the electromechanical behavior of SLS joints confirms the previously described statements. A higher plasticization was observed for 1-month-aged specimens, while a partial recovery of the stiffness was observed after 2 months. By comparing the measured electrical response with simple theoretical calculations, it is possible to obtain the first quantitative idea of damage accumulation and how aging conditions affect the damage evolution. Therefore, this first estimation can be used to better understand the physical mechanisms taking place on CNT-doped adhesive joints under aging conditions subjected to SLS tests.

As a future work, however, it would be necessary to refine the theoretical predictions by taking some effects such as CNT distribution and orientation or the barrier properties of the epoxy matrix into account. This could give more accurate knowledge of the physical behavior of CNT-doped bonded joints under aging conditions.

Author Contributions: X.F.S.-R. conceptualization, methodology, formal analysis, writing—original draft preparation; A.J.-S. conceptualization, writing—review; S.G.P. supervision, formal analysis, writing—review; M.S. conceptualization; A.G. and A.U. funding acquisition. All authors have read and agreed to the published version of the manuscript.

Funding: This research was funded by the Ministerio de Economía y Competitividad of Spanish Government (Project MAT2016-78825-C2-1-R) and Comunidad de Madrid regional government [PROJECT ADITIMAT-CM (S2018/NMT-4411)].

Conflicts of Interest: The authors declare no conflict of interest.

References

1. Hussey, R.; Wilson, J. *Structural Adhesives: Directory and Databook*; Springer Science & Business Media: Berlin/Heidelberg, Germany, 1996.
2. Palaniappan, J.; Wang, H.; Ogin, S.L.; Thorne, A.M.; Reed, G.T.; Crocombe, A.D.; Rech, Y.; Tjin, S.C. Changes in the reflected spectra of embedded chirped fibre Bragg gratings used to monitor disbonding in bonded composite joints. *Compos. Sci. Technol.* **2007**, *67*, 2847–2853. [CrossRef]
3. Palaniappan, J.; Ogin, S.L.; Thorne, A.M.; Reed, G.T.; Crocombe, A.D.; Capell, T.F.; Tjin, S.C.; Mohanty, L. Disbond growth detection in composite-composite single-lap joints using chirped FBG sensors. *Compos. Sci. Technol.* **2008**, *68*, 2410–2417. [CrossRef]
4. Bernasconi, A.; Carboni, M.; Comolli, L. Monitoring of fatigue crack growth in composite adhesively bonded joints using fiber Bragg gratings. *Procedia Eng.* **2011**, *10*, 207–212. [CrossRef]
5. Ruoff, R.S.; Lorents, D.C. Mechanical and thermal properties of carbon nanotubes. *Carbon* **1995**, *33*, 925–930. [CrossRef]
6. Yu, M.F.; Lourie, O.; Dyer, M.J.; Moloni, K.; Kelly, T.F.; Ruoff, R.S. Strength and breaking mechanism of multiwalled carbon nanotubes under tensile load. *Science* **2000**, *287*, 637–640. [CrossRef]

7. De Volder, M.F.L.; Tawfick, S.H.; Baughman, R.H.; Hart, A.J. Carbon Nanotubes: Present and Future Commercial Applications. *Science* **2013**, *339*, 535–539. [CrossRef]
8. Hu, N.; Masuda, Z.; Yan, C.; Yamamoto, G.; Fukunaga, H.; Hashida, T. The electrical properties of polymer nanocomposites with carbon nanotube fillers. *Nanotechnology* **2008**, *19*, 215701. [CrossRef]
9. Wang, M.; Zhang, K.; Dai, X.; Li, Y.; Guo, J.; Liu, H.; Li, G.; Tan, Y.; Zeng, J.; Guo, Z. Enhanced electrical conductivity and piezoresistive sensing in multi-wall carbon nanotubes/polydimethylsiloxane nanocomposites via the construction of a self-segregated structure. *Nanoscale* **2017**, *9*, 11017–11026. [CrossRef]
10. Chen, J.; Han, J.; Xu, D. Thermal and electrical properties of the epoxy nanocomposites reinforced with purified carbon nanotubes. *Mater. Lett.* **2019**, *246*, 20–23. [CrossRef]
11. Plyushch, A.; Lyakhov, D.; Šimėnas, M.; Bychanok, D.; Macutkevič, J.; Michels, D.; Banys, J.; Lamberti, P.; Kuzhir, P. Percolation and Transport Properties in The Mechanically Deformed Composites Filled with Carbon Nanotubes. *Appl. Sci.* **2020**, *10*, 1315. [CrossRef]
12. Ferreira, A.D.B.L.; Nóvoa, P.R.O.; Marques, A.T. Multifunctional Material Systems: A state-of-the-art review. *Compos. Struct.* **2016**, *151*, 3–35. [CrossRef]
13. Mederos-Henry, F.; Mesfin, H.; Danlée, Y.; Jaiswar, R.; Delcorte, A.; Bailly, C.; Hermans, S.; Huynen, I. Smart Nanocomposites for Nanosecond Signal Control: The Nano4waves Approach. *Appl. Sci.* **2020**, *10*, 1102. [CrossRef]
14. Gong, S.; Zhu, Z. Giant piezoresistivity in aligned carbon nanotube nanocomposite: Account for nanotube structural distortion at crossed tunnel junctions. *Nanoscale* **2015**, *7*, 1339–1348. [CrossRef] [PubMed]
15. Pu, J.; Zha, X.; Zhao, M.; Li, S.; Bao, R.; Liu, Z.; Xie, B.; Yang, M.; Guo, Z.; Yang, W. 2D end-to-end carbon nanotube conductive networks in polymer nanocomposites: A conceptual design to dramatically enhance the sensitivities of strain sensors. *Nanoscale* **2018**, *10*, 2191–2198. [CrossRef]
16. Avilés, F.; Oliva, A.I.; Ventura, G.; May-Pat, A.; Oliva-Avilés, A.I. Effect of carbon nanotube length on the piezoresistive response of poly (methyl methacrylate) nanocomposites. *Eur. Polym. J.* **2019**, *110*, 394–402. [CrossRef]
17. Kumar, S.; Falzon, B.G.; Hawkins, S.C. Ultrasensitive embedded sensor for composite joints based on a highly aligned carbon nanotube web. *Carbon* **2019**, *149*, 380–389. [CrossRef]
18. Vietri, U.; Guadagno, L.; Raimondo, M.; Vertuccio, L.; Lafdi, K. Nanofilled epoxy adhesive for structural aeronautic materials. *Compos. Part B Eng.* **2014**, *61*, 73–83. [CrossRef]
19. Gude, M.R.; Prolongo, S.G.; Urena, A. Toughening effect of carbon nanotubes and carbon nanofibres in epoxy adhesives for joining carbon fibre laminates. *Int. J. Adhes. Adhes.* **2015**, *62*, 139–145. [CrossRef]
20. Akpinar, I.A.; Gültekin, K.; Akpinar, S.; Akbulut, H.; Ozel, A. Research on strength of nanocomposite adhesively bonded composite joints. *Compos. Part B Eng.* **2017**, *126*, 143–152. [CrossRef]
21. Akpinar, I.A.; Gültekin, K.; Akpinar, S.; Akbulut, H.; Ozel, A. Experimental analysis on the single-lap joints bonded by a nanocomposite adhesives which obtained by adding nanostructures. *Compos. Part B Eng.* **2017**, *110*, 420–428. [CrossRef]
22. Lim, A.S.; Melrose, Z.R.; Thostenson, E.T.; Chou, T. Damage sensing of adhesively-bonded hybrid composite/steel joints using carbon nanotubes. *Compos. Sci. Technol.* **2011**, *71*, 1183–1189. [CrossRef]
23. Mactabi, R.; Rosca, I.D.; Hoa, S.V. Monitoring the integrity of adhesive joints during fatigue loading using carbon nanotubes. *Compos. Sci. Technol.* **2013**, *78*, 1–9. [CrossRef]
24. Kang, M.; Choi, J.; Kweon, J. Fatigue life evaluation and crack detection of the adhesive joint with carbon nanotubes. *Compos. Struct.* **2014**, *108*, 417–422. [CrossRef]
25. Sandler, J.; Shaffer, M.; Prasse, T.; Bauhofer, W.; Schulte, K.; Windle, A. Development of a dispersion process for carbon nanotubes in an epoxy matrix and the resulting electrical properties. *Polymer* **1999**, *40*, 5967–5971. [CrossRef]
26. Prolongo, S.G.; Buron, M.; Gude, M.R.; Chaos-Moran, R.; Campo, M.; Urena, A. Effects of dispersion techniques of carbon nanofibers on the thermo-physical properties of epoxy nanocomposites. *Compos. Sci. Technol.* **2008**, *68*, 2722–2730. [CrossRef]
27. Chakraborty, A.K.; Plyhm, T.; Barbezat, M.; Necola, A.; Terrasi, G.P. Carbon nanotube (CNT)–epoxy nanocomposites: a systematic investigation of CNT dispersion. *J. Nanoparticle Res.* **2011**, *13*, 6493–6506. [CrossRef]

28. Fernández Sánchez-Romate, X.X.; Molinero, J.; Jiménez-Suárez, A.; Sánchez, M.; Güemes, A.; Ureña, A. Carbon Nanotube Doped Adhesive Films for Detecting Crack Propagation on Bonded Joints: A Deeper Understanding of Anomalous Behaviors. *ACS Appl. Mater. Interfaces* **2017**, *9*, 43267–43274. [CrossRef]
29. Sánchez-Romate, X.F.; Baena, L.; Jiménez-Suárez, A.; Sánchez, M.; Güemes, A.; Ureña, A. Exploring the mechanical and sensing capabilities of multi-material bonded joints with carbon nanotube-doped adhesive films. *Compos. Struct.* **2019**, *229*, 111477. [CrossRef]
30. Sánchez-Romate, X.F.; Moriche, R.; Jiménez-Suárez, A.; Sánchez, M.; Güemes, A.; Ureña, A. An approach using highly sensitive carbon nanotube adhesive films for crack growth detection under flexural load in composite structures. *Compos. Struct.* **2019**, *224*, 111087. [CrossRef]
31. Sánchez-Romate, X.F.; Jiménez-Suarez, A.; Molinero, J.; Sánchez, M.; Güemes, A.; Ureña, A. Development of Bonded Joints Using Novel CNT Doped Adhesive Films: Mechanical and Electrical Properties. *Int. J. Adhes. Adhes.* **2018**, *86*, 98–104. [CrossRef]
32. Vaisman, L.; Wagner, H.D.; Marom, G. The role of surfactants in dispersion of carbon nanotubes. *Adv. Colloid Interface Sci.* **2006**, *128*, 37–46. [CrossRef]
33. Rastogi, R.; Kaushal, R.; Tripathi, S.K.; Sharma, A.L.; Kaur, I.; Bharadwaj, L.M. Comparative study of carbon nanotube dispersion using surfactants. *J. Colloid Interface Sci.* **2008**, *328*, 421–428. [CrossRef]
34. Moshammer, K.; Hennrich, F.; Kappes, M.M. Selective suspension in aqueous sodium dodecyl sulfate according to electronic structure type allows simple separation of metallic from semiconducting single-walled carbon nanotubes. *Nano Res.* **2009**, *2*, 599–606. [CrossRef]
35. Zhao, P.; Einarsson, E.; Lagoudas, G.; Shiomi, J.; Chiashi, S.; Maruyama, S. Tunable separation of single-walled carbon nanotubes by dual-surfactant density gradient ultracentrifugation. *Nano Res.* **2011**, *4*, 623–634. [CrossRef]
36. Lawrence, A.; Pearson, J. Electrical properties of soap water amphiphile systems. Electrical conductance and sodium ion activity of aqueous 2% sodium dodecyl sulphate with added homologous n-aliphatic C 2–7 alcohols. *Trans. Faraday Soc.* **1967**, *63*, 495–504. [CrossRef]
37. Song, Y.; Wang, F.; Guo, G.; Luo, S.; Guo, R. Amphiphilic-Polymer-Coated Carbon Nanotubes as Promoters for Methane Hydrate Formation. *ACS Sustain. Chem. Eng.* **2017**, *5*, 9271–9278. [CrossRef]
38. Weiss, J.; Voigt, M.; Kunze, C.; Sanchez, J.E.H.; Possart, W.; Grundmeier, G. Ageing mechanisms of polyurethane adhesive/steel interfaces. *Int. J. Adhes. Adhes.* **2016**, *70*, 167–175. [CrossRef]
39. Mora, V.B.; Mieloszyk, M.; Ostachowicz, W. Model of moisture absorption by adhesive joint. *Mech. Syst. Signal. Process.* **2018**, *99*, 534–549. [CrossRef]
40. Gurtin, M.E.; Yatomi, C. On a model for two phase diffusion in composite materials. *J. Compos. Mater.* **1979**, *13*, 126–130. [CrossRef]
41. Bal, S.; Saha, S. Effect of sea and distilled water conditioning on the overall mechanical properties of carbon nanotube/epoxy composites. *Int. J. Damage Mech.* **2017**, *26*, 758–770. [CrossRef]
42. Kondrashov, S.; Merkulova, Y.I.; Marakhovskii, P.; D'yachkova, T.; Shashkeev, K.; Popkov, O.; Startsev, O.; Molokov, M.; Kurshev, E.; Yurkov, G.Y. Degradation of physicomechanical properties of epoxy nanocomposites with carbon nanotubes upon heat and humidity aging. *Russ. J. Appl. Chem.* **2017**, *90*, 788–796. [CrossRef]
43. Prolongo, S.G.; Gude, M.R.; Urena, A. Water uptake of epoxy composites reinforced with carbon nanofillers. *Compos. Part A Appl. Sci. Manuf.* **2012**, *43*, 2169–2175. [CrossRef]
44. Liu, L.; Tan, S.J.; Horikawa, T.; Do, D.; Nicholson, D.; Liu, J. Water adsorption on carbon-A review. *Adv. Colloid Interface Sci.* **2017**, *250*, 64–78. [CrossRef]
45. Bystrzejewski, M.; Huczko, A.; Lange, H.; Gemming, T.; Büchner, B.; Rümmeli, M.H. Dispersion and diameter separation of multi-wall carbon nanotubes in aqueous solutions. *J. Colloid Interface Sci.* **2010**, *345*, 138–142. [CrossRef]
46. Kastrisianaki-Guyton, E.S.; Chen, L.; Rogers, S.E.; Cosgrove, T.; van Duijneveldt, J.S. Adsorption of sodium dodecylsulfate on single-walled carbon nanotubes characterised using small-angle neutron scattering. *J. Colloid Interface Sci.* **2016**, *472*, 1–7. [CrossRef]
47. Tsutsumi, Y.; Fujigaya, T.; Nakashima, N. Polymer synthesis inside a nanospace of a surfactant–micelle on carbon nanotubes: creation of highly-stable individual nanotubes/ultrathin cross-linked polymer hybrids. *Rsc. Adv.* **2014**, *4*, 6318–6323. [CrossRef]
48. Carter, H.G.; Kibler, K.G. Langmuir-Type Model for Anomalous Moisture Diffusion in Composite Resins. *J. Compos. Mater.* **1978**, *12*, 118–131. [CrossRef]

49. Popineau, S.; Rondeau-Mouro, C.; Sulpice-Gaillet, C.; Shanahan, M.E. Free/bound water absorption in an epoxy adhesive. *Polymer* **2005**, *46*, 10733–10740. [CrossRef]
50. Paul, D.R.; Robeson, L.M. Polymer nanotechnology: Nanocomposites. *Polymer* **2008**, *49*, 3187–3204. [CrossRef]
51. Lavagna, L.; Massella, D.; Pantano, M.F.; Bosia, F.; Pugno, N.M.; Pavese, M. Grafting carbon nanotubes onto carbon fibres doubles their effective strength and the toughness of the composite. *Compos. Sci. Technol.* **2018**, *166*, 140–149. [CrossRef]
52. Thostenson, E.T.; Chou, T. Carbon nanotube networks: Sensing of distributed strain and damage for life prediction and self healing. *Adv. Mater.* **2006**, *18*, 2837–2841. [CrossRef]
53. Li, C.; Thostenson, E.T.; Chou, T. Dominant role of tunneling resistance in the electrical conductivity of carbon nanotube-based composites. *Appl. Phys. Lett.* **2007**, *91*, 223114. [CrossRef]
54. Sánchez-Romate, X.F.; Artigas, J.; Jiménez-Suárez, A.; Sánchez, M.; Güemes, A.; Ureña, A. Critical parameters of carbon nanotube reinforced composites for structural health monitoring applications: Empirical results versus theoretical predictions. *Compos. Sci. Technol.* **2019**, *171*, 44–53. [CrossRef]
55. Simmons, J.G. Generalized formula for the electric tunnel effect between similar electrodes separated by a thin insulating film. *J. Appl. Phys.* **1963**, *34*, 1793–1803. [CrossRef]
56. Mao, H.; Mahadevan, S. Fatigue damage modelling of composite materials. *Compos. Struct.* **2002**, *58*, 405–410. [CrossRef]
57. Pham, K.; Amor, H.; Marigo, J.; Maurini, C. Gradient Damage Models and Their Use to Approximate Brittle Fracture. *Int. J. Damage Mech.* **2011**, *20*, 618–652. [CrossRef]
58. Fiedler, B.; Hojo, M.; Ochiai, S.; Schulte, K.; Ando, M. Failure behavior of an epoxy matrix under different kinds of static loading. *Compos. Sci. Technol.* **2001**, *61*, 1615–1624. [CrossRef]
59. Sbarufatti, C.; Sánchez-Romate, X.F.; Scaccabarozzi, D.; Cinquemani, S.; Libonati, F.; Güemes, A.; Ureña, A. On the dynamic acquisition of electrical signals for structural health monitoring of carbon nanotube doped composites. In Proceedings of the Structural Health Monitoring 2017, Real-Time Material State Awareness and Data-Driven Safety Assurance-Proceedings of the 11th International Workshop on Structural Health Monitoring, Stanford, CA, USA, 12–14 September 2017; pp. 1921–1928.

Article

Decomposition of Flavonols in the Presence of Saliva

Malgorzata Rogozinska and Magdalena Biesaga *

Faculty of Chemistry, University of Warsaw, Pasteura 1, 02-093 Warsaw, Poland; mgwiazdon@chem.uw.edu.pl
* Correspondence: mbiesaga@chem.uw.edu.pl

Received: 23 September 2020; Accepted: 22 October 2020; Published: 26 October 2020

Abstract: In this study, the LC-MS/MS was applied to explore the stability of four common dietary flavonols, kaempferol, quercetin, isorhamnetin, and myricetin, in the presence of hydrogen peroxide and saliva. In addition, the influence of saliva on the representative quercetin glycosides, rutin, quercitrin, hyperoside, and spiraeoside was examined. Our study showed that, regardless of the oxidative agent used, flavonols stability decreases with increasing B-ring substitution. The decomposition of analyzed compounds was based on their splitting by the opening the heterocyclic C-ring and realizing more simple aromatic compounds. The dead-end products corresponded to different benzoic acid derivatives derived from B-ring. Kaempferol, quercetin, isorhamnetin, and myricetin were transformed into 4-hydroxybeznoic acid, protocatechuic acid, vanillic acid, and gallic acid, respectively. Additionally, for quercetin and myricetin, two intermediate depsides and 2,4,6-trihydroxybenzoic acid derived from A-ring were detected. All analyzed glycosides were resistant to hydrolysis in the presence of saliva. Based on our data, saliva was proven to be a next oxidative agent which leads to the formation of corresponding phenolic acids. Hence, studies on flavonols' metabolism should take into consideration that the flavonols decomposition starts in the oral cavity; hence, in subsequent parts of the human digestive tract, they could be present not in their parent form but as phenolic acids. Further analyses of the influence of saliva on flavonols glycosides need to be performed due to the possible interindividual fluctuations.

Keywords: flavonols; H_2O_2; saliva; metabolism; oxidation; LC-MS/MS

---

1. Introduction

Flavonoids are phenolic compounds widely present in plants and food of plant origin. Both clinical and epidemiological studies show the correlation between the dietary polyphenols intake and the reduction of risk of some chronic diseases such as cardiovascular diseases, cancer, and diabetes, as well as aging [1–4]. These beneficial effects are associated with their antioxidant activity [5–7]. Flavonols, the major class of flavonoids present in the human diet, and among them quercetin, kaempferol, myricetin, and isorhamnetin (Figure 1), are well known to act as antioxidants in vitro and show protective effects against free radicals, reactive oxygen species, and other oxidation agents [8–10]. However, the biological properties of antioxidants such as flavonols depend on their bioavailability and metabolism in the human body. The study of biological responses due to dietary intake of polyphenols cannot be carried out without taking into consideration polyphenols–saliva interactions. Although phenolics are normally ingested through the mouth as elements of food, very little is known about their metabolism in the oral cavity. One of the main goals assigned to saliva is participation in glycosides hydrolysis [11,12], which delivers biologically active aglycones that can be absorbed more effectively in the human digestive tract. Since flavonols mainly occur in food as glycosides, much of the research focused on the metabolism in oral cavity concerns glycosides. However, some reports show the presence of flavonols aglycones in the food of plant origin, for instance in eucalyptus and unifloral types of honey [13]. Moreover, the fermentation carried out during the manufacturing of food could results in the hydrolysis of glycosides to aglycones. Although it was reported that flavonoid glycosides can be absorbed intact via the sodium-dependent glucose transporter [14], it was also shown that many glycosides are not absorbed due to efficient efflux transport by intestinal efflux protein pumps [15].

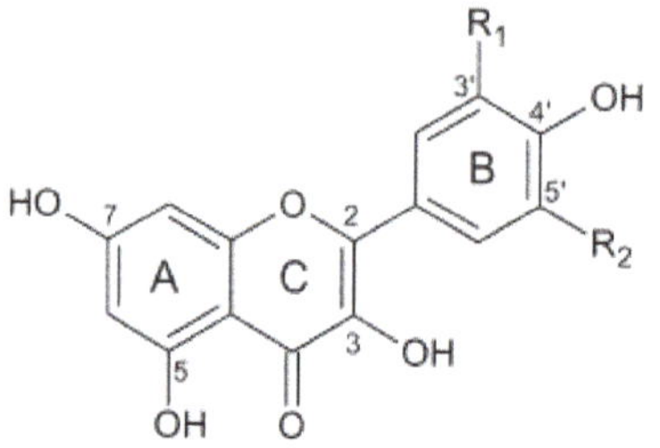

Flavonol	R_1	R_2
Kaempferol	H	H
Quercetin	OH	H
Isorhamnetin	OCH_3	H
Myricetin	OH	OH

Figure 1. Structures of kaempferol, quercetin, isorhamnetin, and myricetin.

It is well established that the oral cavity harbors numerous and diverse microorganisms, which can hydrolyze flavonols glucosides to the aglycones by glucosidases excreted from the bacteria [11,16]. All of these features indicate that the effective absorption of polyphenols in the human digestive tract strongly depends on their deglycosylation, and, as far as we know, study on the metabolism of aglycones in the oral cavity are limited. Interaction of saliva components with bioactive compounds from food occurs due to various reasons. Firstly, chewing food allows its components to stay in the oral cavity for a while, which ensures the contact of saliva and oral mucosa with food. Secondly, during chewing, flavonols are dissolved in saliva, which facilitates their interaction with oral mucosa and salivary proteins [17–20]. It is worth noting that Quijada-Morín et al. [17] outlined interactions between flavan-3-ols and salivary proteins not only as a precipitation issue as it has been usually studied but also as a more complex interaction, which involves the formation of soluble and insoluble complexes.

All of these mechanisms increase lipophilic polyphenols assimilation and causes their retention in the oral cavity over time [21–23].

The human oral cavity contains numerous and diverse microorganisms as commensals [24,25]. Approximately 280 bacterial species from the oral cavity have been isolated in culture and formally named. The possible reason for the decomposition of flavonols in the presence of saliva is the fact that bacteria and leucocytes presented in the oral cavity are able to generate H_2O_2; thus, flavonols could be easily oxidized [26,27]. While there is a lack of reports showing the oxidation of flavonols in the presence of saliva, other oxidizing conditions have been well established, mainly for quercetin. Quercetin is degraded in air conditions with the formation of different depsides, as intermediates in the degradation pathway. Further decomposition results in the formation of 2-(3,4-dihydroxyphenyl)-2-oxoacetic acid, 2,4,6-trihydroxybenzoic acid, and 3,4-dihydroxybenzic acid [28]. Additionally, Maini et al. [29] proposed another degradation products of quercetin after its exposure to UVA radiation: 1,3,5-trihydroxybenzene and 2,4,6-trihydroxybenzaldehyd. In general, oxidation of quercetin by various methods: air, electrochemical, enzymatic, and free radical oxidation may yield, more or less, the same set of oxidized products [30].

For further elucidation of the oxidation processes of flavonols in biological systems, we investigated the stability of four wide-spread flavonols: quercetin, kaempferol, myricetin, and isorhamnetin in the presence of saliva. This work also provides comparative oxidative studies of flavonols using H_2O_2 solution and whole saliva as oxidation agents. We chose flavonols that differ in the number of hydroxyl substituents in the B-ring. Additionally, in the case of isorhamnetin, one hydroxyl group is replaced with a methoxy group, which allows for the blocking of interactions from two adjacent hydroxyl groups. To determine the various degradation products of flavonols, LC-MS/MS system was applied.

2. Materials and Methods

2.1. Chemicals and Reagents

The commercial standards of flavonols and phenolic acids, as well as the rest of the chemicals, were purchased from Sigma-Aldrich (Steinheim, Germany). Glycosides rutin (quercetin 3-rhamnoglucoside), quercitrin (quercetin 3-rhamnoside), hyperoside (quercetin 3-galactoside), and spiraeoside (quercetin 4′-glucoside) were purchased from Extrasynthese (Genay, France). Methanol was obtained from Merck (Darmstadt, Germany). In all experiments, ultrapure water from a Milli-Q system with an electrical resistivity of 18 MΩ × cm was used. Stock solutions of flavonoids were prepared in methanol. Diluted standards were prepared in 25% methanol with a final concentration of 50 mg/L. All solutions were filtered through 0.45 µm membranes (Millipore) and degassed prior to use.

2.2. Collection of Saliva Samples

Five healthy human volunteers (20–30 years old) were recruited to this study based on restricted criteria. All volunteers were self-reported to be in good general health without any chronic diseases, not taking antibiotics, and no history of drinking and smoking habits. Volunteers were requested to refrain from eating for at least 10 h and they could drink only mineral water before collection. Saliva samples were collected in the morning 2 h after brushing teeth with toothpaste free of polyphenols. Before saliva collection, the oral cavity was rinsed with ultrapure water to remove dead skin cells. Saliva was mechanically stimulated by chewing a plastic tube and collected into the plastic containers. Then, individual saliva samples (n = 5) were diluted with water (1:1, w:w) and incubated at 37 °C in a water bath until analysis.

2.3. Influence of Saliva/H_2O_2 Solution on Stability of Flavonols

Standards of kaempferol, isorhamnetin, quercetin, and myricetin were mixed with prepared saliva or H_2O_2 solution at 25 mg/L in 12.5% of MeOH final concentration and incubated for 24 h at 37 °C. Samples were prepared to obtain the ratio 1:1 v:v of saliva or equimolar H_2O_2 to the standard solution. The ratio of saliva and standard solution was evaluated to obtain optimal flavonols decomposition time, which allowed detecting unstable intermediates [31]. Since the non-glycosylated flavonols did not easily dissolve in the water, addition of MeOH had to be used, and the minimum concentration of MeOH was 25% in stock solution (12.5% in sample solution). Additionally, each flavonol was mixed with the same volume of the water, as a control sample. All samples were filtered through 0.45-µm PTFE membranes and analyzed at different times of incubation over 24 h. A long incubation period allowed us to find the dead-end products of flavonols degradation. The pH value of all solutions was controlled and measured at the beginning and end of the reaction.

2.4. Influence of Microbiota

To check if the decomposition of flavonols is the result of the presence of microbiota, we decided to incubate quercetin (as a representative flavonol) with saliva filtered using sterile polyethersulfone (PES) membrane (pore size 0.22 µm) to remove bacteria from the sample. For that purpose, 5 mL of saliva was diluted 1:1 with distilled water and shaken vigorously to reduce viscosity. In parallel, 100 µL of filtered saliva was spread onto 5% blood agar and incubated for 24 h at 37 °C in duplicate to examine if the usage of 0.22 µm membranes ensures the sterility of the sample. At the same time, the experiment with the incubation of quercetin with bot saliva filtered using 0.22 and 0.45 µm membranes was carried out.

2.5. Apparatus

The analytical method used in the presented work was developed in our laboratory and discussed in detail in a previous paper on the study of the phenolic compounds [32]. Chromatographic analyses were carried out using LC-MS/MS system consisted of binary pumps LC20-AD, degasser DGU-20A5, column oven CTO-20AC, and autosampler SIL-20AC, connected to 3200 QTRAP Mass spectrometer (Applied Biosystem/MDS SCIEX). Compounds were separated on Chromolith Performance C18 column (100 × 2mm, Merck) at 30 °C. Formic acid (8 mM, pH 2.8) as eluent A and methanol as eluent B were used. Samples incubated at 37 °C were kept in the same temperature before analysis in an autosampler. The flow rate of mobile phase was 0.2 mL/min and the gradient mode was as follow: 0–3 min, 10% B; 20–25 min, 50% B; 26–40 min, 10% B. LC system was connected to the 3200 QTRAP Mass spectrometer (Applied Biosystem/MDS SCIEX) with electrospray ionization (ESI) working in negative mode. ESI conditions were as follows: capillary temperature of 450 °C, curtain gas at 0.3 MPa, auxiliary gas at 0.3 MPa, and negative ionization mode source voltage of 4.5 kV. Nitrogen was used as curtain and auxiliary gas. Analyst 1.4.2 software was used for data acquisition. LC-MS/MS analysis were carried out by comparing retention time and *m/z* values obtained by MS and MS^2 with the mass spectra of standards obtained under the same conditions. Because some degradation products such as depsides are not commercially available, the presence of these compounds was confirmed by comparison of retention times, masses, and fragmentation spectra of potential oxidation products with literature.

Quantification of compounds was done using the selected reaction monitoring mode (SRM). For each compound, the optimum conditions for SRM mode were determined in infusion mode and two SRM pairs were chosen as representatives (SRM1 and SRM2) (Table 1). Due to the higher intensity of peak obtained using the SRM1 pairs, they were chosen for qualitative analyses. Calibration curves were drawn from the analysis of 5 μL volumes at concentration ranging from 0.5 to 50 mg/L (n = 7) measured in triplicated. Coefficients of linearity (R^2) for the calibration curves were ≥ 0.996. LODs were estimated by decreasing the concentration of the analytes to the smallest detectable peaks, and then its concentration was multiplied by three. LODs ranged 0.1–0.5 mg/L.

Table 1. LC-MS/MS characteristics of phenolic compounds in the negative mode.

Compound	Retention Time, Min	Q1 Mass	Q3 Mass	DP, V	CE, V
Gallic acid	3.0	169	125	−45	−20
			97	−45	−26
Protocatechuic acid	4.4	153	109	−15	−18
			66	−15	−26
4-hydroxybenzoic acid	7.1	137	93	−25	−18
			119	−25	−8
Vanillic acid	9.8	167	152	−30	−16
			107	−30	−24
Hyperoside	16.6	463	300	−80	−32
			271	−80	−56
Spiraeoside	18.4	463	300	−80	−32
			151	−80	−48
Rutin	18.9	609	300	−65	−32
			270	−65	−74
Myricetin	20.1	317	151	−20	−26
			179	−20	−22
Quercitrin	20.4	447	300	−60	−26
			270	−60	−56
Quercetin	22.8	301	151	−40	−30
			179	−40	−24
Kaempferol	25.0	285	151	−45	−25
			117	−45	−53
Isorhamnetin	25.6	315	300	−35	−24
			199	−35	−32

2.6. Statistical Analysis

The statistical analyses of the data were carried out using Microsoft Excel 2016 and Excel's Analysis Toolpak (ANOVA). The one-way analysis of variance (ANOVA) and the significance of differences between sample means were calculated, and p values ≤ 0.05 were taken into account as significant.

3. Results and Discussion

3.1. Oxidation of Flavonols with Hydrogen Peroxide

As compared to the control, a noticeable decrease of quercetin concertation was observed during its incubation in the presence of H_2O_2. As shown in Figure 2, rapid degradation of quercetin could be observed with the simultaneous formation of three other compounds with $[M\text{-}H]^-$ at m/z317 ($t_r = 17.5$ min) $[M\text{-}H]^-$ at m/z =193 ($t_r = 4.4$ min) and $[M\text{-}H]^-$ at m/z 169 ($t_r = 5.0$ min). Oxidation product with $[M\text{-}H]^-$ at m/z 317, identified as depside characteristic for quercetin, was found as an intermediate product, the concentration of which increased in the first 80 min of the experiment. Further incubation led to the decrease of its concentration, whereas new peaks with $[M\text{-}H]^-$ at m/z 193 ($t_r = 4.4$) and $[M\text{-}H]^-$ at m/z 169 ($t_r = 5.0$) appeared. Final degradation products were identified as 2,4,6-trihydroxybenzoic acid (m/z 169) and protocatechuic acid (m/z 193). Such results suggest that the oxidation of quercetin with H_2O_2 is based on its splitting by the opening the heterocyclic C-ring and realizing simpler aromatic compounds. Moreover, oxidation involves the initial oxidative step with subsequent changes in the flavonol skeleton such as the formation of B-ring orthoquinone and rearrangement in the C-ring [33]. Detected oxidation products are similar to those obtained under other conditions such as oxygen [28], UVA and UVB [34], hydroxyl free radical [35], and presence of copper (II) [36,37]. Thus, the hypothesis of Zhou et al. [30] that the oxidation of quercetin using different oxidations agents may yield, more or less, the same set of oxidized products seems to be particularly relevant. Decomposition pattern of myricetin leads throughout the characteristic depside as an intermediate at m/z 321. The LC-MS/MS measurements show that further oxidation led to its decomposition and formation of gallic acid as a corresponding hydroxybenzoic acid derivative at m/z 169 and $t_r = 2.9$ min. Besides, 2,4,6-trihydroxybenzoic acid ($t_r = 4.9$ min) was detected as a degradation product of myricetin. Although it shares the same mass as gallic acid, their separation was obtained in the established method. Unfortunately, neither for isorhamnetin nor for kaempferol corresponding depsides were detected. Nevertheless, transformations which involve initial oxidative steps with subsequent changes in the flavonols skeleton was observed. As a result of kaempferol oxidation, the formation of a compound with $t_r = 7.1$ min and SRM characteristic for 4-hydroxybenzoic acid (137/93) was observed. The breakdown product of isorhamnetin was identified as a vanillic acid with retention time 9.8 min and SRM pair (167/152). These results suggest that, as in the case of quercetin and myricetin, oxidation of kaempferol and isorhamnetin leads to C-ring cleavage in the flavonols' structure. As a result of this reaction, the corresponding hydroxybenzoic acid derivatives are formed. It should be mentioned that two reactive centers in the C-ring were identified as responsible for antioxidant activity in flavonols: the 2,3 double bond in conjugation with the 4-oxo function and the 3- and 5-hydroxyl groups with hydrogen bonding to the same 4-oxo function (for flavonols numbering, see Figure 1) [38,39]. That is why all of the modifications which occur in that area could significantly alter the chemical and biological properties of flavonols.

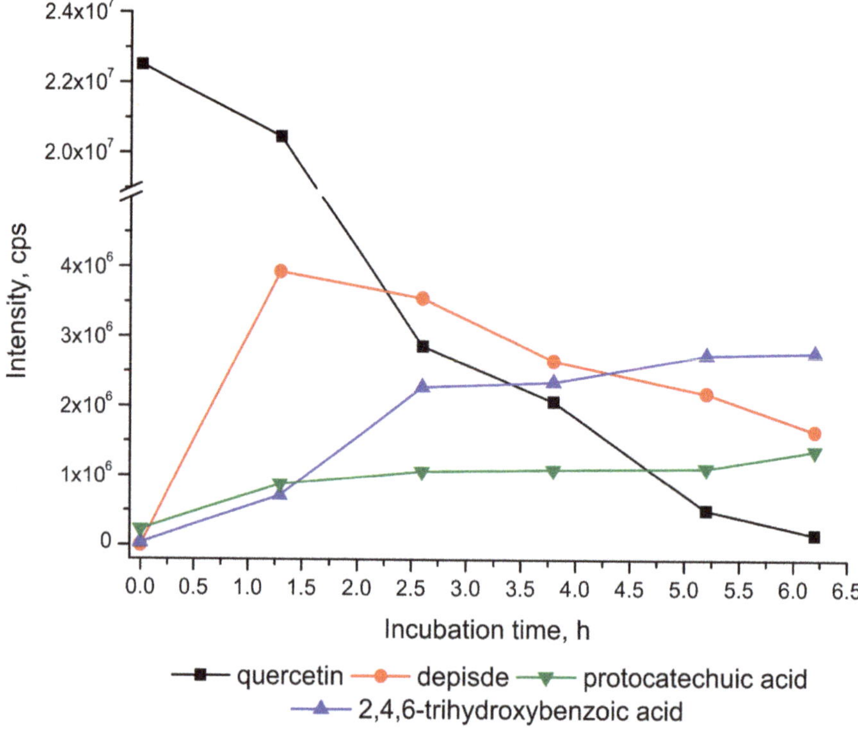

Figure 2. Degradation profile of quercetin and its main products formed during the incubation with H_2O_2 solution.

3.2. Stability of Flavonols in Saliva Solution

Figure 3 presents the loss of the starting amount as percent of remaining kaempferol, quercetin, and isorhamnetin (Figure 3a) as well as myricetin (Figure 3b) for a single representative after 6 h of incubation with saliva. The levels of the examined compounds were significantly different after 6 h of incubation. Kaempferol was the most stable compound in saliva during this time (about 73% left). Quercetin and isorhamnetin were less stable than kaempferol, and the degree of their degradation was 52% and 59%, respectively. Myricetin was the least stable flavonol under these conditions, and its concentration was at trace levels at the end of the experiment. As can be readily seen, myricetin showed rapid degradation and disappeared after 90 min of incubation. The general order of stability of examined flavonols was as follows for all saliva samples collected from five human volunteers: kaempferol > isorhamnetin > quercetin >> myricetin. Generally, the decomposition rate of flavonols increases with an increasing number of hydroxyl groups attached to the B-ring. A similar order of degradation of these flavonols was observed in H_2O_2 solution. These results are in good agreement with observations of Maini et al. for ultraviolet radiation A (UVA) [29]. During the incubation of quercetin with saliva solution, the decomposition of quercetin into compounds with m/z 305 and $m/z = 317$ was observed. These two peaks were identified as the intermediate products of quercetin oxidation (depsides) and their formation under oxidative conditions has already been described [28,29,33]. During further incubation, depsides decomposed with a simultaneous appearance of another peak with the retention time 4.41 min, mass spectrum, and SRM pair (153/109) characteristic for protocatechuic acid (Figure 4). In contrast to the incubation of quercetin in the presence of H_2O_2 mixture, the formation of 2,4,6-trihydroxybenzoic acid as a degradation product was not observed.

Similar results were obtained for myricetin. However, the latter was less stable in the presence of saliva than quercetin. Chromatogram of myricetin incubated with saliva showed a decrease in its peak's intensity, whereas a new peak at m/z = 321 appeared. This peak was considered as depside characteristic for myricetin. Moreover, the formation of a new compound with retention time (t_r = 2.9 min), mass spectrum, and SRM pair characteristic for gallic acid (169/125) was observed. Unfortunately, as mentioned above, among studied flavonols, only for myricetin and quercetin the corresponding depsides were detected. Chromatograms and mass spectrum obtained for isorhamnetin and kaempferol showed their degradation to vanillic and 4-hydroxybenzoic acids, respectively. In an oxidation experiment, Maini et al. [29] suggested that the presence of the corresponding phenolic acid derivative in the absence of any detectable depside concentration is the result of comparable depside formation and hydrolysis rates. Krishnamachari et al. suggested that the presence of both a catechol unit in the B-ring and a free C-3 hydroxyl appears to be a prerequisite for the formation C-ring carbocation or *p*-quinone methide (which formation proceeds predominantly through its tautomer, *o*-quinone) in the oxidative decomposition of flavonols [33,40]. It has been also demonstrated using EPR spectroscopy that the spin distribution during oxidation of quercetin remains entirely on the B-ring, promoting the donation of two electrons leading to the formation of an *o*-quinone [41]. These phenomena explain our observations that myricetin and quercetin decompose rapidly and respective depsides could be observed. Moreover, hydroxyl or methoxy substituents are considered to stabilize the flavonol C-ring carbocation intermediate. As has already been proven, a relatively electron-rich derivative may be more stable and hence more easily formed than its electron-poor analog [42]. Hence, electron-withdrawing and electron-donating groups such as hydroxyl and methoxy groups attached directly to B-ring should influence the rate of *p*-quinone methide formation. In addition, two or more electron-donating groups greatly facilitate its initial generation and stability [42]. Additionally, isorhamnetin due to the presence of a methoxy group greatly enhances the electron-donating properties in the 4′-position [38]. Finally, the presence of corresponding benzoic acid derivatives, as well as 2,4,6-trihydoxybenzoic acid, highlights possible interconversion of the *p*-quinone methide into C-ring carbocation intermediate and its further decomposition. Poorly substituted flavonols such as kaempferol with one hydroxyl group in B-ring skeleton are not able to form *o*-quinone or its tautomeric form *p*-quinone methide species readily. This theory could also explain why any detectable levels of depsides were observed for kaempferol and isorhamnetin.

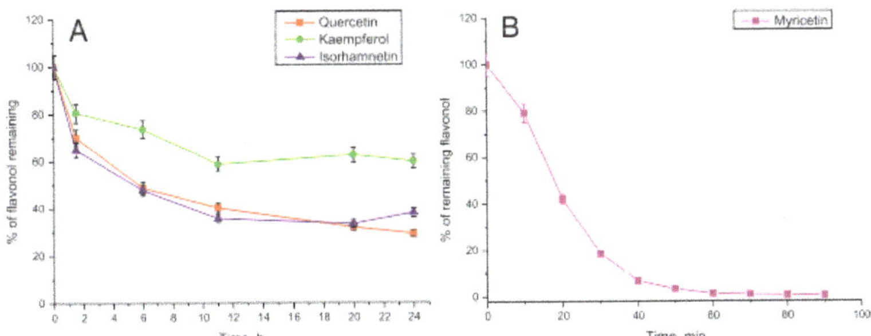

Figure 3. The loss of the starting amount in saliva solution as percent of remaining: (**A**) kaempferol, quercetin, and isorhamnetin; and (**B**) myricetin.

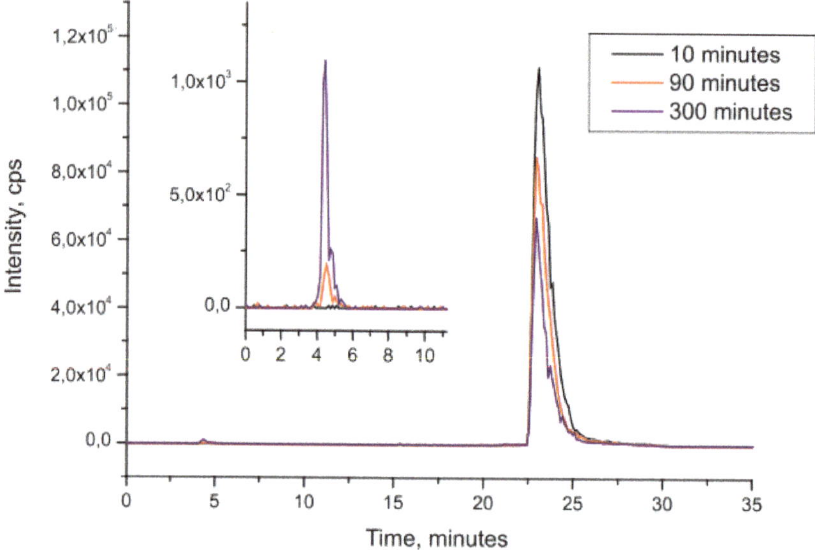

Figure 4. Changes in amount of quercetin (22.8 min) and protocatechuic acid (4.5 min) during incubation of quercetin with saliva.

As mentioned above, saliva was proven to contain phenolics after consumption of phenolics-rich beverages, and these compounds have been found to persist in oral cavity up to 300 min, despite a constant salivary flow [21]. Our research shows that this time is sufficient for the partial decomposition of kaempferol, quercetin, and isorhamnetin and complete decomposition of myricetin. To check if detected compounds may be considered as a dead-end product, we decided to incubate analyzed flavonols for further 18 h. After 24 h of incubation, no additional compounds were detected as degradation products of the analyzed flavonols. However, further decomposition of kaempferol, quercetin, and isorhamnetin was observed. The levels of remaining flavonols calculated as average amount for five volunteers was as follow: 55.42% ± 8.12 of kaempferol, 23.17% ± 5.48 of quercetin, and 39.37% ± 11.11 of isorhamnetin. This indicates that, even after long time of incubation, the order of stability of flavonols remains the same. Since it is known that the stability of flavonoids strongly depends on the pH, we controlled it during the experiment. For all solutions, pH fluctuation ranged from 6.96 ± 0.10 to 7.06 ± 0.15 at the beginning and from 6.98 ± 0.12 to 7.39 ± 0.09 at the end of the experimental period.

Several studies have reported that saliva can hydrolyze flavonoids glycosides, hence we checked four quercetin glycosides: rutin (quercetin 3-rhamnoglucoside), quercitrin (quercetin 3-rhamnoside), hyperoside (quercetin 3-galactoside), and spiraeoside (quercetin 4′-glucoside). Preliminary studies showed that all of the analyzed glycosides were stable in the presence of saliva. The results obtained for rutin and quercitrin are in good agreement with previous studies, which showed that these two glycosides were hydrolyzed very slowly or were resistant to salivary hydrolysis [12,31]. On the other hand, lack of spiraeoside hydrolysis was inconsistent with experiments which suggested that glucose conjugates are rapidly hydrolyzed to corresponding aglycones [11,12]. However, in the same study, a large interindividual variability in hydrolysis rate was also observed. According to the aim of our study, it is interesting to note that sugar moiety attached to flavonol inhibits C-ring decomposition.

The presented studies allow creating the general scheme of flavonols' degradation in the presence of two oxidation agents: H_2O_2 and saliva (Figure 5). Table 2 presents the summarized products detected in samples after oxidation with H_2O_2 and saliva.

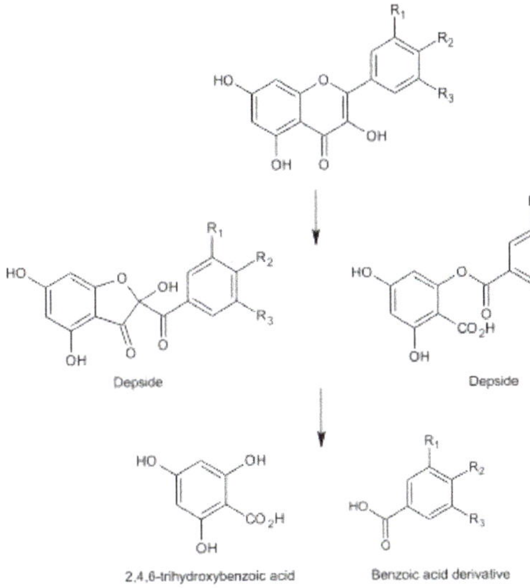

Flavonol	R_1	R_2	R_3
Kaempferol	H	OH	H
Quercetin	OH	OH	H
Isorhamnetin	OCH_3	OH	H
Myricetin	OH	OH	OH

Figure 5. General scheme of flavonols' degradation in the presence of saliva.

Table 2. Comparison of products detected with H_2O_2 and saliva oxidation.

	Kaempferol		**Quercetin**		**Isorhamnetin**		**Myricetin**	
	H_2O_2	**Saliva**	H_2O_2	**Saliva**	H_2O_2	**Saliva**	H_2O_2	**Saliva**
Depside 1	–	–	+	+	–	–	–	–
Depside 2	–	–	–	+	–	–	+	+
2,4,6−trihydroxybenzoic acid	–	–	+	–	–	–	+	–
Benzoic acid derivative	+	+	+	+	+	+	+	+

– not detected; + detected.

Overall, decomposition of flavonols is based on the cleavage of heterocyclic C-ring, with no changes in hydroxyl and methoxy substituents in A-ring and B-ring. Decomposition of myricetin and quercetin leads to the formation of gallic and protocatechuic acid, respectively, which exhibit high redox potential due to the presence of adjacent hydroxyl groups attached to the aromatic ring [43]. Contrarily, vanillic and 4-hydroxybenzoic acids were found to be less efficient in radical neutralization reaction [43].

As mentioned above, the human oral cavity contains numerous and diverse microorganisms as commensals. It is known that microbiota can split the flavonoids by opening the heterocyclic ring and releasing simpler aromatic compounds, such as hydroxyphenylacetic acids from flavonols, which could be further metabolized to derivatives of benzoic acid [44]. Taking that into consideration, we decided to incubate quercetin, as a representative flavonol, with saliva containing microorganisms (filtration through 0.45 μm filters) and without microorganisms (filtration through 0.22 μm sterile filters). In both cases, quercetin was degraded to the same extent during its incubation to protocatechuic acid, as a dead-end product. This indicates that the decomposition of quercetin is independent of the presence of microbiota.

4. Conclusions

As noted in this study, the stability of flavonols in the presence of saliva solution strongly depends on the number of hydroxyl groups attached to the B-ring. Indeed, flavonol stability decreases with increasing B-ring substitution. This proves that saliva is a next oxidative agent, besides UVA and UVB radiation, air, enzymes, and free radicals, which leads to the formation of corresponding phenolic acids as dead-end products of flavonols. The obtained results indicate that myricetin is the most effective flavonol in the process of neutralizing free radicals formed in the oral cavity, due to its easy oxidation caused by the presence of three hydroxyl groups in the B-ring of this compound. Quercetin and isorhamnetin can also be regarded as quite effective antioxidants, capable of oxidizing in the presence of saliva. Nevertheless, their degradation products such as gallic acid and protocatechuic acid still possess reducing potential and are well-known antioxidant units. Rutin (quercetin 3-rhamnoglucoside) and quercitrin (quercetin 3-rhamnoside), as well as, surprisingly, hyperoside (quercetin 3-galactoside) and spiraeoside (quercetin 4′-glucoside), were resistant to hydrolysis in the presence of saliva. Nevertheless, due to the interindividual fluctuations, further analyses should be elucidated on that issue. However, the most intriguing seems to be the relatively short residence time of most foods and their bioactive compounds in the oral cavity. Even so, it should be noticed that this time is sufficient for flavonols decomposition, which clearly shows the process of flavonols metabolic transformation starts in the oral cavity. Hence, studies on flavonols metabolism should take under consideration that, in subsequent parts of the human digestive tract, they could be present not in their parent form but as phenolic acids. Consequently, information on the bioavailability and metabolic pathway of such dietary bioactive compounds is a key part in understanding their beneficial influence on human health.

Author Contributions: Conceptualization, M.R. and M.B.; methodology, M.R., M.B.; validation, M.R., M.B. investigation, M.R., M.B.; writing—original draft preparation, M.R.; writing—review and editing, M.B.; supervision—M.B.; All authors have read and agreed to the published version of the manuscript.

Funding: This research received no external funding.

Acknowledgments: The presented studies were carried out with the approval of Ethics and Bioethics Committee (KEIB–5/2016, Cardinal Stefan Wyszynski University in Warsaw). LC-MS/MS measurements were performed at the Laboratory of Structural Research, Faculty of Chemistry, University of Warsaw, which was established under the European Regional Development Grant WPK_1/1.4.3./2004/72/72/165/2005/U. The authors thank prof. Dorota Korsak for her help in the microbiological part of the presented manuscript and prof. Krystyna Pyrzyńska for her helpful advice on various technical issues examined in this paper.

Conflicts of Interest: The authors declare no conflict of interest.

References

1. Dai, Q.; Borenstein, A.R.; Wu, Y.; Jackson, J.C.; Larson, E.B. Fruit and Vegetable Juices and Alzheimer's Disease: The Kame Project. *Am. J. Med.* **2006**, *119*, 751–759. [CrossRef] [PubMed]
2. Knekt, P.; Kumpulainen, J.; Järvinen, R.; Rissanen, H.; Heliövaara, M.; Reunanen, A.; Hakulinen, T.; Aromaa, A. Flavonoid intake and risk of chronic diseases. *Am. J. Clin. Nutr.* **2002**, *76*, 560–568. [CrossRef] [PubMed]
3. García-Lafuente, A.; Guillamón, E.; Villares, A.; Rostagno, M.A.; Martínez, J.A. Flavonoids as anti-inflammatory agents: Implications in cancer and cardiovascular disease. *Inflamm. Res.* **2009**, *58*, 537–552. [CrossRef] [PubMed]
4. Renaud, S.; de Lorgeril, M. Wine, alcohol, platelets, and the French paradox for coronary heart disease. *Lancet* **1992**, *339*, 1523–1526. [CrossRef]
5. Peng, I.W.; Kuo, S.M. Research Communication: Flavonoid Structure Affects the Inhibition of Lipid Peroxidation in Caco-2 Intestinal Cells at Physiological Concentrations. *J. Nutr.* **2003**, *133*, 2184–2187. [CrossRef]
6. Lotito, S.B.; Frei, B. Relevance of apple polyphenols as antioxidants in human plasma: Contrasting in vitro and in vivo effects. *Free. Radic. Biol. Med.* **2004**, *36*, 201–211.
7. Filipe, P.; Morlière, P.; Patterson, L.K.; Hug, G.L.; Mazière, J.C.; Mazière, C.; Freitas, J.P.; Fernandes, A.; Santus, R. Repair of Amino Acid Radicals of Apolipoprotein B100 of Low-Density Lipoproteins by Flavonoids. A Pulse Radiolysis Study with Quercetin and Rutin†. *Biochemistry* **2002**, *41*, 11057–11064. [CrossRef]
8. Perez-Vizcaino, F.; Duarte, J. Flavonols and cardiovascular disease. *Mol. Asp. Med.* **2010**, *31*, 478–494. [CrossRef]
9. Liao, W.; Chen, L.; Ma, X.; Jiao, R.; Li, X.; Wang, Y. Protective effects of kaempferol against reactive oxygen species-induced hemolysis and its antiproliferative activity on human cancer cells. *Eur. J. Med. Chem.* **2016**, *114*, 24–32. [CrossRef]

10. Boots, A.W.; Haenen, G.R.; Bast, A. Health effects of quercetin: From antioxidant to nutraceutical. *Eur. J. Pharmacol.* **2008**, *585*, 325–337. [CrossRef]
11. Hirota, S.; Nishioka, T.; Shimoda, T.; Miura, K.; Ansai, T.; Takahama, U. Quercetin Glucosides are Hydrolyzed to Quercetin in Human Oral Cavity to Participate in Peroxidase-Dependent Scavenging of Hydrogen Peroxide. *Food Sci. Technol. Res.* **2001**, *7*, 239–245. [CrossRef]
12. Walle, T.; Browning, A.M.; Steed, L.L.; Reed, S.G.; Walle, U.K. Flavonoid Glucosides are Hydrolyzed and Thus Activated in the Oral Cavity in Humans. *J. Nutr.* **2005**, *135*, 48–52. [CrossRef] [PubMed]
13. Truchado, P.; Ferreres, F.; Tomas-Barberan, F. Liquid chromatography–Tandem mass spectrometry reveals the widespread occurrence of flavonoid glycosides in honey, and their potential as floral origin markers. *J. Chromatogr. A* **2009**, *1216*, 7241–7248. [CrossRef] [PubMed]
14. Walle, T.; Walle, U.K. The β-D-glucoside and sodium-dependent glucose transporter 1 (SGLT1)-inhibitor phloridzin is transported by both sglt1 and multidrug resistance-associated proteins 1/2. *Drug Metab. Dispos.* **2003**, *31*, 1288–1291. [CrossRef]
15. Walgren, R.A.; Karnaky, K.J.; Lindenmayer, G.E.; Walle, T. Efflux of Dietary Flavonoid Quercetin 4′-β-Glucoside across Human Intestinal Caco-2 Cell Monolayers by Apical Multidrug Resistance-Associated Protein-2. *J. Pharmacol. Exp. Ther.* **2000**, *294*, 830–836.
16. Tenovuo, J.O. *Human Saliva: Clinical Chemistry and Microbiology*; CRC Press: Boca Raton, FL, USA, 1998; pp. 55–91.
17. Quijada-Morín, N.; Crespo-Expósito, C.; Rivas-Gonzalo, J.C.; García-Estévez, I.; Escribano-Bailón, M. Effect of the addition of flavan-3-ols on the HPLC-DAD salivary-protein profile. *Food Chem.* **2016**, *207*, 272–278. [CrossRef]
18. Cala, O.; Fabre, S.; Pinaud, N.; Dufourc, E.J.; Laguerre, M.; Fouquet, E.; Pianet, I. Towards a Molecular Interpretation of Astringency: Synthesis, 3D Structure, Colloidal State, and Human Saliva Protein Recognition of Procyanidins. *Planta Med.* **2011**, *77*, 1116–1122. [CrossRef]
19. Zhou, S.; Seo, S.; Alli, I.; Chang, Y.W. Interactions of caseins with phenolic acids found in chocolate. *Food Res. Int.* **2015**, *74*, 177–184. [CrossRef]
20. Pascal, C.; Poncet-Legrand, C.; Imberty, A.; Gautier, C.; Sarni-Manchado, P.; Cheynier, V.; Vernhet, A. Interactions between a non Glycosylated Human Proline-Rich Protein and Flavan-3-ols are Affected by Protein Concentration and Polyphenol/Protein Ratio. *J. Agric. Food Chem.* **2007**, *55*, 4895–4901. [CrossRef]
21. Ginsburg, I.; Koren, E.; Shalish, M.; Kanner, J.; Kohen, R. Saliva increases the availability of lipophilic polyphenols as antioxidants and enhances their retention in the oral cavity. *Arch. Oral Biol.* **2012**, *57*, 1327–1334. [CrossRef]
22. Yang, C.S.; Lee, M.J.; Chen, L. Human salivary tea catechin levels and catechin esterase activities: Implication in human cancer prevention studies. *Cancer Epidemiol. Biomark. Prev.* **1999**, *8*, 83–89.
23. Siebert, K.J.; Maekawa, A.A.; Lynn, P. The effects of green tea drinking on salivary polyphenol concentration and perception of acid astringency. *Food Qual. Prefer.* **2011**, *22*, 157–164. [CrossRef]
24. Mager, D.L.; Ximenez-Fyvie, L.A.; Haffajee, A.D.; Socransky, S.S. Distribution of selected bacterial species on intraoral surfaces. *J. Clin. Periodontol.* **2003**, *30*, 644–654. [CrossRef] [PubMed]
25. Aas, J.A.; Paster, B.J.; Stokes, L.N.; Olsen, I.; Dewhirst, F.E. Defining the Normal Bacterial Flora of the Oral Cavity. *J. Clin. Microbiol.* **2005**, *43*, 5721–5732. [CrossRef] [PubMed]
26. Yamamoto, M.; Saeki, K.; Utsumi, K. Isolation of human salivary polymorphonuclear leukocytes and their stimulation-coupled responses. *Arch. Biochem. Biophys.* **1991**, *289*, 76–82. [CrossRef]
27. Thomas, E.L.; Pera, K.A. Oxygen metabolism of Streptococcus mutans: Uptake of oxygen and release of superoxide and hydrogen peroxide. *J. Bacteriol.* **1983**, *154*, 1236–1244. [CrossRef]
28. Ramešová, Š.; Sokolova, R.; Degano, I.; Bulíčková, J.; Žabka, J.; Gál, M. On the stability of the bioactive flavonoids quercetin and luteolin under oxygen-free conditions. *Anal. Bioanal. Chem.* **2011**, *402*, 975–982. [CrossRef]
29. Maini, S.; Hodgson, H.L.; Krol, E.S. The UVA and Aqueous Stability of Flavonoids is Dependent on B-Ring Substitution. *J. Agric. Food Chem.* **2012**, *60*, 6966–6976. [CrossRef]
30. Zhou, A.; Sadik, O.A. Comparative Analysis of Quercetin Oxidation by Electrochemical, Enzymatic, Autoxidation, and Free Radical Generation Techniques: A Mechanistic Study. *J. Agric. Food Chem.* **2008**, *56*, 12081–12091. [CrossRef]

31. Gwiazdon, M.; Biesaga, M. Stability of Quercetin and its Glycosides in the Presence of Saliva. In *Quercetin: Food Sources, Antioxidant Properties and Health Effects*; Nova Science Publishers: Hauppauge, NY, USA, 2015; pp. 209–222. ISBN 978-1-63483-595-4.
32. Biesaga, M.; Pyrzynska, K. Liquid chromatography/tandem mass spectrometry studies of the phenolic compounds in honey. *J. Chromatogr. A* **2009**, *1216*, 6620–6626. [CrossRef]
33. Krishnamachari, V.; Levine, L.H.; Paré, P.W. Flavonoid Oxidation by the Radical Generator AIBN: A Unified Mechanism for Quercetin Radical Scavenging. *J. Agric. Food Chem.* **2002**, *50*, 4357–4363. [CrossRef] [PubMed]
34. Fahlman, B.M.; Krol, E.S. UVA and UVB radiation-induced oxidation products of quercetin. *J. Photochem. Photobiol. B Biol.* **2009**, *97*, 123–131. [CrossRef] [PubMed]
35. Makris, D.P.; Rossiter, J.T. Hydroxyl Free Radical-Mediated Oxidative Degradation of Quercetin and Morin: A Preliminary Investigation. *J. Food Compos. Anal.* **2002**, *15*, 103–113. [CrossRef]
36. Jungbluth, G.; Rühling, I.; Ternes, W. Oxidation of flavonols with Cu(II), Fe(II) and Fe(III) in aqueous media. *J. Chem. Soc. Perkin Trans. 2* **2000**, 1946–1952. [CrossRef]
37. Pękal, A.; Biesaga, M.; Pyrzynska, K. Interaction of quercetin with copper ions: Complexation, oxidation and reactivity towards radicals. *BioMetals* **2010**, *24*, 41–49. [CrossRef]
38. Pannala, A.S.; Chan, T.S.; O'Brien, P.J.; Rice-Evans, C. Flavonoid B-Ring Chemistry and Antioxidant Activity: Fast Reaction Kinetics. *Biochem. Biophys. Res. Commun.* **2001**, *282*, 1161–1168. [CrossRef]
39. Jovanovic, S.V.; Steenken, S.; Hara, Y.; Simic, M.G. Reduction potentials of flavonoid and model phenoxyl radicals. Which ring in flavonoids is responsible for antioxidant activity? *J. Chem. Soc. Perkin Trans. 2* **1996**, 2497–2504. [CrossRef]
40. Krishnamachari, V.; Levine, L.H.; Zhou, C.; Paré, P.W. In Vitro Flavon-3-ol Oxidation Mediated by a B Ring Hydroxylation Pattern. *Chem. Res. Toxicol.* **2004**, *17*, 795–804. [CrossRef]
41. Van Acker, S.A.B.E.; de Groot, M.J.; Berg, D.J.V.D.; Tromp, M.N.J.L.; Kelder, G.D.O.D.; van der Vijgh, A.W.J.F.; Bast, A. A Quantum Chemical Explanation of the Antioxidant Activity of Flavonoids. *Chem. Res. Toxicol.* **1996**, *9*, 1305–1312. [CrossRef]
42. Weinert, E.E.; Dondi, R.; Colloredo-Melz, S.; Frankenfield, K.N.; Mitchell, C.H.; Freccero, M.; Rokita, S.E. Substituents on Quinone Methides Strongly Modulate Formation and Stability of Their Nucleophilic Adducts. *J. Am. Chem. Soc.* **2006**, *128*, 11940–11947. [CrossRef]
43. Mathew, S.; Abraham, T.E.; Zakaria, Z.A. Reactivity of phenolic compounds towards free radicals under in vitro conditions. *J. Food Sci. Technol.* **2015**, *52*, 5790–5798. [CrossRef] [PubMed]
44. Manach, C.; Scalbert, A.; Morand, C.; Rémésy, C.; Jiménez, L. Polyphenols: Food sources and bioavailability. *Am. J. Clin. Nutr.* **2004**, *79*, 727–747. [CrossRef]

Publisher's Note: MDPI stays neutral with regard to jurisdictional claims in published maps and institutional affiliations.

MDPI
St. Alban-Anlage 66
4052 Basel
Switzerland
Tel. +41 61 683 77 34
Fax +41 61 302 89 18
www.mdpi.com

Applied Sciences Editorial Office
E-mail: applsci@mdpi.com
www.mdpi.com/journal/applsci

www.ingramcontent.com/pod-product-compliance
Lightning Source LLC
LaVergne TN
LVHW071244170726
843515LV00006BA/1326

* 9 7 8 3 0 3 6 5 1 1 1 4 6 *